IMPACT MATHEMATICS

Algebra and More

Course 1

IMPACT MATHEMATICS

Algebra and More

Course 1

Developed by
Education Development Center, Inc.

Senior Project Director: Cynthia J. Orrell

Senior Curriculum Developers: Michelle Manes, Sydney Foster, Daniel Lynn Watt, Ricky Carter, Joan Lukas, Kristen Herbert

Curriculum Developers: Haim Eshach, Phil Lewis, Melanie Palma, Peter Braunfeld, Amy Gluckman, Paula Pace

Special Contributors: Faye Nisonoff Ruopp, Elizabeth D. Bjork

 **Glencoe**

New York, New York   Columbus, Ohio   Chicago, Illinois   Peoria, Illinois   Woodland Hills, California

Glencoe

The algebra content for *Impact Mathematics* was adapted from the series, *Access to Algebra*, by Neville Grace, Jayne Johnston, Barry Kissane, Ian Lowe, and Sue Willis. Permission to adapt this material was obtained from the publisher, Curriculum Corporation of Level 5, 2 Lonsdale Street, Melbourne, Australia.

Send all inquiries to:
Glencoe/McGraw-Hill
8787 Orion Place
Columbus, OH 43240

ISBN: 0-07-860909-7

7 8 9 10 079/055 12 11 10 09 08 07 06

Impact Mathematics Project Reviewers

Education Development Center appreciates all the feedback from the curriculum specialists and teachers who participated in review and testing.

Special thanks to:

Peter Braunfeld
Professor of Mathematics Emeritus
University of Illinois

Sherry L. Meier
Assistant Professor of Mathematics
Illinois State University

Judith Roitman
Professor of Mathematics
University of Kansas

....

Marcie Abramson
Thurston Middle School
Boston, Massachusetts

Alan Dallman
Amherst Middle School
Amherst, Massachusetts

Steven J. Fox
Bendle Middle School
Burton, Michigan

Denise Airola
Fayetteville Public Schools
Fayetteville, Arizona

Sharon DeCarlo
Sudbury Public Schools
Sudbury, Massachusetts

Kenneth L. Goodwin Jr.
Middletown Middle School
Middletown, Delaware

Chadley Anderson
Syracuse Junior High School
Syracuse, Utah

David P. DeLeon
Preston Area School
Lakewood, Pennsylvania

Fred E. Gross
Sudbury Public Schools
Sudbury, Massachusetts

Jeanne A. Arnold
Mead Junior High
Elk Grove Village, Illinois

Jacob J. Dick
Cedar Grove School
Cedar Grove, Wisconsin

Penny Hauben
Murray Avenue School
Huntingdon, Pennsylvania

Joanne J. Astin
Lincoln Middle School
Forrest City, Arkansas

Sharon Ann Dudek
Holabird Middle School
Baltimore, Maryland

Jean Hawkins
James River Day School
Lynchburg, Virginia

Jack Beard
Urbana Junior High
Urbana, Ohio

Cheryl Elisara
Centennial Middle School
Spokane, Washington

Robert Kalac
Butler Junior High
Frombell, Pennsylvania

Chad Cluver
Maroa-Forsyth Junior High
Maroa, Illinois

Patricia Elsroth
Wayne Highlands Middle School
Honesdale, Pennsylvania

Robin S. Kalder
Somers High School
Somers, New York

Robert C. Bieringer
Patchogue-Medford School Dist.
Center Moriches, New York

Dianne Fink
Bell Junior High
San Diego, California

Darrin Kamps
Lucille Umbarge Elementary
Burlington, Washington

Susan Coppleman
Nathaniel H. Wixon Middle School
South Dennis, Massachusetts

Terry Fleenore
E.B. Stanley Middle School
Abingdon, Virginia

Sandra Keller
Middletown Middle School
Middletown, Delaware

Sandi Curtiss
Gateway Middle School
Everett, Washington

Kathleen Forgac
Waring School
Massachusetts

Pat King
Holmes Junior High
Davis, California

Kim Lazarus
San Diego Jewish Academy
La Jolla, California

Ophria Levant
Webber Academy
Calgary, Alberta
Canada

Mary Lundquist
Farmington High School
Farmington, Connecticut

Ellen McDonald-Knight
San Diego Unified School District
San Diego, California

Ann Miller
Castle Rock Middle School
Castle Rock, Colorado

Julie Mootz
Ecker Hill Middle School
Park City, Utah

Jeanne Nelson
New Lisbon Junior High
New Lisbon, Wisconsin

DeAnne Oakley-Wimbush
Pulaski Middle School
Chester, Pennsylvania

Tom Patterson
Ponderosa Jr. High School
Klamath Falls, Oregon

Maria Peterson
Chenery Middle School
Belmont, Massachusetts

Lonnie Pilar
Tri-County Middle School
Howard City, Michigan

Karen Pizarek
Northern Hills Middle School
Grand Rapids, Michigan

Debbie Ryan
Overbrook Cluster
Philadelphia, Pennsylvania

Sue Saunders
Abell Jr. High School
Midland, Texas

Ivy Schram
Massachusetts Department of Youth
Services
Massachusetts

Robert Segall
Windham Public Schools
Willimantic, Connecticut

Kassandra Segars
Hubert Middle School
Savannah, Georgia

Laurie Shappee
Larson Middle School
Troy, Michigan

Sandra Silver
Windham Public Schools
Willimantic, Connecticut

Karen Smith
East Middle School
Braintree, Massachusetts

Kim Spillane
Oxford Central School
Oxford, New Jersey

Carol Struchtemeyer
Lexington R-5 Schools
Lexington, Missouri

Kathy L. Terwelp
Summit Public Schools
Summit, New Jersey

Laura Sosnoski Tracey
Somerville, Massachusetts

Marcia Uhls
Truesdale Middle School
Wichita, Kansas

Vendula Vogel
Westridge School for Girls
Pasadena, California

Judith A. Webber
Grand Blanc Middle School
Grand Blanc, Michigan

Sandy Weishaar
Woodland Junior High
Fayetteville, Arkansas

Tamara L. Weiss
Forest Hills Middle School
Forest Hills, Michigan

Kerrin Wertz
Haverford Middle School
Havertown, Pennsylvania

Anthony Williams
Jackie Robinson Middle School
Brooklyn, New York

Deborah Winkler
The Baker School
Brookline, Massachusetts

Lucy Zizka
Best Middle School
Ferndale, Michigan

CONTENTS

Chapter Three

Working with Fractions and Decimals 152

Chapter Four

Making Sense of Percents 224

Chapter Five

Exploring Graphs 276

Chapter Six

Analyzing Data 340

Chapter Seven

Variables and Rules 408

Chapter Eight

Geometry and Measurement 464

Chapter Nine

Chapter Ten

CHAPTER 1

All about Patterns

Real-Life Math

A Bee Tree Although a female honeybee has two parents, a male honeybee has only a mother. The family tree of a male honeybee's ancestors reveals an interesting pattern of numbers.

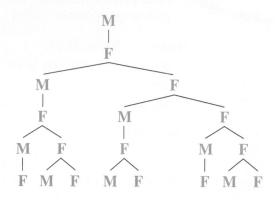

M	1 male bee
F	1 parent
M F	2 grandparents
F M F	3 great-grandparents
M F F M F	5 great-great-grandparents
F M F M F F M F	8 great-great-great-grandparents

The numbers of bees in the generations—1, 1, 2, 3, 5, 8, and so on—form a famous list of numbers known as the *Fibonacci sequence.*

Think About It Can you discover a pattern in the family tree or the list of numbers that will help you find the next two or three numbers in the Fibonacci sequence?

Family Letter

Dear Student and Family Members,

Our class is about to begin an exciting year of mathematics. Don't worry—mathematics is more than just adding and subtracting numbers. Mathematics has been called the "science of patterns." Recognizing and describing patterns and using patterns to make predictions are important mathematical skills.

We'll begin by looking for patterns in Pascal's triangle, a number triangle containing many patterns.

```
                    1
                 1     1
              1     2     1
           1     3     3     1
        1     4     6     4     1
     1     5    10    10     5     1
```

Can you describe any patterns in the triangle? Try to predict the numbers in the next row of the table. Don't worry if you can't find any patterns yet. We'll be learning lots about this triangle in the next few days.

We'll also be looking for shape patterns. For example, the surface of a honeycomb, like the one shown here in the background, is made up of a pattern of hexagons that fit together with no overlaps. Can you make a similar pattern with squares? How about with triangles?

Vocabulary
Along the way, you'll be learning about these new terms:

angle

concave polygon

line symmetry

order of operations

polygon

regular polygon

sequence

term

triangle inequality

vertex

What can you do at home?
During the next few weeks, your student may show interest in patterns and rules. Ask him or her to think about common occurrences of patterns and rules, such as this rule to estimate how many miles you are from a lightning strike: *Count the number of seconds between seeing the lightning and hearing the thunder, and then divide by 5.*

Looking for Patterns

Patterns are everywhere! You can see patterns in wallpaper, fabric, buildings, flowers, and insects. You can hear patterns in music and song lyrics and even in the sound of a person's voice. You can follow patterns to catch a bus or a train or to locate a store with a particular address.

Patterns are an important part of mathematics. You use them every time you read a number, perform a mathematical operation, interpret a graph, or identify a shape. In this lesson, you will search for, describe, and extend many types of patterns.

Explore

In this diagram, you can begin at "Start" and trace a path, following the arrows, to any of the letters.

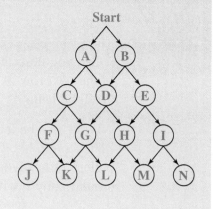

How many different paths are there from Start to A? Describe each path.

How many paths are there from Start to D? Describe each path.

How many paths are there from Start to G? Describe each path.

There are four paths from Start to K. Describe all four.

Add another row of circles to a copy of the diagram, following the pattern of arrows and letters. How many paths are there from Start to S? Describe them.

On a new copy of the diagram, replace each letter with the number of paths from Start to that letter. For example, replace A with 1 and K with 4.

The triangle of numbers you just created is quite famous. You will learn more about the triangle and the patterns it contains in Investigation 1.

Investigation ▶ 1 Pascal's Triangle and Sequences

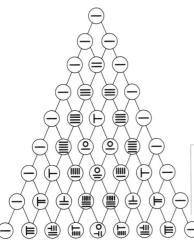

The number triangle as it appears in *Precious Mirror of the Four Elements*, written by Chinese mathematician Chu Shih-Chieh in 1303

The number triangle you created in the Explore has fascinated mathematicians for centuries because of the many patterns it contains. Chinese and Islamic mathematicians worked with the triangle as early as A.D. 1100. Blaise Pascal, a French mathematician who studied it in 1653, called it the *arithmetic triangle*. It is now known as *Pascal's triangle* in his honor.

Problem Set A

Below is a copy of the diagram you made in the Explore. The word "Start" has been replaced by the numeral 1, and the rows have been labeled.

```
              1              Row 0
            1   1            Row 1
          1   2   1          Row 2
        1   3   3   1        Row 3
      1   4   6   4   1      Row 4
    1   5  10  10   5   1    Row 5
```

There are many patterns in this triangle. For example, each row reads the same forward as it does backward.

1. Describe as many patterns in the triangle as you can.

2. To add more rows to the triangle, you could count paths as you did in the Explore—but that might take a lot of time. Instead, use some of the patterns you found in Problem 1 to extend the triangle to Row 7. You may not be able to figure out all the numbers, but fill in as many as you can.

3. One way to add new rows to the triangle is to consider how each number is related to the two numbers just above it to the left and right. Look at the numbers in Rows 3 and 4. Describe a rule for finding the numbers in Row 4 from those in Row 3. Does your rule work for other rows of the triangle as well?

4. Use your rule from Problem 3 to complete the triangle to Row 9.

Pascal's triangle has many interesting patterns in it. You have probably worked with other patterns in the form of puzzles like these:

> Fill in the blanks.
>
> **Puzzle A:** 2, 5, 8, 11, __, __ , __
>
> **Puzzle B:** 16, 8, 4, 2, __, __, __
>
> **Puzzle C:** 3, 2, 3, 2, __, __, __
>
> **Puzzle D:** ★, ✳, ★, ✳, __, __, __

VOCABULARY
sequence
term

To solve these puzzles, you need to find a pattern in the part of the list given and use it to figure out the next few items. Ordered lists like these are called **sequences.** Each item in a sequence is called a **term.** Terms may also be referred to as *stages*.

Think Discuss

Here is Puzzle A. Describe a rule you can follow to get from one term to the next.

$$2, 5, 8, 11, \_\_, \_\_ , \_\_$$

According to your rule, what are the next three terms?

Now look at Puzzle B. Describe the pattern you see.

$$16, 8, 4, 2, \_\_, \_\_, \_\_$$

According to the pattern you described, what are the next three terms?

What pattern do you see in Puzzle C: 3, 2, 3, 2, __, __, __?

According to the pattern, what are the next three terms?

Sequences don't always involve numbers. Look at Puzzle D, for example.

$$★, ✳, ★, ✳, \_\_, \_\_, \_\_$$

Describe the pattern, and give the next three terms.

In Puzzles A and B, each term is found by applying a rule to the term before it. In Puzzles C and D, the terms follow a repeating pattern. In the next problem set, you will explore more sequences of both types.

Problem Set B

1. The sequences in Parts a–e follow a repeating pattern. Give the next three terms or stages of each sequence.

   a.

   b. 3, 6, 9, 3, 6, 9, 3, 6, . . .

   c.

   d. 7, 1, 1, 7, 1, 1, 7, 1, 1, . . .

   e. $\frac{1}{2}, \frac{2}{3}, \frac{1}{2}, \frac{2}{3}, \frac{1}{2}, \frac{2}{3}, \ldots$

2. In Parts a–e, each term in the sequence is found by applying a rule to the term before it (the *preceding* term). Give the next three terms of each sequence.

   a. 3, 6, 9, 12, . . .

   b.

   c. 100, 98.5, 97, . . .

   d. 3, 5, 8, 12, . . .

   e. $\frac{1}{2}, \frac{1}{3}, \frac{1}{4}, \frac{1}{5}, \ldots$

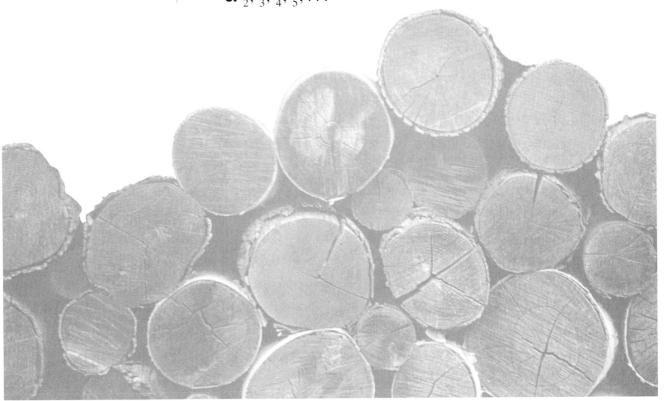

3. Below are two sequences, one made with toothpicks and the other with counters. You and your partner should each choose a different sequence. Do Parts a–c on your own using your sequence.

Sequence A

Sequence B

a. Make or draw the next three terms of your sequence.

b. How many toothpicks or counters will be in the tenth term? Check by making or drawing the tenth term.

c. Give a number sequence that describes the number of toothpicks or counters in each term of your pattern.

d. Compare your answers to Parts a–c with your partner's. What is the same about your answers? What is different?

4. Describe the pattern in each number sequence, and use the pattern to fill in the missing terms.

a. 5, 12, 19, 26, ___ , ___ , ___

b. 0, 9, 18, 27, ___ , ___ , ___

c. 125, 250, ___ , 1,000, ___ , ___ , 8,000

d. 1, 0.1, ___, 0.001, ___, 0.00001, ___

e. 4, 6, 9, 11, 14, 16, 19, ___, ___, ___

5. Consider this sequence of symbols:

$$\Delta, \Delta, \Delta, \Omega, \Omega, \Delta, \Delta, \Delta, \Omega, \Omega, \Delta, \Delta, \Delta, \Omega, \Omega, \ldots$$

a. If this repeating pattern continues, what will the next six terms be?

b. What will the 30th term be?

c. How could you find the 100th term without drawing 100 symbols? What will the 100th term be?

Just the facts

The symbols in Problem 5 are letters of the Greek alphabet. Δ is the letter *delta*, and Ω is the letter *omega*. Greek letters are used frequently in physics and advanced mathematics.

Just the facts

The *Fibonacci numbers*—the numbers in the sequence—can be found in the arrangements of leaves and flowers on plants and of scales on pine cones and pineapples.

6. The sequence below is known as the *Fibonacci sequence* after the mathematician who studied it. The Fibonacci sequence is interesting because it appears often in both natural and manufactured things.

$$1, 1, 2, 3, 5, 8, 13, \ldots$$

a. Study the sequence carefully to see whether you can discover the pattern. Give the next three terms of the sequence.

b. Write instructions for continuing the Fibonacci sequence.

Share & Summarize

1. The diagram from the Explore on page 4 is repeated below. How is Pascal's triangle related to the number of paths from Start to each letter in this diagram?

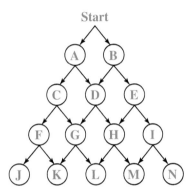

2. You discovered that each number in Pascal's triangle is the sum of the two numbers just above it. Explain what this means in terms of the number of paths to a particular letter in the diagram above.

3. Describe some strategies you use when searching for a pattern in a sequence.

On Your Own Exercises

1. Here are the first few rows of Pascal's triangle:

			1			Row 0
		1		1		Row 1
	1		2		1	Row 2
1		3		3		1	Row 3

    a. How many numbers are in each row shown?

    b. How many numbers are in Row 4? In Row 5? In Row 6?

    c. If you are given a row number, how can you determine how many numbers are in that row?

    d. In some rows, every number appears twice. Other rows have a middle number that appears only once. Will Row 10 have a middle number? Will Row 9? How do you know?

2. A certain row of Pascal's triangle has 252 as the middle number and 210 just to the right of the middle number.

    $\cdots$?   ?   ?   ?   252   210   ?   ?   ? $\cdots$

    a. What is the number just to the left of the middle number? How do you know?

    b. What is the middle number two rows later? How do you know?

Describe the pattern in each sequence, and use the pattern to find the next three terms.

    3. 3, 12, 48, 192, ___ , ___ , ___

    4. 0.1, 0.4, 0.7, 1.0, ___ , ___ , ___

    5. 2, 5, 4, 7, 6, 9, ___, ___, ___

    6. Δ, ∞, Δ, Δ, ∞, Δ, Δ, Δ, ∞, ___, ___, ___

    7. $^-5$, $^-4$, $^-3$, $^-2$, ___ , ___, ___

    8. a, c, e, g, ___, ___, ___

Just the facts

In mathematics, the symbol Δ represents the amount of change in a quantity, and the symbol ∞ represents infinity.

impactmath.com/self_check_quiz

9. Some patterns in Pascal's triangle appear in unexpected ways. For example, look at the pattern in the sums of the rows.

$$1$$
$$1 + 1$$
$$1 + 2 + 1$$
$$1 + 3 + 3 + 1$$
$$1 + 4 + 6 + 4 + 1$$
$$1 + 5 + 10 + 10 + 5 + 1$$

Row 0   Sum = 1
Row 1   Sum = 2
Row 2   Sum = 4

a. Find the sum of each row shown above.

b. Describe the pattern in the row sums.

10. The pattern below involves two rows of numbers. If the pattern were continued, what number would be directly to the right of 98? Explain how you know.

	3		6		9		12		15		18
1	2	4	5	7	8	10	11	13	14	16	17

11. Look at this pattern of numbers. If it were continued, what number would be directly below 100?

```
              1
           2  3  4
        5  6  7  8  9
    10  11  12  13  14  15  16
```

12. For this problem, you may want to draw the shapes on graph paper.

a. Find the next term in this sequence:

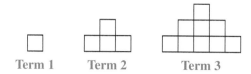

Term 1   Term 2   Term 3

b. This table shows the number of squares in the bottom rows of Terms 1 and 2. Copy and complete the table to show the number of squares in the bottom rows of the next two terms.

Term	Squares in Bottom Row
1	1
2	3
3	
4	

c. Look at your table carefully. Describe the pattern of numbers in the second column. Use your pattern to extend the table to show the number of squares in the bottom rows of Terms 5 and 6.

d. Predict the number of squares in the bottom row of Term 30.

e. Now make a table to show the *total number of squares* in each of the first five terms.

Term	Total Number of Squares
1	1
2	4
3	
4	
5	

f. Look for a pattern in your table from Part e. Use the pattern to predict the total number of squares in Term 10.

In your **own words**

What is a pattern? Is every sequence of numbers a pattern? Is every sequence of shapes a pattern? Explain your answers.

13. Imagine that an ant is standing in the square labeled A on the grid below. The ant can move horizontally or vertically, with each step taking him one square from where he started.

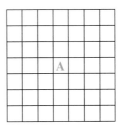

   a. On a copy of the grid, color each square (except the center square) according to the least number of steps it takes the ant to get there. Use one color for all squares that are one step away, another color for all squares that are two steps away, and so on.

   b. What shapes are formed by squares of the same color? How many squares of each color are there? What other patterns do you notice?

Mixed Review

Find each sum or difference without using a calculator.

14. $5,853 - 788$ **15.** $1,054 + 1,492$ **16.** $47,745 - 2,943$

17. Write *thirty-two thousand, five hundred sixty-three* in standard form.

18. Write *fourteen million, three hundred two thousand, two* in standard form.

19. Write 324 in words.   **20.** Write 12,640 in words.

Remember
Writing a number in *standard form* means writing it using digits. For example, standard form for *seventeen* is 17.

21. **Geometry** Imagine that you have 12 square tiles, each measuring 1 inch on a side.

1 in.

1 in.

   a. In how many different ways can you put all 12 tiles together to make a rectangle? Sketch each possible rectangle.

   b. Which of your rectangles has the greatest perimeter? What is its perimeter?

   c. Which of your rectangles has the least perimeter? What is its perimeter?

Remember
The *perimeter* of a figure is the distance around the figure.

1.2 Following Rules

Jing drew a rectangle. She then wrote a rule for generating a sequence of shapes starting with her rectangle.

Starting rectangle:

Rule: Draw a rectangle twice the size of the preceding rectangle.

By following Jing's rule, Caroline drew this sequence:

Jahmal followed Jing's rule and drew this sequence:

Think & Discuss

Could both sequences above be correct? Explain your answer.

Rosita also followed Jing's rule. The sequence she drew was different from both Caroline's and Jahmal's. What might Rosita's sequence look like?

Rewrite the rule so that Caroline's sequence is correct but Jahmal's is not. Try to make your rule clear enough that anyone following it would get the sequence Caroline did.

Investigation ▶ Sequences and Rules

You have just seen how three students could follow the same rule yet draw three different sequences. This is because Jing's rule is *ambiguous*—it can be interpreted in more than one way. In both mathematics and everyday life, it is often important to state rules in such a way that everyone will get the same result.

Problem Set A

1. Create a sequence of shapes in which each shape can be made by applying a rule to the preceding shape. On a blank sheet of paper, draw the first shape of your sequence and write the rule. Try to make your rule clear enough that anyone following it will get the sequence you have in mind.

2. Exchange starting shapes and rules with your partner. Follow your partner's rule to draw at least the next three shapes in the sequence.

3. Compare the sequence you drew in Problem 2 with your partner's original sequence. Are they the same? If not, describe how they are different and why. If either your rule or your partner's rule is ambiguous, work together to rewrite it to make it clear.

Rules are often used to describe how two quantities are related. For example, a rule might tell you how to calculate one quantity from another.

> ### EXAMPLE
>
> An adult dose of SniffleLess cold medicine is 2 ounces. The dose for a child under 12 can be calculated by using this rule:
>
> *Divide the child's age by 12, and multiply the result by 2 ounces.*
>
> This rule tells how a child's dose is related to the age of the child. If you know the child's age, you can use the rule to calculate the dose.
>
> You can apply the rule to calculate the dose for a 3-year-old child:
>
> $$\begin{aligned} \text{dose} &= 3 \div 12 \times 2 \text{ ounces} \\ &= 0.25 \times 2 \text{ ounces} \\ &= 0.5 \text{ ounce} \end{aligned}$$

In Problem Set B, you will look at some common rules for finding one quantity from another.

Problem Set B

1. You can use this rule to estimate how many miles away a bolt of lightning struck:

   Count the seconds between seeing the lightning flash and hearing the thunder, and divide by 5.

   Use the rule to estimate how far away a bolt of lightning struck if you counted 15 seconds between the flash and the thunder.

2. Hannah's grandmother uses this rule to figure out how many spoonfuls of tea to put in her teapot:

   Use one spoonful for each person, and then add one extra spoonful.

   a. How much tea would Hannah's grandmother use for four people?

   b. Hannah's grandfather thinks this rule makes the tea too strong. Make up a rule he might like better.

   c. Using your rule from Part b, how many spoonfuls of tea are needed for four people?

   d. Hannah's cousin Amy likes her tea much stronger than Hannah's grandmother does. Make up a rule Amy might like, and use it to figure out how many spoonfuls of tea would be needed for four people.

3. A cookbook gives this rule for roasting beef:

   Cook for 20 minutes at 475°F. Then lower the heat to 375°F and cook for 15 minutes per pound. If you like rare beef, remove the roast from the oven. If you like it medium, cook it an additional 7 minutes. If you like it well done, cook it an additional 14 minutes.

   What is the total cooking time for a 4-pound beef roast if you like it medium?

Just the facts

It is believed that people first began drinking hot tea more than 5,000 years ago in China. Iced tea wasn't introduced until 1904 at the St. Louis World's Fair, when an Englishman named Richard Blechynden added ice to the drink because no one was buying his hot tea.

The rules in Problem Set B are fairly simple. Many rules involve more complicated calculations. If you don't need to find an exact value, you can sometimes use a simpler rule to find an approximation.

Problem Set

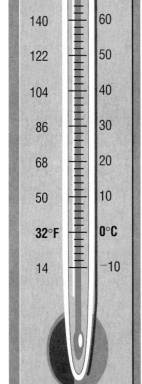

1. You can convert temperatures from degrees Celsius (°C) to degrees Fahrenheit (°F) using this rule:

   Multiply the degrees Celsius by 1.8 and add 32.

   Copy and complete the table to show some Celsius temperatures and their Fahrenheit equivalents.

Degrees Celsius	Degrees Fahrenheit
0	32
10	50
20	
30	
40	
50	

2. The Lopez family spent their summer vacation in Canada. Ms. Lopez used this rule to convert Celsius temperatures to Fahrenheit temperatures:

   Multiply the degrees Celsius by 2 and add 30.

   This rule makes it easy to do mental calculations, but it gives only an *approximation* of the actual Fahrenheit temperature.

   a. Complete this table to show the approximate Fahrenheit temperatures this rule gives for the listed Celsius temperatures.

Degrees Celsius	Approximate Degrees Fahrenheit
0	30
10	50
20	
30	
40	
50	

   b. For which Celsius temperature do the two rules above give the same result?

   c. For which Celsius temperatures in the table does Ms. Lopez's rule give a Fahrenheit temperature that is too high?

Toronto, the capital of the province of Ontario

3. One day the Lopez family flew from Toronto, where the temperature was 37°C, to Winnipeg, where the temperature was 23°C.

   a. Use the rule from Problem 1 to find the exact Fahrenheit temperatures for the two cities.

   b. Use Ms. Lopez's rule from Problem 2 to find the approximate Fahrenheit temperatures for the two cities.

   c. For which city did Ms. Lopez's rule give the more accurate temperature?

4. Look back at your answers to Problems 2 and 3. What happens to the Fahrenheit approximation as the Celsius temperature increases?

Share & Summarize

1. Below are the first term and a rule for a sequence.

   First term: 20

   Rule: Write the number that is 2 units from the preceding number on the number line.

   a. Give the first few terms of two sequences that fit the rule.

   b. Rewrite the rule so that only one of your sequences is correct.

2. At the corner market, bananas cost 49¢ a pound. Write a rule for calculating the cost of a bunch of bananas.

Investigation **2** Order of Operations

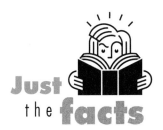

Just the facts

Conventions are not unchangeable like the physical law "When you drop an object, it falls to the ground." People can agree to change a convention and do something different.

A *convention* is a rule people have agreed to follow because it is helpful or convenient for everyone to do the same thing. The rules "When you drive, keep to the right" and "In the grocery store, wait in line to pay for your selections" are two conventions.

Reading across the page from left to right is a convention that English-speaking people have adopted. When you see the words "dog bites child," you know to read "dog" then "bites" then "child" and not "child bites dog." Not all languages follow this convention. For example, Hebrew is read across the page from right to left, and Japanese is read down the page from left to right.

To do mathematics, you need to know how to read mathematical expressions. For example, how would you read this expression?

$$5 + 3 \times 7$$

There are several possibilities:

- *Left to right:* Add 5 and 3 to get 8, and then multiply by 7. The result is 56.

- *Right to left:* Multiply 7 and 3 to get 21, and then add 5. The result is 26.

- *Multiply and then add:* Multiply 3 and 7 to get 21, and then add 5. The result is 26.

VOCABULARY
order of operations

To communicate in the language of mathematics, people follow a convention for reading and evaluating expressions. The convention, called the **order of operations,** says that expressions should be evaluated in this order:

- Evaluate any expressions inside parentheses.

- Do multiplications and divisions from left to right.

- Do additions and subtractions from left to right.

Remember

Evaluating a mathematical expression means finding its value.

To evaluate $5 + 3 \times 7$, you multiply first and then add:

$$5 + 3 \times 7 = 5 + 21 = 26$$

If you want to indicate that the addition should be done first, you would use parentheses:

$$(5 + 3) \times 7 = 8 \times 7 = 56$$

These calculations follow the order of operations:

$$15 - 3 \times 4 = 15 - 12 = 3$$
$$1 + 4 \times (2 + 3) = 1 + 4 \times 5 = 1 + 20 = 21$$
$$3 + 6 \div 2 - 1 = 3 + 3 - 1 = 6 - 1 = 5$$

Another convention in mathematics involves the symbols used to represent multiplication. You are familiar with the $\times$ symbol. An asterisk or a small dot between two numbers also means to multiply. So, each of these expressions means "three times four":

$$3 \times 4 \qquad 3 \cdot 4 \qquad 3 * 4$$

Problem Set D

In Problems 1–4, use the order of operations to decide which of the expressions are equal.

1. $8 \cdot 4 + 6$ $\qquad$ $(8 \cdot 4) + 6$ $\qquad$ $8 \times (4 + 6)$

2. $2 + 8 \cdot 4 + 6$ $\qquad$ $(2 + 8) \times (4 + 6)$ $\qquad$ $2 + (8 \cdot 4) + 6$

3. $(10 - 4) \times 2$ $\qquad$ $10 - (4 * 2)$ $\qquad$ $10 - 4 * 2$

4. $24 \div 6 * 2$ $\qquad$ $(24 \div 6) \times 2$ $\qquad$ $24 \div (6 \cdot 2)$

5. Make up a mathematical expression with at least three operations, and calculate the result. Then write your expression on a separate sheet of paper, and swap expressions with your partner. Evaluate your partner's expression, and have your partner check your result.

6. Most modern calculators follow the order of operations.

   a. Use your calculator to compute $2 + 3 \times 4$. What is the result? Did your calculator follow the order of operations?

   b. Use your calculator to compute $1 + 4 \times 2 + 3$. What is the result? Did your calculator follow the order of operations?

Problem Set E

Mr. Conte gets electricity and gas from the Smallville Power Company. The company uses this rule to calculate a customer's bill:

Charge $0.1205 per kilowatt-hour (kwh) of electricity used and $0.657 per therm of gas used.

1. This month, Mr. Conte's household used 726 units of electricity and 51.7 units of gas. How much should his bill be? Give your answer in dollars and cents.

2. The computer system at Smallville Power crashed, so the clerks have to use calculators to determine the bills. The calculators do *not* use the order of operations. Instead, they evaluate the operations in the order they are entered. To figure out Mr. Conte's bill, the clerk enters the expression below. Will the result be correct, too little, or too much? Explain.

$$726 \times \$0.1205 + 51.7 \times \$0.657$$

3. Suppose the clerk enters the calculation below instead. Will the result be correct, too much, or too little? Explain.

$$\$0.1205 \times 726 + \$0.657 \times 51.7$$

A fraction bar is often used to indicate division. For example, these expressions both mean "divide 10 by 2" or "10 divided by 2":

$$10 \div 2 \qquad \frac{10}{2}$$

Sometimes a fraction bar is used in more complicated expressions:

$$\frac{2 + 3}{4 + 4}$$

In expressions such as this, the bar not only means "divide," it also acts as a grouping symbol—grouping the numbers and operations above the bar and grouping the numbers and operations below the bar. It is as if the expressions above and below the bar are inside parentheses.

The expression $\frac{2 + 3}{4 + 4}$ means "Add 2 + 3, then add 4 + 4, and divide the results." So this expression means $\frac{5}{8}$, or 0.625.

This more complete order of operations includes the fraction bar:

- Evaluate expressions inside parentheses and above and below fraction bars.

- Do multiplications and divisions from left to right.

- Do additions and subtractions from left to right.

Problem Set F

Find the value of each expression.

1. $\frac{2 + 2}{1 + 1}$

2. $2 + \frac{2}{1 + 1}$

3. Your calculator does not have a fraction bar as a grouping symbol, so you have to be careful when entering expressions like $\frac{2 + 2}{1 + 1}$.

  a. What result does your calculator give if you enter 2 + 2/1 + 1 (or 2 + 2 ÷ 1 + 1)? Can you explain why you get that result?

  b. What should you enter to evaluate $\frac{2 + 2}{1 + 1}$?

Share & Summarize

Why is it important to learn mathematical conventions such as the order of operations?

On Your Own Exercises

Use the first term and rule given to create a sequence. Tell whether your sequence is the only one possible. If it isn't, give another sequence that fits the rule.

1. *First term:* 40
   Rule: Divide the preceding term by 2.

2. *First term:*
   Rule: Draw a shape with one more side than the preceding shape.

3. Starting with a closed geometric figure with straight sides of equal lengths, you can use the rule below to create a design.

   Find the midpoint (middle point) of each side of the figure. Connect the midpoints, in order, to make a new shape. (It will be the same shape as the original, but smaller.)

   a. Copy this square. Follow the rule three times, each time starting with the figure you drew the previous time.

   b. Copy this triangle. Follow the rule three times, each time starting with the figure you drew the previous time.

   c. Draw your own shape, and follow the rule three times to make a design.

4. **Measurement** Luis is making a dessert that requires three eggs for each cup of flour.

   a. How many eggs does he need for three cups of flour?

   b. For a party, Luis made a large batch of his dessert using a dozen eggs. How much flour did he use?

5. Economics Althea uses this rule to figure out how much to charge for baby-sitting:

Charge $5 per hour for one child, plus $2 per hour for each additional child.

a. Last Saturday she watched the Newsome twins for 3 hours. How much money did she earn? Explain how you found your answer.

b. Mr. Foster hires Althea to watch his three children for 2 hours. How much will she charge?

c. Does Althea earn more for watching two children for 3 hours or three children for 2 hours?

d. Althea hopes to earn $25 next weekend to buy her sister a birthday present. Describe two ways she could earn at least $25 baby-sitting.

6. Measurement You can convert speeds from kilometers per hour to miles per hour by using this rule:

Multiply the number of kilometers per hour by 0.62.

a. Convert each kilometers-per-hour value in the table below to miles per hour.

Kilometers per Hour	Miles per Hour
50	
60	
70	
80	
90	
100	
110	
120	

b. As part of his job, Mr. Lopez does a lot of driving in Canada. He uses this rule to approximate the speed in miles per hour from a given speed in kilometers per hour:

Divide the number of kilometers per hour by 2 and add 10.

Use Mr. Lopez's rule to convert each kilometers-per-hour value in the table to an approximate miles-per-hour value.

c. For which kilometers-per-hour values from the tables are the results for the two rules closest?

d. For which kilometers-per-hour values in the table does Mr. Lopez's rule give a value that is too high?

Evaluate each expression.

7. $3 + 3 \cdot 2 + 2$

8. $(3 + 3) \times (2 + 2)$

9. $(3 + 3) + 2 \div 2$

10. $\dfrac{7 + 6 - 2 \cdot 6}{11 - 5 \cdot 2}$

Tell whether each expression was evaluated correctly using the order of operations. If not, give the correct result.

11. $10 \times (1 + 5) - 7 = 8$

12. $54 - 27 \div 3 = 45$

13. $(16 - 4 \cdot 2) - (14 \div 2) = 5$

14. $100 - 33 \cdot 2 - (4 + 8) = 22$

Connect & Extend

15. You can produce a sequence of numbers by applying this rule to each term:

If the number is even, get the next number by dividing by 2. If the number is odd, get the next number by multiplying by 3 and adding 1.

a. Use this rule to produce a sequence with 1 as the first term. Describe the pattern in the sequence.

b. Now use the rule to produce a sequence with 8 as the first term. Keep finding new terms until you see a pattern in the sequence. Describe what happens.

c. Use the rule to generate two more sequences. Keep finding new terms until you see a pattern.

d. Using your calculator and the rule, generate a sequence with 331 as its first term. Again, keep finding new terms until you see a pattern.

e. Describe what you discovered in Parts a–d.

16. Measurement One mile is about 1.6 kilometers.

    a. Which is the greater distance, 1 mile or 1 kilometer?

    b. Los Angeles and New York City are about 2,460 miles apart. How many kilometers apart are they?

    c. If the speed limit on a road in Canada is 50 kilometers per hour, what is the speed limit in miles per hour?

    d. In Investigation 2, you learned that lightning is 1 mile away for every 5 seconds you count between the lightning and the following clap of thunder. About how many seconds would it take you to hear the thunder if the lightning were 1 kilometer away?

17. Economics Calls on a particular pay telephone are charged according to this rule:

Charge 25 cents for the call, plus 10 cents for every 3 minutes, or part of 3 minutes, after the first 3 minutes.

    a. How much would it cost you to make a 10-minute call?

    b. If you have $1.15 in change, how long can you talk if you make a single call?

In Exercises 18–21, tell whether each rule is

    • a convention, or

    • a rule we can't change

In your **own words**

Explain the mathematical convention that tells you how to read a three-digit whole number like 645 and know that it is different from a number like 546.

18. Nine times a number is equal to the difference between 10 times the number and the number.

19. In an expression involving only addition and multiplication and no parentheses, such as $2 \cdot 3 + 4 \cdot 5 + 6$, do the multiplication first.

20. $4 + 3 = 7$

21. Use a decimal point to separate the integer part of a number from the fractional part.

22. This computation gives the same result whether you compute correctly (using order of operations) or whether you do the computations from left to right:

$$16 - 6 \cdot 2 - 15 \div 5$$

a. Find the value of the expression both ways, and show that you get the same result.

b. Find another computation that you should not evaluate from left to right, but that gives the correct result if you do.

Mixed Review

Find each sum or difference without using a calculator.

23. $73.97 - 12.43$ **24.** $4.642 - 2.1$ **25.** $37.13 - 16.4$

26. $194.5 + 73.94$ **27.** $54.32 + 45.68$ **28.** $73.7654 - 5$

29. Lucita drew this grid:

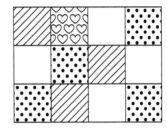

a. What fraction of the squares contain dots?

b. What percent of the squares are striped?

c. What fraction of the squares have hearts?

d. Describe how Lucita could fill in the blank squares to create a grid in which 50% of the squares contain dots, $\frac{1}{4}$ have hearts, and 25% have stripes.

e. Describe how Lucita could fill in the blank squares to create a grid in which $\frac{2}{3}$ of the squares have the same pattern.

30. What number is halfway between 1.8 and 3.2 on the number line?

1.3 Writing Rules for Patterns

There is a pattern in the page numbers of a newspaper. The following activity will help you discover it.

MATERIALS

a section of newspaper

Explore

Each person in your group should take one sheet from the same section of a newspaper.

Notice that your sheet contains four printed pages, two on each side. Write down the pair of page numbers on one side of the sheet and the pair of page numbers on the other side.

Compare the two page numbers on one side of your sheet with the two on the other side. Describe any patterns you notice.

Now compare your two pairs of numbers with those of the other students in your group. Describe any patterns that fit every pair of numbers.

Now work with your group to solve this problem:

A section of a newspaper has 48 pages (numbered from 1 to 48). What is the sum of all the page numbers in the section?

Explain how you found your answer. (Try to find your answer without adding all the numbers.)

Investigation 1. Finding Rules

A fun way to practice recognizing patterns and finding rules is to play a game called *What's My Rule?* In this game, one player thinks of a rule about numbers, and the other players try to guess the rule.

EXAMPLE

Hannah, Jahmal, and Miguel were playing *What's My Rule?*

Now you will have a chance to play *What's My Rule?* As you play, try to come up with some strategies for finding the rule quickly.

Problem Set A

1. Play *What's My Rule?* at least six times with your group. Take turns making up the rule. Do the following for each game you play:

   • Write down the name of the person who made up the rule.

   • Make a table showing the numbers the players guess and the results the rule gives for those numbers.

   • After a player correctly guesses the rule, write it down.

2. Work with your group to create a list of strategies for playing *What's My Rule?*

In *What's My Rule?* you try to guess a rule that another student made up. Now you will play a rule-guessing game that doesn't require a partner.

To play, imagine that a machine has taken some *input* numbers, applied a rule to each one, and given the resulting *output* numbers. Your job is to guess the rule the machine used.

Think & Discuss

Here are the outputs one machine gave for the inputs 6, 3, 10, and 11. What rule did the machine use?

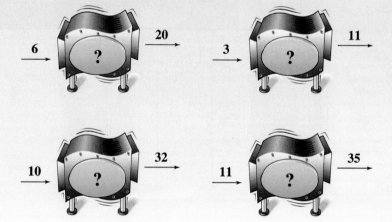

Problem Set B

Each table shows the outputs a particular machine produced for the given inputs. Find a rule the machine could have used. Check to make sure your rule works for all the inputs listed.

1.

Input	3	5	8	4	1
Output	2	4	7	3	0

2.

Input	4	7	10	3	0
Output	2	3.5	5	1.5	0

3.

Input	10	6	3	4	0	100
Output	23	15	9	11	3	203

Share & Summarize

1. In one game of *What's My Rule?* the first clue was "2 gives 4." Write at least two rules that fit this clue.

2. The next clue in the same game was "3 gives 9." Write at least two rules that fit this clue. Do any of the rules you wrote for the first clue work for this clue as well?

3. The third clue was "10 gives 100." Give a rule that fits all three clues. How did you find the rule?

4. Describe some strategies you use to find a rule for an input/output table.

Investigation 2 ▶ Connecting Numbers

In Investigation 1, you found rules relating input and output numbers. Now you will write rules connecting pairs of numbers in a pattern of toothpicks. You will discover that finding a rule can help you figure out the number of toothpicks in any part of the pattern without building every step along the way.

MATERIALS

toothpicks (optional)

Problem Set C

Look at this sequence of toothpick figures.

Term 1   Term 2   Term 3   Term 4

In this sequence, there are 4 toothpicks in Term 1, 7 toothpicks in Term 2, and 10 toothpicks in Term 3.

Term	Toothpicks
1	4
2	7
3	10

1. How many toothpicks are in Term 4? If you continued the pattern, how many toothpicks would you need to make Term 5? Term 6? Term 10?

2. It would take a long time to build or draw Term 100. Describe a shortcut for finding the number of toothpicks in Term 100.

3. Could you use your shortcut to find the number of toothpicks for *any* term of the sequence? Write a rule for finding the number of toothpicks needed for any term, and explain why it works. (Hint: It is not enough to show that your rule works in a few specific cases. Try to explain why it works based on how the terms are built.)

Ms. Washington asked her students to write reports about how they found their rules in Problem Set C.

Here is Rosita, Conor, and Marcus' report.

○ Report by Rosita, Conor, and Marcus

Term 1 is a square with 4 toothpicks. Three more toothpicks are added at each term to make another square. So, Term 2 has 4 plus one group of 3. Term 3 has 4 plus two groups of 3, and so on.

Term 1   Term 2   Term 3

We checked our rule for Term 8: it should have 4 toothpicks plus 7 groups of 3. That's $4 + 7 \times 3 = 25$ toothpicks altogether. When we drew the shape, it took 25 toothpicks.

○

Term 8

We found two ways to write our rule.

One way is this:

To find the number of toothpicks for any term, start with 4 toothpicks, and add the number of the term minus 1, times 3.

A shorter way is this:

number of toothpicks = 4 + (term number − 1) × 3

A second group found their rule a different way.

Here is Luke, Althea, and Miguel's report.

Report by Luke, Althea, and Miguel

We made a table to show the number of toothpicks at each term.

Term	Toothpicks
1	4
2	7
3	10
4	13

The pattern in the second column is 4, 7, 10, 13, ... The numbers go up by 3 for each new term. It would take a long time to do this up to Term 100, so we looked for a connection between the term number and the number of toothpicks.

We guessed "add 3 to the term number," but that worked for only the first row (1 + 3 = 4). It didn't work for the second row (2 + 3 ≠ 7).

Then we noticed that the rule "multiply the term number by 3 and add 1" worked for the first two rows. We checked, and it worked for the other rows too.

We drew diagrams to show why the rule works.

Term 1   Term 2   Term 3   Term 4

We can see that our rule will work for any term because each new term needs 3 new toothpicks, and there is always 1 extra toothpick.

In Problem Set D, you will practice writing rules for sequences and explaining why your rules work. You may find it helpful to build the patterns with toothpicks.

Problem Set D

Do Parts a–d for each sequence of shapes below.

a. Figure out how many toothpicks are in each of the first five terms. Record your results in a table.

b. Tell how many toothpicks are in Term 100.

c. Write a rule that connects the number of toothpicks to the term number. Use your rule to predict the number of toothpicks in Terms 6 and 7, and check your predictions by building or drawing those terms. If your rule doesn't work, revise it until it does.

d. Explain why your rule will work for any term number.

Term	Toothpicks
1	6
2	11
3	16
4	21
5	26

100   501

1.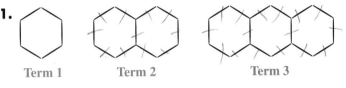

Term 1   Term 2   Term 3

2.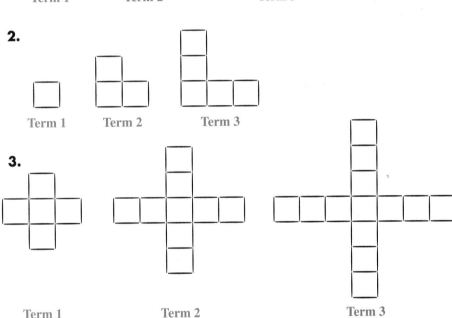

Term 1   Term 2   Term 3

3.

Term 1   Term 2   Term 3

Share & Summarize

Caroline asked, "How can I know that a rule for a toothpick sequence is correct unless I check it for every term?" Write a sentence or two answering Caroline's question.

On Your Own Exercises

Find a rule that works for all the pairs in each input/output table. Use your rule to find the missing outputs.

1.

Input	0	1	2	5	8	12
Output	4	5	6	9		

2.

Input	3	24	36	12	45	60
Output	1	8	12	4		

3.

Input	2	10	16	22	32	44
Output	0	4	7	10		

4.

Input	1	2	3	4	6	10
Output	9	19	29	39		

5. Consider this sequence of figures:

Term 1   Term 2   Term 3

a. Sketch the next two terms in the sequence.

b. Complete the table to show the number of toothpicks in each term.

Term	1	2	3	4	5
Toothpicks	6				

c. Predict the number of toothpicks in Term 100.

impactmath.com/self_check_quiz

6. Conor and Althea both found a rule for predicting the number of toothpicks in each term of this sequence:

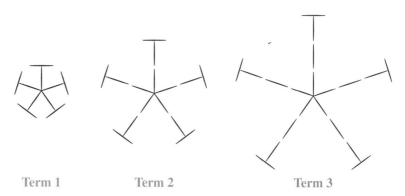

Term 1   Term 2   Term 3

Conor's rule was "Add 1 to the term number, and multiply what you get by 5."

Althea's rule was "Multiply the term number by 5, and add 5 more."

a. Do both rules fit the three terms shown?

b. Use both rules to predict the number of toothpicks in Term 100. Do the rules give the same result?

c. Choose one of the rules, and explain why it is correct.

7. This sequence of figures is made from stars:

Term 1   Term 2   Term 3   Term 4

a. Figure out how many stars are in each of the first five terms, and record your results in a table.

b. How many stars are needed for Term 100?

c. Write a rule that connects the number of stars to the term number. Use your rule to predict the number of stars in Terms 6 and 7, and check your predictions by building or drawing those terms. If your rule doesn't work, revise it until it does.

d. Explain why your rule will work for any term number.

8. This sequence of figures is made from flowers:

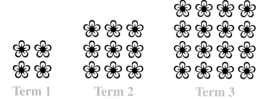

Term 1   Term 2   Term 3

a. Figure out how many flowers are in each of the first five terms, and record your results in a table.

b. How many flowers are needed for Term 100?

c. Write a rule that connects the number of flowers to the term number. Use your rule to predict the number of flowers in Terms 6 and 7, and check your predictions by building or drawing those terms. If your rule doesn't work, revise it until it does.

d. Explain why your rule will work for any term number.

9. Not all input/output tables involve numbers. In this table, the inputs are words and the outputs are letters.

Input	Alice	Justin	Kiran	Marcus	Jimmy	Sarah
Output	i	s	r	r		

a. Complete the last two columns of the table.

b. What would be the output for your name?

c. Describe a rule for finding the output letter for any input word.

d. Are there input words that have no outputs? Explain your answer.

10. In this input/output table, the inputs are numbers and the outputs are letters.

Input	1	2	3	4	5	6
Output	O	T	T	F	F	S

a. What would be the outputs for the inputs 7 and 8?

b. Describe a rule for finding the output letter for any input number.

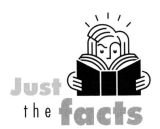

Just the facts

An *anagram* is a word or phrase formed by reordering the letters of another word or phrase. Anagrams were popular in 17th-century France. King Louis XIII even had a Royal Anagrammatist who worked full time creating anagrams to amuse Louis and his guests.

11. Rosita was trying to find a relationship between the number of letters in a word and the number of different ways the letters can be arranged. She considered only words in which all the letters are different.

Number of Letters	Example	Number of Arrangements
1	A	1 (A)
2	OF	2 (OF, FO)
3	CAT	6 (CAT, CTA, ACT, ATC, TAC, TCA)

a. Continue Rosita's table, finding the number of arrangements of four different letters. (You could use MATH as your example, since it has four different letters.)

b. Challenge Predict the number of arrangements of five different letters. Explain how you found your answer.

12. Life Science Geese often fly in a V-shaped pattern. Below is a sequence of such patterns.

Term 1   Term 2   Term 3

a. Draw Terms 4 and 5. Use dots or other shapes to represent each bird.

b. How many geese are in Term 100?

c. Find a rule relating the number of geese to the term number.

d. Can such a V-pattern have exactly 41,390,132 geese? Explain.

Just the facts

Flying in a V-shaped pattern is an efficient way to travel. As each goose flaps its wings, it creates an "uplift" for the birds behind it. When the lead goose tires, it moves out of position, allowing another bird to take its place.

13. Consider this sequence of stars:

Term 1   Term 2   Term 3   Term 4

Could there be a term in this sequence with exactly 12,239 stars? Explain.

14. You can create a sequence of squares of increasing size by arranging identical copies of a small square. The first three terms in such a sequence are shown here.

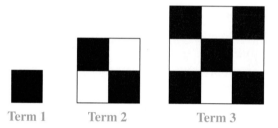

Term 1   Term 2   Term 3

a. Draw the next two terms in the sequence.

b. The numbers of small squares in the terms of this sequence are called *square numbers*. The first square number is 1, the second square number is 4, and so on. Give the third, fourth, and fifth square numbers.

c. Without drawing a picture, find the 25th square number.

d. Write a rule for finding the square number for any term in the sequence.

15. You can create a sequence of triangles of increasing size by arranging identical copies of a small triangle. The first three terms in such a sequence are shown here.

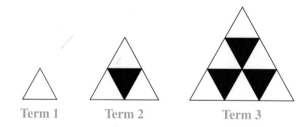

Term 1   Term 2   Term 3

a. Draw the next two terms in the sequence.

b. How many small black and white triangles do you need for Term 1? For Term 2? For Term 3?

c. Compare the number of small triangles in each term of this sequence with the number of squares in each term of the sequence in Problem 14. What do you notice?

d. The numbers of small *white* triangles in the terms of this sequence are called *triangular numbers*. The first triangular number is 1, the second triangular number is 3, and so on. What are the third, fourth, and fifth triangular numbers?

e. Try to find the 20th triangular number without drawing a picture.

f. Challenge Write a rule for finding the triangular number for any term in the sequence.

Find each sum or difference.

16. $\frac{3}{7} + \frac{1}{7}$ **17.** $\frac{13}{32} + \frac{13}{32} + \frac{6}{32}$ **18.** $\frac{1}{2} + \frac{1}{4} + \frac{1}{4}$

19. $\frac{9}{5} - \frac{6}{5}$ **20.** $\frac{12}{15} - \frac{1}{15} - \frac{1}{15}$ **21.** $\frac{5}{7} - \frac{2}{7} - \frac{3}{7}$

Earth Science The symbols in Exercises 22–24 are used in *meteorology*, the study of weather. Copy each symbol, and draw all its lines of symmetry.

22. violent rain showers   **23.** ice pellets   **24.** hurricane

Evaluate each expression.

25. $14 - 12 \div 2$ **26.** $5 + 10 \div 5 \cdot 2$ **27.** $16 + 16 \cdot 4 - 32$

Give the next four terms in each sequence.

28. 64, 32, 16, 8, . . .   **29.** 4, 6, 5, 7, 6, 8, 7, . . .

Patterns in Geometry

Throughout this chapter, you have explored patterns. You have looked at patterns in Pascal's triangle, in sequences, in toothpick designs, and in everyday life. Now you will investigate patterns in geometric shapes.

Explore

How many squares are in this design?
(Hint: The answer is more than 16!)

Investigation 1 ▶ Polygons

VOCABULARY
polygon

Polygons are flat (two-dimensional) geometric figures that have these characteristics:

- They are made of straight line segments.
- Each segment touches exactly two other segments, one at each of its endpoints.

These shapes are polygons:

These shapes are not polygons:

Think & Discuss

Look at the shapes above that are not polygons. Explain why each of these shapes does not fit the definition of a polygon.

In Greek, *poly* means "many" and *gon* means "angle." With the exception of *quadrilateral*, which means "four sides," polygon names refer to the number of angles. For example, *pentagon* means "five angles" and *octagon* means "eight angles."

Polygons can be classified according to the number of sides they have. You have probably heard many of these names before.

Name	Sides	Examples
Triangle	3	
Quadrilateral	4	
Pentagon	5	
Hexagon	6	
Heptagon	7	
Octagon	8	
Nonagon	9	
Decagon	10	

Most polygons with more than 10 sides have no special name. A polygon with 11 sides is described as an *11-gon,* a polygon with 12 sides is a *12-gon,* and so on. Each of the polygons below is a 17-gon.

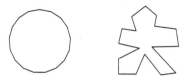

Each corner of a polygon, where two sides meet, is called a **vertex.** The plural of vertex is *vertices.* Labeling vertices with capital letters makes it easy to refer to a polygon by name.

EXAMPLE

This figure contains two triangles and one quadrilateral:

To name one of the polygons in the figure, list its vertices in order as you move around it in either direction. One name for the green triangle is △*ABC.* Other names are possible, including *BCA* and *ACB.* One name for the white triangle is △*ADC.*

The quadrilateral in the figure could be named quadrilateral *ABCD,* or *BCDA,* or *DCBA,* or *DABC.* All of these names list the vertices in order as you move around the quadrilateral. The name *ACBD* is *not* correct.

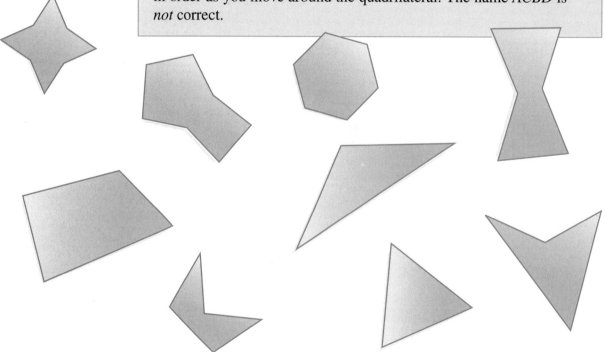

Problem Set A

You will now search for polygons in given figures. Each figure has a total score that is calculated by adding

- 3 points for each triangle,
- 4 points for each quadrilateral,
- 5 points for each pentagon, and
- 6 points for each hexagon.

As you work, try to discover a systematic way to find and list all the polygons in a figure. Be careful to give only one name for each polygon.

Record your work for each problem in a table like this one, which has already been started for Problem 1.

Polygon	Names	Score
Triangle	*ABC, ADC*	6
Quadrilateral		
Pentagon		
Hexagon		
	Total Score	

1.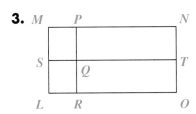

2.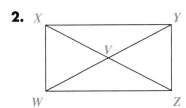

3.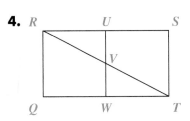

4.

5. Now create your own figure that is worth at least 30 points. Label the vertices. List each of the triangles, quadrilaterals, pentagons, and hexagons in your figure.

Share & Summarize

1. Draw two polygons. Also draw two shapes that are not polygons. Explain why the shapes that are not polygons do not fit the definition of a polygon.

2. In Problem Set A, you had to find ways to list all the polygons in a figure without repeating any. Describe one strategy you used.

Investigation 2 ▶ Angles

You probably already have a good idea about what an angle is. You may think about an angle as a rotation (or a turn) about a point, like an arm bending at the elbow or hinged boards snapping shut at the start of a movie scene.

You may also think about an angle as two sides that meet at a point, like the hands of a clock or the vanes of a windmill.

Or you may think of an angle as a wedge, like a piece of cheese or a slice of pizza.

In mathematics, an **angle** is defined as two *rays* with the same endpoint. A ray is straight, like a line. It has an endpoint where it starts, and it goes forever in the other direction.

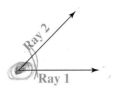

Angles can be measured in *degrees*. Below are some angles with measures you may be familiar with.

- The angle at the vertex of a square measures 90°. You can think of a 90° angle as a rotation $\frac{1}{4}$ of the way around a circle.

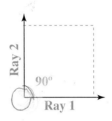

- Two rays pointing in opposite directions form a 180° angle. A 180° angle is a rotation $\frac{1}{2}$ of the way around a circle.

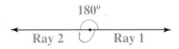

- A 360° angle is a rotation around a complete circle. In a 360° angle, the rays point in the same direction.

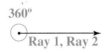

You can use 90°, 180°, and 360° angles to help estimate the measures of other angles. For example, the angle below is about a third of a 90° angle, so it has a measure of about 30°.

Think & Discuss

Copies of the polygon at right can be arranged to form a star.

What is the measure of the angle that is marked in the star? How do you know?

MATERIALS

paper polygons or pattern blocks

Problem Set B

You will be given several copies of each polygon below. Your job is to determine the angle measures for each polygon.

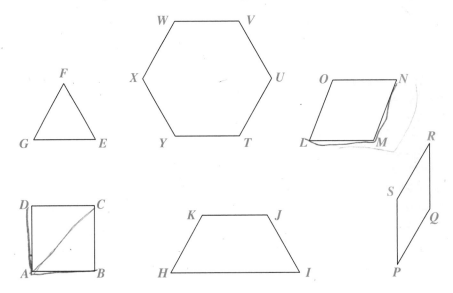

To find the measures of the angles, you can use 90°, 180°, and 360° angles as guides, and you can compare the angles of the polygons with one another.

Your answers should be a record of each vertex, *A–Y,* and the measure of the angle at that vertex. (Note that for many of the polygons, two or more of the angles are identical, so you only have to find the measure of one of them.)

You will now use the angles you found in Problem Set B to help estimate the measures of other angles.

Problem Set C

Estimate the measure of each angle. To help make your estimates, you can compare the angles to 90°, 180°, and 360° angles and to the angles of the polygons in Problem Set B. For each angle, explain how you made your estimate.

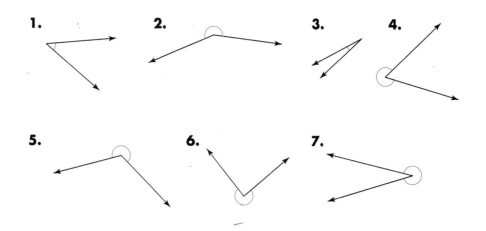

1.   **2.**   **3.**   **4.**

5.   **6.**   **7.**

Share & Summarize

1. Describe how you can estimate the measure of an angle.

2. Marcus said the angles below have the same measure. Hannah said Angle 2 is larger than Angle 1. Who is correct? Explain.

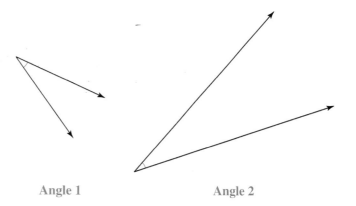

Angle 1   Angle 2

Investigation 3 Classifying Polygons

Polygons can be divided into groups according to certain properties.

VOCABULARY
concave polygon

Concave polygons look like they are "collapsed" or have a "dent" on one or more sides. Any polygon with an angle measuring more than 180° is concave. These are concave polygons:

 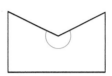

The polygons below are not concave. Such polygons are sometimes called *convex polygons*.

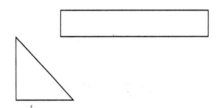

VOCABULARY
regular polygon

Regular polygons have sides that are all the same length and angles that are all the same size. These polygons are regular:

The polygons below are not regular. Such polygons are sometimes referred to as *irregular*.

   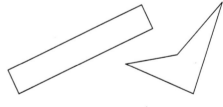

VOCABULARY
line symmetry

A polygon has **line symmetry,** or *reflection symmetry*, if you can fold it in half along a line so the two halves match exactly. The "folding line" is called the *line of symmetry*.

The polygons below have line symmetry. The lines of symmetry are shown as dashed lines. Notice that three of the polygons have more than one line of symmetry.

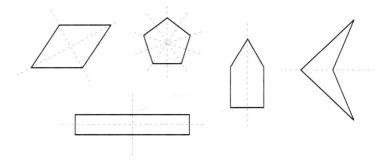

These polygons do not have line symmetry:

Think & Discuss

Consider the polygons below.

This diagram shows how these four polygons can be grouped into the categories *concave* and *not concave:*

 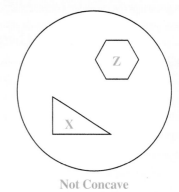

Concave   Not Concave

Now make a diagram to show how the polygons can be grouped into the categories *line symmetry* and *not concave.* Use a circle to represent each category.

- set of polygons and category labels
- large Venn diagram

Problem Set D

You will now play a polygon-classification game with your group. Your group will need a set of polygons and category labels and a large Venn diagram.

Here are the polygons used for the game:

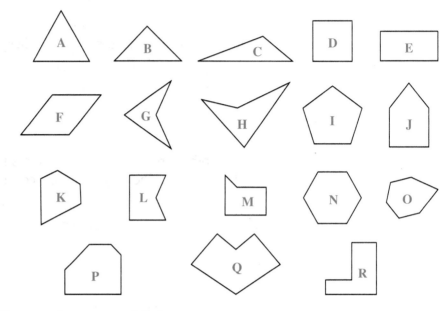

Here are the category labels:

Regular	Concave	Triangle
Not Regular	Not Concave	Not Triangle
Quadrilateral	Pentagon	Hexagon
Not Quadrilateral	Not Pentagon	Not Hexagon
Line Symmetry	No Line Symmetry	

And here is the Venn diagram:

Remember

A Venn diagram uses circles to represent relationships among sets of objects.

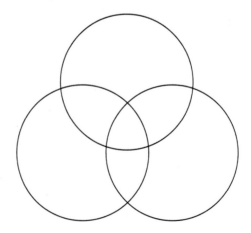

1. As a warm-up for the game, put one of the labels *Regular, Concave,* and *Triangle* next to each of the circles on the diagram. Work with your group to place each of the polygons in the correct region of the diagram.

   To record your work, sketch the three-circle diagram, label each circle, and record the polygons you placed in each region of the diagram. (Just record the letters; you don't need to draw the polygons.)

2. Now you are ready to play the game. Choose one member of your group to be the leader, and follow these rules:

   • The leader selects three category cards and looks at them *without showing them to the other group members.*

   • The leader uses the cards to label the regions, placing one card *face down* next to each circle.

   • The other group members take turns selecting a polygon, and the leader places the polygon in the correct region of the diagram.

   • After a player's shape has been placed in the diagram, he or she may guess what the labels are. The first player to guess all three labels correctly wins.

   At the end of each game, work with your group to place the remaining shapes, and then copy the final diagram. Take turns being the leader until each member of the group has had a chance.

3. Work with your group to create a diagram in which no polygons are placed in an overlapping region (that is, no polygon belongs to more than one category).

4. Work with your group to create a diagram in which all of the polygons are placed either in the overlapping regions or outside the circles (that is, in which no polygon belongs to just one category).

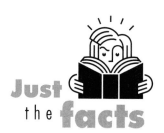

Just the facts

Venn diagrams are named after John Venn (1834–1923) of England, who made them popular. Venn, a priest and historian, published two books on logic in the 1880s.

Share & Summarize

1. Determine what the labels on this diagram must be. Use the category labels from Problem Set D.

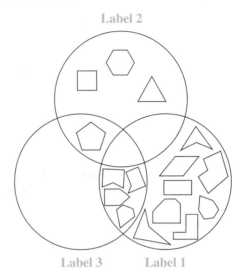

Label 2

Label 3   Label 1

2. Explain why there are no polygons in the overlap of the Label 1 circle and the Label 2 circle.

3. Explain why there are no polygons in the Label 3 circle that are not also in one of the other circles.

Investigation 4 Triangles

In many ways, triangles are the simplest polygons. They are the polygons with the fewest sides, and any polygon can be split into triangles. For this reason, learning about triangles can help you understand other polygons as well.

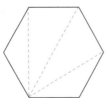 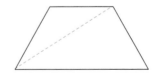 

In the next problem set, you will build triangles from linkage strips. The triangles will look like those below. The sides of this triangle are 2, 3, and 4 units long. Notice that a "unit" is the space between two holes.

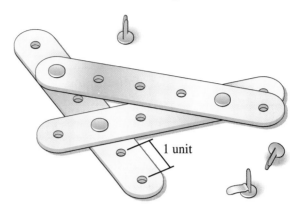

1 unit

Do you think *any* three segments can be joined to make a triangle? You will investigate this question in the next problem set.

MATERIALS

linkage strips
and fasteners

Problem Set E

1. Copy the table below, and then do the following for each row:

   • Try to build a triangle with the given side lengths.

   • In the "Triangle?" column, enter "yes" if you could make a triangle and "no" if you could not.

   • If you could make a triangle, try to make a *different* triangle from the same side lengths. (For two triangles to be different, they must have different shapes.) In the "Different Triangle?" column, enter "yes" if you could make another triangle and "no" if you could not.

Side 1	Side 2	Side 3	Triangle?	Different Triangle?
4 units	4 units	4 units		
4 units	4 units	3 units		
4 units	4 units	2 units		
4 units	4 units	1 unit		
4 units	3 units	1 unit		
4 units	2 units	2 units		
3 units	3 units	3 units		
3 units	3 units	1 unit		
3 units	2 units	2 units		
3 units	2 units	1 unit		
3 units	1 unit	1 unit		

2. Do you think you could make a triangle with segments 4, 4, and 10 units long? Explain your answer.

3. Do you think you could make a triangle with segments 10, 15, and 16 units long? Explain your answer.

4. Describe a rule you can use to determine whether three given segments will make a triangle. Test your rule on a few cases different from those in the table until you are convinced it is correct.

5. Do you think you can make more than one triangle with the same set of side lengths? Explain.

VOCABULARY
triangle inequality

Your work in the last problem set can help you understand a famous mathematical property called the **triangle inequality.**

Triangle Inequality

The sum of the lengths of any two sides of a triangle is greater than the length of the third side.

Think & Discuss

The triangle inequality says that the sum of the lengths of *any* two sides of a triangle is greater than the length of the third side. However, to determine whether three given segments will form a triangle, you only need to compare the sum of the lengths of the two shorter segments with the length of the longest segment. Explain why.

Caroline said, "I know that any three segments that are the same length can form a triangle. I don't even need to check." Is Caroline correct? Explain.

The word *triangle* means "three angles." You can see that any triangle has three angles, one at each vertex.

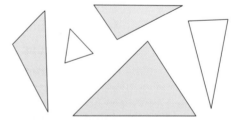

In the next problem set, you will look for a rule relating the angle measures of a triangle.

Problem Set F

1. Draw your own triangle. One triangle is shown here, but yours doesn't have to look like this one. Use a ruler or other straightedge so that your triangle is neat.

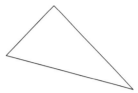

2. Tear off the three vertices of your triangle. It is important to tear them, not cut them, so you can tell which part of each piece was a vertex.

Arrange the three vertices as shown below. What is the measure of the angle they form?

3. Compare your answer for Problem 2 with the answers of others in your class. Did everyone get the same result?

4. What do you think is the rule relating the measures of the angles of a triangle?

Share & Summarize

1. Give lengths of three segments you know *will not* form a triangle. Explain how you know your answer is correct.

2. Give a set of three lengths you know *will* form a triangle. Explain how you know your answer is correct.

3. Suppose a triangle has vertices *A, B,* and *C*. What is the sum of the measures of the angles at these vertices?

Investigation ► Polygons to Polyhedra

MATERIALS

- paper polygons
- tape

You have been working with two-dimensional shapes. In this lab investigation, you will explore three-dimensional shapes.

A *polyhedron* is a closed, three-dimensional figure made of polygons. The shapes below are polyhedra. You have probably seen some of these shapes before.

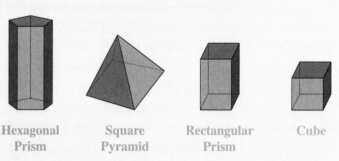

| Hexagonal Prism | Square Pyramid | Rectangular Prism | Cube |

The polygons that make up a polyhedron are called *faces.* The segments where the faces meet are called *edges.* The corners are called *vertices.*

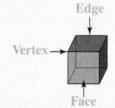

In a *regular polyhedron,* the faces are identical regular polygons, and the same number of faces meet at each vertex. The cube shown above is a regular polyhedron. It has identical square faces, and three faces meet at each vertex. None of the other shapes above is a regular polyhedron. Can you see why?

There is an infinite number of regular polygons—you can always make one with more sides. However, there is a very small number of regular polyhedra. In this lab investigation, you will find them all.

Remember

A *regular polygon* has sides that are all the same length and angles that are all the same size.

Construct the Polyhedra

1. Start with the equilateral triangles, and follow these steps.

> ***Step 1:*** Tape three triangles together around a vertex as shown.

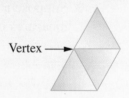

Vertex →

Just the facts

The word *hedron* is Greek for "face," so a polyhedron is a figure with "many faces." Polyhedra are named for the number of faces they have. A cube, for example, can also be called a *hexahedron*, for "six faces."

Step 2: Bring the two outside triangles together and tape them in place, creating a three-dimensional shape.

Vertex ⟶

Step 3: Notice that, at one of the vertices, three triangles meet. At the other vertices, only two triangles meet. At a vertex with only two triangles, add another triangle, so the vertex now has three triangles around it. Now see if you can create a closed shape with three triangles at each vertex. If not, continue to add triangles until you can create a closed shape.

2. Repeat the process you used in Question 1, but start with four triangles around a vertex. Add triangles until the figure closes and there are four triangles around each vertex.

Vertex ⟶

Vertex ⟶

3. Repeat the process again, starting with five triangles around a vertex.

4. Repeat the process once again, starting with six triangles around a vertex. What happens?

5. Now start with squares. Create a polyhedron with three squares around each vertex. What polyhedron did you make?

6. Try to create a polyhedron with four squares around each vertex. What happens?

7. Now start with pentagons. Try to create a polyhedron with three regular pentagons around each vertex. Can you do it?

8. Try to create a polyhedron with four regular pentagons around each vertex. What happens?

9. Now start with hexagons. Try to create a polyhedron with three regular hexagons around each vertex. What happens?

10. What happens when you try to create a polyhedron from regular heptagons?

You have just created all the regular polyhedra.

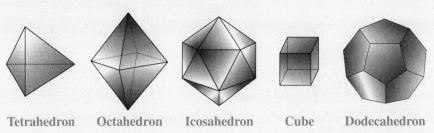

Tetrahedron　Octahedron　Icosahedron　Cube　Dodecahedron

An interesting pattern relates the number of faces, edges, and vertices of all polyhedra. Looking at the regular polyhedra you created can help you find that pattern.

Find a Pattern

11. On each of your polyhedra, find the number of faces, the number of vertices, and the number of edges. This may take some clever counting! Record your results in a table.

Polyhedron	Faces	Vertices	Edges
Tetrahedron			
Octahedron			
Icosahedron			
Cube			
Dodecahedron			

12. Can you find a way to relate the number of faces and vertices to the number of edges?

What Did You Learn?

13. Use what you learned while building the polyhedra to explain why there are only five regular polyhedra.

Just the **facts**

Regular polyhedra are also called *Platonic solids* for the Greek philosopher Plato, who believed they were the building blocks of nature. He thought fire was made from tetrahedra, earth from cubes, air from octahedra, water from icosahedra, and planets and stars from dodecahedra.

On Your Own Exercises

Practice & **Apply**

1. How many triangles are in this figure? (Don't just count the smallest triangles!)

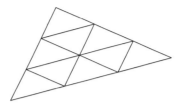

2. Look at the figure in Exercise 1.

   a. Copy the figure, and label each vertex with a capital letter.

   b. In your figure, find at least one of each of the following polygons:

   • quadrilateral

   • pentagon

   • hexagon

   Use your vertex labels to name each shape.

   c. Find the polygon with the maximum number of sides in your figure. Use the vertex labels to name the shape.

3. List all the polygons in the figure below. Compute the figure's score by adding

   • 3 points for each triangle

   • 4 points for each quadrilateral

   • 5 points for each pentagon

   • 6 points for each hexagon

   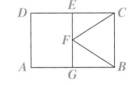

   Record your work in a table like the one below.

Polygon	Names	Score
Triangle		
Quadrilateral		
Pentagon		
Hexagon		
	Total Score	

In Exercises 4–7, several identical angles have the same vertex. Find the measure of the marked angle, and explain how you found it.

4.

5.

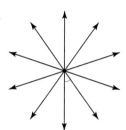

6.

7.

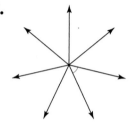

8. A 180° angle is sometimes called a *straight angle*. Explain why that name makes sense.

9. You know that a 360° rotation is one complete rotation around a circle. Find the degree measures for each of these rotations.

   a. half a rotation

   b. two complete rotations

   c. $1\frac{1}{2}$ rotations

10. Draw two angles that each measure more than 90°. Explain how you know they measure more than 90°.

11. Draw two angles that each measure less than 90°. Explain how you know they measure less than 90°.

12. The diagram shows the result of one round of the game you played in Problem Set D.

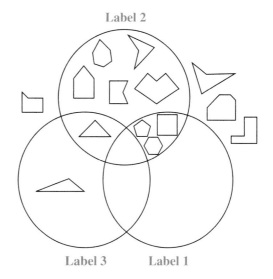

a. Figure out what the labels must be. Use the category labels from Problem Set D.

b. Where would you place each of these shapes?

In Exercises 13–16, draw a polygon that fits the given description, if possible. If it is not possible, say so.

13. a regular polygon with four sides

14. a concave polygon with a line of symmetry

15. a concave triangle

16. a triangle with just one line of symmetry

Tell whether it is possible to make a triangle with the given side lengths.

17. 1, 1, 1   **18.** 1, 1, 2

19. 3, 4, 5   **20.** 25, 25, 200

Tell whether the given measures could be the angle measures of a triangle.

21. 10°, 30°, 30°   **22.** 90°, 90°, 90°

23. 60°, 90°, 30°   **24.** 45°, 45°, 45°

25. 72°, 72°, 36°   **26.** 45°, 55°, 80°

If the given measures could be the measures of two angles of a triangle, give the measure of the third angle. If not, explain why.

27. 10°, 30°   **28.** 90°, 90°

29. 60°, 60°   **30.** 45°, 45°

31. A *diagonal* of a polygon is a segment that connects two vertices but is not a side of the polygon. In each polygon below, the dashed segment is one of the diagonals.

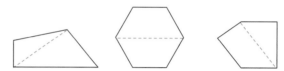

The number of diagonals you can draw from a vertex of a polygon depends on the number of vertices the polygon has.

a. Copy each of these regular polygons. On each polygon, choose a vertex and draw every possible diagonal from that vertex.

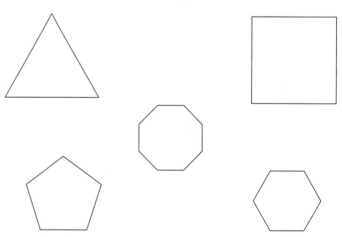

b. Copy and complete the table.

Polygon	Vertices	Diagonals from a Vertex
	3	
Quadrilateral		
	5	
Hexagon		
Heptagon	7	
Octagon		

c. Describe a rule that connects the number of vertices a polygon has to the number of diagonals that can be drawn from each vertex.

d. Explain how you know your rule will work for polygons with any number of vertices.

e. Challenge Describe a rule for predicting the *total number of diagonals* you can draw if you know the number of vertices in a polygon, and explain how you found your rule. Add a column to your table to help you organize your thinking.

Total Diagonals
0
2

32. Look for polygons in your home or school or in books from other classes. Describe at least three different polygons you find, and tell where you found them.

33. Find three angles in your home or school with measures equal to 90°, three with measures less than 90°, and three with measures greater than 90°. Describe where you found each angle.

34. Order the angles below from smallest to largest.

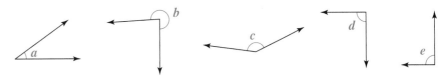

35. Statistics In a survey for the school yearbook, students were asked to name their favorite class. Conor made a circle graph to display the results, but he forgot to label the wedges.

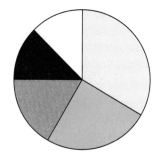

a. Of the students surveyed, $\frac{1}{3}$ liked math best. Which color wedge represents these students? What is the angle measure of that wedge?

b. About $\frac{1}{4}$ of the students liked their foreign language class best. Which wedge represents these students? What is the angle measure of that wedge?

c. Conor remembers that he used light blue to represent students who like science best. What fraction of the students surveyed chose science as their favorite subject?

d. Drama and English tied, with $\frac{1}{8}$ of the students choosing each. Which wedges represent drama and English? What is the angle measure of each wedge?

In Exercises 36–38, describe a rule for creating each shape based on the preceding shape, and draw the next two shapes in the sequence.

36.

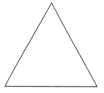 

37.

38.

 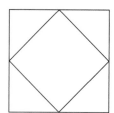 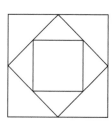

39. Circle diagrams, like those you used to classify polygons, are sometimes used to solve logic puzzles like this one:

Camp Poison Oak offers two sports, soccer and swimming. Of 30 campers, 24 play soccer, 20 swim, and 4 play no sport at all. How many campers both swim and play soccer?

The diagram below includes a circle for each sport. The 4 outside the circles represents the four campers that do not play either sport. Use the diagram to help you solve the logic puzzle.

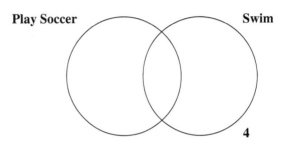

40. Consider these triangles:

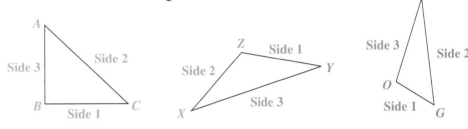

a. For each triangle, tell which side is longest and which angle has the greatest measure.

b. In another triangle, △*PQR*, the angle at vertex *R* has the greatest measure. Where is the longest side? Answer in words or by drawing a picture.

c. Challenge Suppose △*CAT* has two 80° angles, one at vertex *C* and one at vertex *T*. Where is the longest side of the triangle? Explain your thinking. It may help to draw a picture.

In y o u r **own words**

Explain what each of the following words means, and give at least two facts related to each word:

• polygon
• angle
• triangle

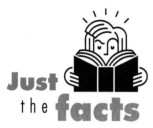

Just the facts

Triangles are the only polygons that are *rigid* in the way described in Exercise 42. If you use linkage strips to build a polygon with more than three sides, you can push on the sides or vertices to create an infinite number of different shapes.

Mixed Review

41. The introduction to Investigation 4 said that every polygon can be split into triangles.

  a. Copy each polygon below, and see if you can split it into triangles.

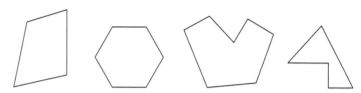

  b. Draw two polygons of your own, and show how to split each of them into triangles.

42. In Problem Set E, you found that when you built a triangle, you could not push or pull on the sides or vertices to change the triangle into a different triangle. Because of this property, triangles are often used as supports in buildings, bridges, and other structures. Look around your home and neighborhood for examples of triangles used as supports. Describe at least two examples you find.

Write each decimal as a fraction.

43. 0.25   **44.** 0.017   **45.** 0.040   **46.** 0.10203

Find each quantity.

47. $\frac{1}{5}$ of 200   **48.** $\frac{2}{6}$ of 120   **49.** $\frac{3}{4}$ of 28   **50.** $\frac{1}{4}$ of 0.4

51. $\frac{1}{2}$ of 1   **52.** $\frac{1}{2}$ of $\frac{1}{2}$ **53.** $\frac{1}{2}$ of $\frac{1}{4}$ **54.** $\frac{1}{2}$ of $\frac{1}{8}$

55. Economics Jing and Caroline went to a restaurant for lunch. Their bill is shown below.

  a. Jing said that she decides how much tip to leave by doubling the tax. Use Jing's rule to determine the tip.

  b. Caroline said that she figures the tip by moving the decimal point on the pretax total one place to the left and doubling the result. Use Caroline's rule to determine the tip.

Bill	
Tuna Melt	$3.95
Veggie Club	3.00
Milk	0.80
Orange Juice	1.25
Pretax Total	$9.00
8% Tax	0.72
Grand Total	$9.72

  c. The girls decided to use Caroline's rule. To figure out how much they each had to pay, they added the tip to the grand total and divided the result in half. How much did each girl pay?

Chapter Summary

In this chapter, you explored patterns and rules. You began by searching for patterns in Pascal's triangle and in sequences, and finding ways to describe and extend the patterns.

You then followed common rules and rules for creating sequences, and you wrote rules for others to follow. You also learned about the *order of operations,* a convention for evaluating and writing mathematical expressions.

Next you focused on writing rules connecting two quantities, such as the term number and the number of toothpicks in the term, and the inputs and outputs in a game of *What's My Rule?*

Finally, you focused on patterns in geometry. You learned to identify, name, and classify polygons. You also discovered some important properties about the side lengths and angle measures of triangles.

Strategies and Applications

The questions in this section will help you review and apply the important ideas and strategies developed in this chapter.

Recognizing, describing, and extending patterns

1. Use your calculator to help you complete this table.

Number of 3s	Expression	Product
1	3	3
2	3 · 3	9
3	3 · 3 · 3	
4	3 · 3 · 3 · 3	
5	3 · 3 · 3 · 3 · 3	
6	3 · 3 · 3 · 3 · 3 · 3	
7	3 · 3 · 3 · 3 · 3 · 3 · 3	
8	3 · 3 · 3 · 3 · 3 · 3 · 3 · 3	

a. Look at the ones digits of the products. What pattern do you see?

b. Predict the ones digits of the product of nine 3s and the product of ten 3s. Use your calculator to check your predictions.

c. What is the ones digit of the product of twenty-five 3s? Explain how you found your answer.

2. Lakita works as a word processor. She charges customers according to this rule:

Charge $7.50 for the project plus $2 per page.

a. Kashi hired Lakita to word process an 8-page term paper. How much did Lakita charge him?

b. Ms. Thompson hired Lakita to word process a business report. Lakita charged her $67.50 for the job. How many pages were in the report?

c. Lakita thinks she might get more customers if she doesn't charge the fixed rate of $7.50. She decides to use this new rule:

Charge $2.50 per page.

How much more or less would Kashi and Ms. Thompson have been charged if Lakita had used this new rule?

3. Consider this starting term and rule:

Starting term: ▲

Rule: Add three triangles to the preceding term.

a. Give the first four terms of two sequences that fit this rule.

b. Rewrite the rule so that only one of your sequences is correct.

Applying the order of operations

4. Start with this string of numbers:

3   4   6   2   4   3

a. Copy the string of numbers. Create a mathematical expression by inserting operation symbols $(+, -, \times, \div)$ and parentheses between the numbers. Evaluate your expression.

b. Copy the string two more times. Create and evaluate two more mathematical expressions so that each of your three expressions gives a different result.

Writing rules that connect two quantities

5. Here are the first three terms of a sequence made from squares:

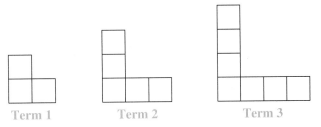

Term 1   Term 2   Term 3

a. Figure out how many squares are in each of the first five terms. Record your results in a table.

b. How many squares are needed for Term 100?

c. Write a rule that connects the number of squares to the term number. Use your rule to predict the number of squares in Terms 6 and 7, and check your predictions by drawing those terms. If your rule doesn't work, revise it until it does.

d. Explain why your rule will work for any term number.

Identifying, naming, and classifying polygons

Tell whether each figure is a polygon. If it isn't, explain why.

6.    **7.**    **8.**

Draw a polygon that fits the given description, if possible. If it is not possible, say so.

9. a concave hexagon with line symmetry

10. a regular quadrilateral without line symmetry

11. a concave pentagon with no line symmetry

Understanding and applying properties of triangles

12. Explain how you can tell whether three segments can be joined to form a triangle. Give the lengths of three segments that can form a triangle and the lengths of three segments that cannot form a triangle.

13. If you know the measures of two angles of a triangle, how can you find the measure of the third angle? Explain why your method works.

Demonstrating Skills

Describe a rule for creating each sequence, and give the next three terms.

14. 2, 5, 8, 11, 14, . . .

15. 1, 2, 1, 2, 2, 1, 2, 2, 2, 1, . . .

16. 512, 256, 128, 64, . . .

17. 1, 4, 2, 5, 3, 6, 4, . . .

Evaluate each expression.

18. $6 + 4 - 5 \div 5$

19. $5 * (4 + 5) + 3$

20. $2 + \frac{7 \cdot 4}{5 + 2}$

21. $2 * 3 + 2 * 3 + 2$

22. $15 - 12 \div 3 - 9$

23. $3 \cdot 2 \div 3 \cdot 3 \div 2$

In Questions 24–26, refer to this figure:

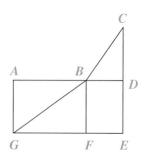

24. Name all the triangles in the figure.

25. Name all the quadrilaterals in the figure.

26. Name all the pentagons in the figure.

In Questions 27–32, tell which of these terms describe each polygon. List all terms that apply.

triangle   pentagon   concave   line symmetry
quadrilateral   hexagon   regular

27.

28.

29.

30.

31.

32.

Estimate the measure of each angle.

33.

34.

35.

Tell whether it is possible to make a triangle with the given side lengths.

36. 5, 6, 7   **37.** 11, 4, 15   **38.** 21, 14, 11

Tell whether the given measures could be the angle measures of a triangle.

39. 45°, 45°, 45°   **40.** 80°, 40°, 80°   **41.** 54°, 66°, 60°

CHAPTER 2

All about Numbers

Market Place Values *Buy low, sell high!* You may have heard this piece of wisdom about stock investing. Stocks allow people to own parts of companies—from fast-food chains to software developers to retail stores. Stockowners hope the value of their stock will rise over time, allowing them to sell their stocks at a higher price than they paid for them.

Think About It Stock-market reports use positive and negative decimals to describe how a stock is doing. How might you show *a loss of $3* using a negative number? How about *a loss of $1.25*?

Family Letter

Dear Student and Family Members,

Our next chapter in mathematics extends the ideas of patterns to look at patterns in factors, multiples, fractions, decimals, and negative numbers—numbers you use every day.

You'll begin by imagining that you work at a factory that uses machines to stretch things like gum, strings, and spaghetti by the number shown on the front of the machine. For example, if you want a piece of licorice to be five times its original length, you could use a ×5 machine. You can replace some machines with a hookup of machines to show the *factor pairs* of a number. For example, instead of a ×10 machine, you could use a hookup of two machines. The first is a ×2 machine. Can you guess what the second machine is?

There are many number patterns found in fractions and decimals. Here's a pattern shown by fractions with the same denominator. Try to predict the numbers in the next column.

Fraction	$\frac{1}{5}$	$\frac{2}{5}$	$\frac{3}{5}$	$\frac{4}{5}$	$\frac{5}{5}$	$\frac{6}{5}$	$\frac{7}{5}$	?
Decimal	0.2	0.4	0.6	0.8	1.0	1.2	1.4	?

Knowing these patterns and the decimal equivalents of common fractions will make it easier for you to calculate with fractions and decimals.

Vocabulary
This chapter also introduces many terms which may be new to you. Here are some of them:

absolute value	**greatest common factor**	**prime factorization**
composite number	**least common multiple**	**prime number**
equivalent fractions	**mixed number**	**relatively prime**
exponent	**multiple**	**repeating decimal**
factor	**opposites**	

What can you do at home?
Fractions and decimals are everywhere. Ask your student to note the many ways fractions and decimals (other than money) are used in his or her day-to-day life.

2.1 Factors and Multiples

Imagine that you work at a stretching factory. The factory has an incredible set of machines that will stretch almost anything.

To double the length of a stick of gum, for example, you can put it through a ×2 machine.

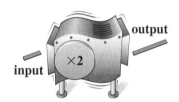

To stretch a caterpillar to five times its length, you can use a ×5 machine.

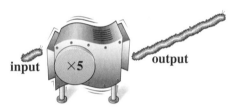

The factory has machines for every whole-number stretch up to 200. This means there is a ×1 machine, a ×2 machine, a ×3 machine, and so on, up to ×200.

Just the facts

Licorice, the candy, is flavored with juice from the licorice herb. The herb is native to southern Europe and is often used to mask the taste of bitter medicines.

Think & Discuss

- Your supervisor has asked you to fill the following orders. Tell which machine you would use to perform each stretch.

  Order 1: Stretch a 1-foot chain into a 5-foot chain.

  Order 2: Stretch a 2-inch pencil into a 16-inch pencil.

  Order 3: Stretch a 3-inch wire into a 21-inch wire.

- Your co-worker, Winnie, sends a 1-foot piece of licorice through the ×2 machine five times. How does this change the length of the licorice?

- Another co-worker, Bill, puts a 1-foot piece of licorice through the ×1 machine five times. How does this change the length of the licorice?

- This afternoon you will be using the ×2 machine and the ×3 machine to stretch 1-foot strands of spaghetti. Give 10 different output lengths you can make using one or both of these machines.

Investigation 1 ▶ Factors

Machines in the stretching factory can be hooked together. When machines are hooked together, the output of one machine becomes the input for the next. For example, when a 1-foot piece of licorice passes through the hookup below, the ×2 machine stretches it into a 2-foot piece. Then the ×3 machine stretches the 2-foot piece into a 6-foot piece.

Problem Set A

1. Winnie is using this hookup to stretch rope:

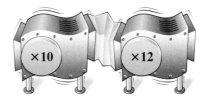

  a. What happens to a 1-foot length of rope put through this hookup?

  b. What single machine would do the same stretch as Winnie's hookup?

  c. Would Winnie get the same result if she used this hookup? Explain.

2. Bill wants to use a hookup of five ×1 machines to stretch a rope to five times its length. Winnie says this won't work. Who is right? Why?

3. What happens when a 1-foot piece of rope is sent through a hookup of two ×5 machines?

When a stretching machine breaks down, you can use your knowledge of hookups to replace it.

Problem Set B

1. If a ×12 machine breaks down, what machines can you hook up to replace it?

2. If a ×51 machine breaks down, what machines can you hook up to replace it?

3. Suppose a ×64 machine breaks down.

   a. What two-machine hookups will replace it?

   b. What three-machine hookups will replace it?

   c. What four-machine hookups will replace it?

4. If a ×81 machine breaks down, what hookups will replace it?

5. If a ×101 machine breaks down, what hookups will replace it?

When you figure out what hookups will replace a particular machine, you are thinking about *factors* of the machine's number. A **factor** of a whole number is another whole number that divides into it without a remainder. For example, 1, 2, 3, 4, 6, 8, 12, and 24 are factors of 24.

When you found two-machine hookups, you were finding *factor pairs*. A **factor pair** for a number is two factors whose product equals that number. Here are the factor pairs for 24:

| 1 and 24 | 2 and 12 | 3 and 8 | 4 and 6 |

Order doesn't matter in a factor pair—2 and 12 is the same pair as 12 and 2.

Problem Set C

List the factors of each number.

1. 21   2. 36   3. 37   4. 63

5. List all the factors of 150. Then list all the factor pairs for 150.

6. List all the factors of 93. Then list all the factor pairs for 93.

Remember

A number is divisible by:

- 2 if the ones digit is divisible by 2.
- 3 if the sum of the digits is divisible by 3.
- 4 if the number formed by the last two digits is divisible by 4.
- 5 if the ones digit is 0 or 5.
- 6 if the number is divisible by both 2 and 3.
- 8 if the number formed by the last three digits is divisible by 8.
- 9 if the sum of the digits is divisible by 9.
- 10 if the ones digit is 0.

Share & Summarize

1. Explain how finding a hookup to replace a broken machine involves finding factors.

2. Describe some methods you use for finding factors of a number.

Investigation Prime Numbers

You, Winnie, and Bill plan to open your own stretching factory. Since stretching machines are expensive, you decide to buy only machines with stretches up to ×50. You also decide not to purchase any machine that can be replaced by hooking other machines together or by using the same machine more than once.

MATERIALS

copy of the stretching-machine table

Problem Set D

1. The table lists stretching machines from ×1 to ×50. On a copy of the table, cross out all the machines that can be replaced by other machines. Circle the machines that cannot be replaced. When you finish, compare your table with your partner's.

×1	×2	×3	×4	×5	×6	×7	×8	×9	×10
×11	×12	×13	×14	×15	×16	×17	×18	×19	×20
×21	×22	×23	×24	×25	×26	×27	×28	×29	×30
×31	×32	×33	×34	×35	×36	×37	×38	×39	×40
×41	×42	×43	×44	×45	×46	×47	×48	×49	×50

2. The machines you crossed out can be replaced by using the machines you circled.

   a. Explain how the ×24 machine can be replaced.

   b. Explain how the ×49 machine can be replaced.

   c. Choose three more machines you crossed out, and explain how each can be replaced by using the machines you circled.

3. Use the idea of factors to explain why the circled machines cannot be replaced.

4. Why is ×2 the only even-numbered machine you need?

5. Winnie points out that you can save more money by not purchasing a ×1 machine. Bill says, "There is no hookup that can replace the ×1 machine. We must buy it!"

What do *you* think? Does your factory need the ×1 machine? Explain. If you think it is not needed, cross it out in your table.

VOCABULARY
prime number
composite number

The numbers on the machines that cannot be replaced (excluding the ×1 machine) are *prime numbers*. A **prime number** is a whole number greater than 1 with only two factors: itself and 1.

The numbers on the machines that can be replaced are *composite numbers*. A whole number greater than 1 with more than two factors is a **composite number.** The number 1 is neither prime nor composite.

Any composite number can be written as a product of prime numbers. Here are some examples:

$$10 = 5 \cdot 2 \qquad 117 = 3 \cdot 3 \cdot 13 \qquad 693 = 3 \cdot 3 \cdot 7 \cdot 11$$

VOCABULARY
prime factorization

When you write a composite number as a product of prime numbers, you are finding its **prime factorization.** You can use a *factor tree* to find the prime factorization of any number.

EXAMPLE

Below are the steps for making a factor tree for the number 20. Two possible factor trees are shown.

- Find any factor pair for 20. Draw a "branch" leading to each factor.

- If a factor is prime, circle it. If a factor is not prime, find a factor pair for it.

- Continue until you can't factor any further.

Both of these factor trees show that the prime factorization of 20 is 2 · 2 · 5. (Remember, when you multiply numbers, order doesn't matter.)

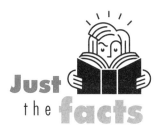
When a factor occurs more than once in a prime factorization, you can use *exponents* to write the factorization in a shorter form. An **exponent** is a small, raised number that tells how many times a factor is multiplied.

3 is an exponent.

2^3

- 2^3 is read "2 to the third power."
- The exponent tells you to multiply three factors of 2: $2 \cdot 2 \cdot 2 = 8$.

You can use exponents to rewrite the prime factorization of 20 as $2^2 \cdot 5$.

Problem Set E

1. Use a factor tree or another method to find the prime factorization of 24.

2. Use a factor tree or another method to find the prime factorization of 49.

3. Explain how your answers to Problems 1 and 2 are related to the machines needed to replace a ×24 machine and a ×49 machine in your stretching factory.

Use a factor tree to find the prime factorization of each number. Write the prime factorization using exponents when appropriate.

4. 100　　　**5.** 99　　　**6.** 5,050　　　**7.** 111,111

Share Summarize

1. Describe the difference between a prime number and a composite number. Give three examples of each.

2. Find the prime factorization of each composite number you listed in Question 1.

Investigation 3 ▶ Common Factors

Business has been good at your factory. You have earned enough to buy one machine for every prime-number stretch from 2 to 101. However, there still don't seem to be enough machines to go around! You find that you and your co-workers often need to use the same machine at the same time.

Explore

Here are the orders you and Winnie are scheduled to fill this morning:

Time	Your Orders	Winnie's Orders
10:00	stretch $\times 32$	stretch $\times 16$
10:30	stretch $\times 15$	stretch $\times 21$
11:00	stretch $\times 99$	stretch $\times 6$
11:30	stretch $\times 49$	stretch $\times 14$

Which machines do you and Winnie need to use at the same time?

Rearrange the schedule so both you and Winnie can fill your orders.

VOCABULARY
common factor

The scheduling problems in the Explore occurred when the stretch numbers had a *common factor*. A **common factor** of two or more numbers is a number that is a factor of all of the numbers.

Problem Set F

In Problems 1–3, list the factors of each number. Then find the common factors of the numbers.

1. 100 and 75

2. 33 and 132

3. 36, 84, and 112

4. Is it possible for two numbers to have 8 as a common factor but not 2? If so, give an example. If not, explain why not.

5. Is it possible for two numbers to have no common factors? If so, give an example. If not, explain why not.

6. Is it possible for two numbers to have only one common factor? If so, give an example. If not, explain why not.

The **greatest common factor** of two or more numbers is the greatest of their common factors. The abbreviation **GCF** is often used for greatest common factor. Two or more numbers are **relatively prime** if their only common factor is 1.

EXAMPLE

Look at the common factors of 12 and 16.

Factors of 12: 1, 2, 3, 4, 6, 12   Factors of 16: 1, 2, 4, 8, 16

The common factors of 12 and 16 are 1, 2, and 4. Therefore, 12 and 16 are *not* relatively prime. The GCF of 12 and 16 is 4.

Now consider the common factors 8 and 15.

Factors of 8: 1, 2, 4, 8   Factors of 15: 1, 3, 5, 15

The only common factor of 8 and 15 is 1. Therefore, 8 and 15 are relatively prime. The GCF of 8 and 15 is 1.

The game in Problem Set G will give you practice finding greatest common factors and determining whether numbers are relatively prime.

Problem Set G

Here are the rules for the *GCF* game:

GCF Game Rules

For the game board, create a 4 × 4 grid of consecutive numbers. One possible board is shown here. Then follow this procedure:

12	13	14	15
16	17	18	19
20	21	22	23
24	25	26	27

- Player A circles a number.
- Player B circles a second number. Player B's score for this turn is the GCF of the two circled numbers. Cross out the circled numbers.
- Player B circles any number that is not crossed out.
- Player A circles another number that has not been crossed out. Player A's score for the turn is the GCF of the two numbers. Cross out the circled numbers.
- Take turns until all numbers have been crossed out. The player with the higher score at the end of the game wins.

1. Play *GCF* four times with your partner. As you play, think about strategies you can use to score the most points.

2. Describe the strategies you used when playing *GCF*. Use the terms *relatively prime* and *greatest common factor* in your explanation.

Share & Summarize

1. Here is the board for a game of *GCF*. Your partner has just circled 15. Which number would you circle? Why?

12	ⓧ13	14	⑮
16	ⓧ17	18	ⓧ19
ⓧ20	21	22	ⓧ23
24	ⓧ25	26	27

2. You make the first move in this round of *GCF*. Which number would you circle? Why?

ⓧ6	ⓧ7	8	9
10	ⓧ11	12	ⓧ13
ⓧ14	15	16	ⓧ17
18	ⓧ19	ⓧ20	21

Investigation 4 ▶ Multiples

In the previous investigations, you found factors of a given number. Now you will find numbers that have a given number as a factor.

For example, think about numbers that have 20 as a factor. You can easily generate such numbers by multiplying 20 by whole numbers.

$$20 \times 1 = 20$$
$$20 \times 2 = 40$$
$$20 \times 3 = 60$$
$$20 \times 4 = 80$$

VOCABULARY
multiple

The products 20, 40, 60, and 80 are *multiples* of 20. A **multiple** of a whole number is the product of that number and another whole number.

Problem Set H

1. List four multiples of 20 that are greater than 1,000.

2. Is there a limit to how many multiples a whole number greater than 0 can have? Explain.

3. Is there a limit to how many factors a whole number greater than 0 can have? Explain.

4. Is a number its own factor? Is a number its own multiple? Explain.

5. How can you determine whether 1,234 is a multiple of 7?

VOCABULARY
common multiple
least common multiple, LCM

A **common multiple** of a set of numbers is a multiple of all the numbers. The **least common multiple** is the smallest of these. The abbreviation **LCM** is often used for least common multiple.

EXAMPLE

Consider the multiples of 6 and 9.

Multiples of 6: 6, 12, 18, 24, 30, 36, 42, 48, 54, 60, 66, 72, . . .

Multiples of 9: 9, 18, 27, 36, 45, 54, 63, 72, 81, . . .

The numbers 18, 36, 54, and 72 are common multiples of 6 and 9. The LCM is 18.

Problem Set ▌

1. Use a strip of paper labeled with the whole numbers from 1 to 40.

| 1 2 3 4 5 6 7 8 9 10 11 12 13 14 15 16 17 18 19 20 21 22 23 24 25 26 27 28 29 30 31 32 33 34 35 36 37 38 39 40 |

a. Circle all the multiples of 2 between 1 and 40, and draw a square around all the multiples of 3.

b. List the common multiples of 2 and 3 between 1 and 40. What do these numbers have in common?

c. What are the next three common multiples of 2 and 3?

d. What is the LCM of 2 and 3?

2. Use a new strip of paper labeled from 1 to 40.

| 1 2 3 4 5 6 7 8 9 10 11 12 13 14 15 16 17 18 19 20 21 22 23 24 25 26 27 28 29 30 31 32 33 34 35 36 37 38 39 40 |

a. Circle all the multiples of 4 between 1 and 40, and draw a square around all the multiples of 6.

b. List the common multiples of 4 and 6 between 1 and 40. What do these numbers have in common?

c. What are the next three common multiples of 4 and 6?

d. What is the LCM of 4 and 6?

3. Look at your third numbered strip.

| 1 2 3 4 5 6 7 8 9 10 11 12 13 14 15 16 17 18 19 20 21 22 23 24 25 26 27 28 29 30 31 32 33 34 35 36 37 38 39 40 |

a. Which two numbers greater than 1 have the most common multiples between 1 and 40?

b. What is the LCM of these two numbers?

c. How do you know these two numbers have the most common multiples of all the pairs you could choose?

Use the idea of LCMs to help solve the next set of problems.

Problem Set J

Ileana often attends family reunions in the summer.

- Her mother's family, the Colóns, has a reunion every third summer.

- Her father's family, the Kanes, has a reunion every fourth summer.

- Her stepfather's family, the Ashbys, has a reunion every sixth summer.

This summer Ileana attended all three reunions!

1. Think about the Colón and Kane reunions.

   a. How many years will it be until the Colón and Kane reunions occur in the same summer again? Explain how you found your answer.

   b. How many years will it be until the second time both reunions happen again?

   c. Will both reunions be held 48 years from now? Will both be held 70 years from now? How do you know?

2. Now think about the Colón and Ashby reunions.

   a. How many years will it be until the Colón and Ashby reunions occur in the same summer again? Explain how you found your answer.

   b. How many years will it be until the second time both reunions happen again?

   c. Will both reunions be held 40 years from now? How do you know?

3. Now consider the Ashby and Kane reunions.

   a. How many years will it be until the Ashby and Kane reunions are held in the same summer again? Explain how you found your answer.

   b. How many years will it be until the second time both reunions happen again?

   c. Will both reunions be held 72 years from now? How do you know?

4. How many years will it be until all three reunions occur in the same summer again?

Share & Summarize

Explain how your answers to the problems in Problem Set J are related to the ideas of common multiples and least common multiples.

Lab
Investigation ▶ A Locker Problem

The students in Ms. Dolce's homeroom have lockers numbered 1–30, located down a long hallway. One day the class did an experiment. The first student, Student 1, walked down the hall and opened every locker.

Student 2 closed Locker 2 and every second locker after it.

Student 3 closed Locker 3 and *changed the state* of every third locker after it. This means that if the locker was open, Student 3 closed it; if the locker was closed, Student 3 opened it.

Student 4 changed the state of every fourth locker, starting with Locker 4. The students continued this pattern until all 30 students had had a turn. In this lab investigation, you will consider this question: *Which lockers were open after Student 30 took her turn?*

Try It Out

1. Work with your partner to develop a plan for finding the answer to the question above. Write a few sentences describing your plan.

2. Carry out your plan. Which lockers were open after Student 30 took her turn? What do the numbers on the open lockers have in common?

Analyze What Happened

3. Consider how a student's number is related to the lockers he touched.

   a. Which lockers were touched by both Student 6 and Student 10? By both Student 2 and Student 5? By both Student 8 and Student 10?

   b. Use what you have learned about factors and multiples to explain why your answers to Part a make sense.

4. How many lockers would there have to be for both Student 8 and Student 9 to touch the same locker? Explain.

5. Think about how a locker's number is related to the numbers of the students who touched that locker.

    a. Which students touched both Locker 7 and Locker 12? Both Lockers 3 and 27? Both Lockers 20 and 24?

    b. Use what you have learned about factors and multiples to explain why your answers to Part a make sense.

6. Think about the lockers that were touched by only two students.

    a. Which lockers were touched by only two students? Were these lockers open or closed after Student 30's turn?

    b. What do the numbers on these lockers have in common? Why?

7. Think about the lockers that were touched by exactly three students.

    a. Which lockers were touched by exactly three students? Were these lockers open or closed after Student 30's turn?

    b. Use what you know about factors and multiples to explain why the locker numbers you listed in Part a must be correct.

8. Use what you know about factors and multiples to explain why the lockers you listed in Question 2 were open after Student 30 took her turn.

9. **Challenge** Which lockers were touched the greatest number of times?

What Did You Learn?

10. Suppose the experiment is repeated with 100 lockers and 100 students.

    a. Which lockers will be open after Student 100 takes his turn?

    b. Which lockers will be touched by Student 15?

    c. Which students will touch Locker 91?

    d. Which lockers will be touched by both Student 9 and Student 30?

    e. Which students will touch Lockers 30, 50, and 100?

11. How many lockers would be open if the experiment were repeated with 200 lockers and 200 students? With 500 lockers and 500 students?

12. What is the fewest lockers and students needed for 31 lockers to be open at the end of the experiment?

On Your Own Exercises

Practice  **Apply**

1. One day, the ×18 machine at the stretching factory breaks down.

   a. What two-machine hookups could you use to replace this machine?

   b. What three-machine hookups could you use?

   c. What four-machine hookups could you use?

2. What hookups can be used to replace a ×37 machine?

3. List all the factors of 28. Then list all the factor pairs for 28.

4. List all the factors of 56. Then list all the factor pairs for 56.

In Exercises 5–7, describe how you could fill the order using only prime-number stretching machines.

5. Stretch a piece of rope to 12 times its original length.

6. Stretch a piece of twine to 21 times its original length.

7. Stretch a crayon to 72 times its original length.

In Exercises 8–10, use a factor tree to find the prime factorization of the given number. Use exponents in your answer when appropriate.

   8. 84   **9.** 64   **10.** 1,272

In Exercises 11–14, answer Parts a–c.

   a. What are the common factors of the two numbers?

   b. What is the GCF of the two numbers?

   c. Are the two numbers relatively prime?

11. 13 and 24

12. 50 and 75

13. 144 and 54

14. 24 and 35

 impactmath.com/self_check_quiz

15. Suppose you are playing *GCF* with your partner. Below is the board so far. Your partner has just circled 24. Which number would you circle? Why?

16. Suppose you are playing *GCF* with your partner. Below is the board so far. You circle the first number in this round. Which number should you circle? Why?

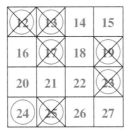

List five multiples of each number.

17. 14   **18.** 66   **19.** 25

20. Is 3,269,283 a multiple of 3? How do you know?

21. Is 1,000 a multiple of 3? How do you know?

22. List all the common multiples of 3 and 5 that are less than 60.

23. List all the common multiples of 7 and 14 that are less than 60.

24. List all the common multiples of 2, 3, and 15 that are less than 100.

25. Rosita earns money by walking dogs after school and on weekends. She walks Madeline every other day, Buddy every fourth day, and Ernie every fifth day. Today she walked all three dogs.

    a. How many days will it be before Rosita walks Madeline and Buddy on the same day again?

    b. How many days will it be before she walks Buddy and Ernie on the same day again?

    c. How many days will it be before she walks Madeline and Ernie on the same day again?

    d. How many days will it be before she walks all three dogs on the same day again?

    e. Which dogs will Rosita walk 62 days from today?

26. Geometry Simone wants to form a rectangle from 12 square tiles.

    a. Sketch every possible rectangle Simone could make.

    b. What are the factor pairs of 12? How are the factor pairs related to the rectangles you sketched in Part a?

    c. Simone wants to form a rectangle from 20 square tiles. Give the dimensions of each possible rectangle she could make, and tell how you know you have found all the possibilities.

27. Daryll said, "I am thinking of a number. The number is less than 50, and 2 and 4 are two of its nine factors. What is my number?"

28. Emma said, "I am thinking of a number. The number is less than 100, 19 is one of its four factors, and the sum of its digits is 12. What is my number?"

29. Jacinta said, "One of the factors of my number is 24. What other numbers must also be factors of my number?"

30. The *proper factors* of a number are all of its factors except the number itself. For example, the proper factors of 6 are 1, 2, and 3. A *perfect number* is a number whose proper factors add to that number. For example, 6 is a perfect number because $1 + 2 + 3 = 6$.

There is one perfect number between 20 and 30. What is it?

31. *Goldbach's conjecture* states that every even number except 2 can be written as the sum of two prime numbers. For example, $10 = 3 + 7$ and $24 = 19 + 5$. Although the conjecture appears to be true, no one has ever been able to prove it.

    a. Write each of the numbers 4, 36, and 100 as the sum of two prime numbers.

    b. Choose two even numbers of your own, and write each as the sum of two prime numbers.

Just the facts

The next greatest perfect number is 496. As of 1999, there were only 38 known perfect numbers—all of which are even. No one knows how many perfect numbers there are or whether there are any odd perfect numbers.

32. A *reversible prime* is a prime number whose digits can be reversed to form another prime number. For example, 13 is a reversible prime because reversing its digits gives 31, another prime number. There are nine reversible primes between 10 and 100. List them all.

33. *Twin primes* are prime numbers that differ by 2. For example, 11 and 13 are twin primes. Find three more pairs of twin primes.

34. History The Greek mathematician and astronomer Eratosthenes created a technique for finding prime numbers. The technique is known as the *Sieve of Eratosthenes* because nonprime numbers are sifted out, like sand through a sieve, leaving the prime numbers.

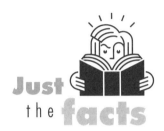
Just the facts

Using only a primitive tool and shadows to measure the angle of the sun's rays, Eratosthenes, who lived more than 2,200 years ago, was able to accurately calculate the circumference of Earth.

a. To create the Sieve of Eratosthenes, start with a grid of numbers from 1 to 100 and follow these steps:

1	2	3	4	5	6	7	8	9	10
11	12	13	14	15	16	17	18	19	20
21	22	23	24	25	26	27	28	29	30
31	32	33	34	35	36	37	38	39	40
41	42	43	44	45	46	47	48	49	50
51	52	53	54	55	56	57	58	59	60
61	62	63	64	65	66	67	68	69	70
71	72	73	74	75	76	77	78	79	80
81	82	83	84	85	86	87	88	89	90
91	92	93	94	95	96	97	98	99	100

Step 1: Cross out 1.

Step 2: Circle 2, and then cross out every multiple of 2 after 2 (that is, 4, 6, 8, 10, and so on).

Step 3: Circle 3—the next number that is not crossed out—and cross out every multiple of 3 (6, 9, 12, 15, and so on) after 3 that hasn't already been crossed out.

Step 4: Circle 5—the next number that is not crossed out—and cross out every multiple of 5 (10, 15, 20, 25, and so on) after 5 that hasn't already been crossed out.

Continue in this manner, circling the next available number and crossing out its multiples, until every number is either crossed out or circled. When you are finished, all the prime numbers between 1 and 100 will be circled.

b. Explain why this technique guarantees that only the prime numbers will be circled.

35. A kindergarten teacher passed out 30 large stickers and 90 small stickers to his students. Each child received the same number of large stickers and the same number of small stickers, and no stickers were left over.

    a. How many children could have been in the class? List all of the possibilities.

    b. How is the greatest number of children that could have been in the class related to the GCF of 30 and 90?

36. In Parts a–d, determine whether each pair of consecutive numbers is relatively prime.

    a. 7 and 8   **b.** 14 and 15   **c.** 20 and 21   **d.** 55 and 56

    e. Choose two more pairs of consecutive numbers, and determine whether they are relatively prime.

    f. Use evidence from Parts a–e to make a *conjecture*—an educated guess—about any pair of consecutive numbers.

37. **Social Studies** U.S. senators are elected for six-year terms, and U.S. presidents are elected for four-year terms. One of the Senate seats in Nebraska was up for election in the year 2000, the same year a president was elected.

    a. What is the next year the same Senate seat will be up for election *and* a presidential election will take place? (Assume the seat isn't vacated early.)

    b. List three other years when this will happen. What do these years have in common?

    c. When was the last year *before* 2000 that this Senate seat was up for election in the same year as a presidential election?

Remember

Two whole numbers are consecutive if their difference is 1.

Just the facts

The Constitution says that a senator must be at least 30 years old, have been a U.S. citizen for 9 years, and, when elected, be a resident of the state from which he or she is chosen.

38. Jing found that the LCM of 3 and 5 is 15 and that the LCM of 2 and 3 is 6. She concluded that she could find the LCM of any two numbers by multiplying them.

a. Copy and complete this table to show the product and the LCM of each pair of numbers.

Numbers	Product	LCM
2 and 3	6	6
2 and 4		
2 and 5		
3 and 4		
3 and 6		
6 and 9		
4 and 10		

b. Is the LCM of two numbers always equal to their product? If not, look for information in the table that will help you predict when the LCM of a pair of numbers *is* the product.

39. To find the LCM of two numbers, Althea first writes their prime factorizations. She then finds the product of the shortest string of primes that includes all primes from both factorizations. For example, she wrote the prime factorizations of 210 and 252.

$$210 = 2 \cdot 3 \cdot 5 \cdot 7 \qquad 252 = 2 \cdot 2 \cdot 3 \cdot 3 \cdot 7$$

The shortest string that includes all the primes is $2 \cdot 2 \cdot 3 \cdot 3 \cdot 5 \cdot 7$, so the LCM is $2 \cdot 2 \cdot 3 \cdot 3 \cdot 5 \cdot 7$, or 1,260.

Try Althea's method on at least two more pairs of numbers.

Mixed Review

Find each sum or difference without using a calculator.

40. $165.7 + 47.5$

41. $3.179 - 0.238$

42. $976,556 + 0.002$

43. $87.78 + 94.76$

44. $10.0101 + 1.101$

45. $9.02 - 7.34$

Geometry Tell what fraction of each figure is shaded.

46.

47.

Patterns in Fractions

You probably know quite a bit about fractions already. A fraction can be used to describe part of a whole or to name a number between two whole numbers.

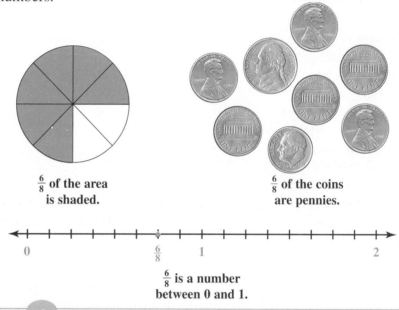

$\frac{6}{8}$ **of the area**
is shaded.

$\frac{6}{8}$ **of the coins**
are pennies.

$\frac{6}{8}$ **is a number**
between 0 and 1.

Think & Discuss

In each representation above—the circle, the collection of coins, and the number line—what does the 8 in $\frac{6}{8}$ represent? What does the 6 represent?

Trace around the circle, and shade in $\frac{3}{4}$ of its area. How does the area you shaded compare to the area shaded above? Explain why your answer makes sense.

Trace the number line, and indicate where $\frac{12}{16}$ is located. How does the location of $\frac{12}{16}$ compare with the location of $\frac{6}{8}$? Explain why your answer makes sense.

In this lesson, you will see how the factor and multiple relationships you studied in Lesson 2.1 can help you think about and work with fractions.

Investigation 1 Visualizing Fractions

Just the facts

The oldest known piece of pottery was made in China in about 7900 B.C. The potter's wheel was invented in China in about 3100 B.C.

In this investigation, you will review some ideas about fractions.

Problem Set A

The students in Mr. Jacob's art class are sitting in four groups. Mr. Jacobs gives each group some bricks of clay to share equally among its members. All the bricks are the same size.

- Group 1 has 5 students and receives 4 bricks.

- Group 2 has 6 students and receives 4 bricks.

- Group 3 has 12 students and receives 9 bricks.

- Group 4 has 5 students and receives 6 bricks.

1. For each group, determine what fraction of a brick each student will get. Explain how you found your answers.

2. Did Mr. Jacobs pass out the clay fairly? Explain your answer.

You should have found that the fraction of clay each member of Group 4 received was greater than 1. The next Example shows how Miguel, Luke, and Hannah thought about dividing six bricks of clay among five students.

EXAMPLE

First I gave each student one brick. I divided the extra brick into fifths & gave $\frac{1}{5}$ to each student. So, each student got $1\frac{1}{5}$ bricks.

I drew 6 bricks and divided each into fifths. Each student got one of the fifths from each brick for a total of $\frac{6}{5}$.

I solved the division problem $6 \div 5$ and found that each student receives $1\frac{1}{5}$ bricks.

The Example on page 97 shows two ways of expressing a fraction greater than 1. Luke's answer, $\frac{6}{5}$, is a fraction in which the numerator is greater than the denominator. Miguel and Hannah's answer, $1\frac{1}{5}$, is a **mixed number**—a whole number and a fraction.

Problem Set B

In Problems 1–4, give your answer as a mixed number and as a fraction.

1. If 12 bricks of clay are divided among 5 students, what portion of a brick will each student receive?

2. Mr. Dante's geese laid 18 eggs. What fraction of a dozen is this?

3. Each grid below has 100 squares. What fraction of a grid is the entire shaded portion?

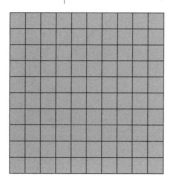 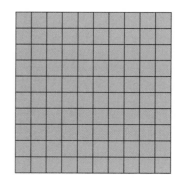 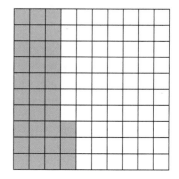

4. What number is indicated by the point?

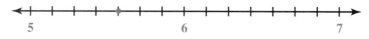

Share & Summarize

Write a word problem, like those in Problems 1 and 2 of Problem Set B, that leads to a fraction greater than 1. Show two ways of expressing the fraction.

Investigation Equivalent Fractions

In this investigation, you will see how different fractions can represent the same part of a whole and how different fractions can represent the same number.

Problem Set C

Caroline, Conor, Miguel, Rosita, and Jahmal baked fruit bars in their home-economics class. Each student cut his or her bar into a different number of equal pieces.

1. Caroline wants to trade some of her lemon bar for an equal portion of Rosita's raspberry bar. She could trade $\frac{1}{3}$ of her bar for $\frac{2}{6}$ of Rosita's. What other fair trades could they make? List all the possibilities. Give your answers as fractions of fruit bars.

2. Conor wants to trade some of his apple bar for an equal portion of Jahmal's peach bar. Describe all the fair trades they could make.

3. Describe all the fair trades Miguel and Jahmal could make.

4. Which pairs of students can trade only whole fruit bars?

5. List some fractions of a fruit bar that are fair trades for $\frac{1}{2}$ of a bar.

In Problem Set C, $\frac{1}{5}$ of Miguel's fruit bar is the same as $\frac{2}{10}$ of Jahmal's fruit bar. Fractions such as $\frac{1}{5}$ and $\frac{2}{10}$ *describe the same portion of a whole* or *name the same number.* Such fractions are called **equivalent fractions.**

You can find a fraction equivalent to a given fraction by multiplying or dividing the numerator and denominator by the same number. Althea worked out an example to convince herself that dividing by the same number gives an equivalent fraction.

I can show $\frac{6}{9}$ by dividing a rectangle into 9 pieces and shading 6 of them.

$\frac{6}{9}$

If I divide both parts of $\frac{6}{9}$ by 3, I get $\frac{2}{3}$. In my picture, this is like turning every three pieces into one big piece. I can see that $\frac{6}{9} = \frac{2}{3}$!

$\frac{6}{9} = \frac{2}{3}$

So dividing both parts of a fraction by a number is like grouping the pieces to make bigger pieces.

Since the shaded amount doesn't change, the fractions are equivalent.

Think & Discuss

Use an argument similar to Althea's to convince yourself that multiplying the numerator and denominator of a fraction by a number gives an equivalent fraction.

A fraction is in **lowest terms** if its numerator and denominator are relatively prime. For example, the fractions $\frac{2}{3}$, $\frac{12}{18}$, and $\frac{20}{30}$ are all equivalent. However, only $\frac{2}{3}$ is in lowest terms, because the only common factor of 2 and 3 is 1.

Remember

Two numbers are relatively prime if their only common factor is 1.

Just the facts

The sizes of wrenches and drill bits are often given as fractions.

Problem Set D

1. Fractions can be grouped into "families" of equivalent fractions. A fraction family is named for the member that is in lowest terms. Here is part of the "$\frac{3}{4}$ fraction family":

$$\frac{3}{4} \quad \frac{6}{8} \quad \frac{9}{12} \quad \frac{12}{16} \quad \frac{15}{20} \quad \frac{18}{24} \quad \frac{21}{28} \quad \frac{24}{32} \quad \frac{27}{36} \quad \frac{30}{40}$$

   a. What do the numerators of these fractions have in common? What do the denominators have in common?

   b. How do you know that all the fractions in this family are equivalent?

   c. Find at least three more fractions in this family, and explain how you found them.

   d. Is $\frac{164}{216}$ in this fraction family? Explain how you know.

2. Now consider the $\frac{3}{8}$ fraction family.

   a. List four members of this family with numerators greater than 15.

   b. List four members of this family with numerators less than 15.

3. The fractions below belong to the same family.

$$\frac{15}{21} \quad \frac{20}{28} \quad \frac{25}{35}$$

   a. Find four more fractions in this family, two with denominators less than 21 and two with denominators greater than 35.

   b. What is the name of this fraction family?

4. How can you determine what family a fraction belongs to?

5. Are $\frac{6}{10}$ and $\frac{16}{20}$ in the same fraction family? Explain how you know.

Share & Summarize

1. In Problem Set C, how did you determine which trades could be made? Give an example to help explain your answer.

2. Explain how you can find fractions equivalent to a given fraction. Demonstrate your method by choosing a fraction and finding four fractions equivalent to it.

3. Describe a method for determining whether two given fractions are equivalent.

Investigation ▶3 Comparing Fractions

In the last investigation, you explored families of equivalent fractions. You saw how you could find fractions equivalent to a given fraction by multiplying or dividing the numerator and denominator by the same number. You will now use what you learned to compare fractions.

Problem Set E

1. Here are some members of the $\frac{1}{7}$ and the $\frac{2}{11}$ fraction families:

The $\frac{1}{7}$ Fraction Family

$$\frac{1}{7} \quad \frac{2}{14} \quad \frac{3}{21} \quad \frac{4}{28} \quad \frac{5}{35} \quad \frac{6}{42} \quad \frac{7}{49} \quad \frac{8}{56} \quad \frac{9}{63} \quad \frac{10}{70} \quad \frac{11}{77} \quad \frac{12}{84}$$

The $\frac{2}{11}$ Fraction Family

$$\frac{2}{11} \quad \frac{4}{22} \quad \frac{6}{33} \quad \frac{8}{44} \quad \frac{10}{55} \quad \frac{12}{66} \quad \frac{14}{77} \quad \frac{16}{88} \quad \frac{18}{99} \quad \frac{20}{110}$$

a. Recall that all the fractions in the $\frac{1}{2}$ fraction family equal $\frac{1}{2}$. Choose a pair of fractions, one from each family above, that you could use to easily compare $\frac{1}{7}$ and $\frac{2}{11}$.

b. Which fraction is greater, $\frac{1}{7}$ or $\frac{2}{11}$? Explain how you know.

2. Consider the fractions $\frac{3}{4}$ and $\frac{7}{12}$.

a. List some members of their fraction families.

b. List two pairs of fractions you could use to compare $\frac{3}{4}$ and $\frac{7}{12}$.

c. Which fraction is greater, $\frac{3}{4}$ or $\frac{7}{12}$?

In Problem Set E, you probably found that you can compare two fractions by finding members of their fraction families with a *common denominator* or with a *common numerator*.

The diagram on the left shows how fractions with a common denominator of 10 compare. The diagram on the right shows how fractions with a common numerator of 7 compare.

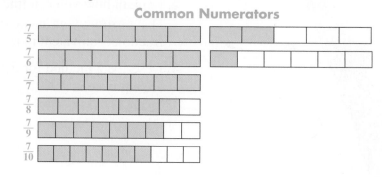

Think & Discuss

If two fractions have the same denominator but different numerators, how can you tell which of the fractions is greater? Explain why your method works.

If two fractions have the same numerator but different denominators, how can you tell which of the fractions is greater? Explain why your method works.

When you are given two fractions to compare, how can you quickly find equivalent fractions with a common denominator? With a common numerator?

Problem Set F

1. Rewrite $\frac{4}{17}$ and $\frac{3}{10}$ with a common denominator, and then tell which fraction is greater.

2. Rewrite $\frac{4}{9}$ and $\frac{8}{15}$ with a common numerator, and then tell which fraction is greater.

3. Consider the fractions $\frac{5}{8}$ and $\frac{7}{10}$.

   a. Rewrite the fractions with the *least* common denominator.

   b. Which fraction is greater, $\frac{5}{8}$ or $\frac{7}{10}$?

   c. What is the relationship between the least common denominator and the multiples of the original denominators, 8 and 10?

Replace each ● with $<$, $>$, or $=$ to make a true statement.

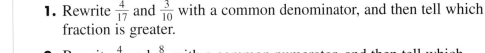

4. $\frac{3}{4}$ ● $\frac{7}{12}$ **5.** $\frac{8}{13}$ ● $\frac{12}{19}$ **6.** $\frac{48}{120}$ ● $\frac{12}{39}$

7. $\frac{17}{11}$ ● $\frac{11}{7}$ **8.** $\frac{13}{12}$ ● $\frac{6}{5}$ **9.** $\frac{19}{36}$ ● $\frac{10}{24}$

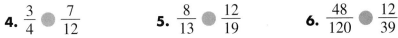

Remember

- $>$ means "is greater than"
- $<$ means "is less than"
- These statements mean the same thing:
  $$3 < 4 \qquad 4 > 3$$

Share & Summarize

Order these fractions from least to greatest using any method you like.

$$\frac{1}{3} \qquad \frac{2}{9} \qquad \frac{5}{3} \qquad \frac{5}{4} \qquad \frac{7}{9} \qquad \frac{1}{2} \qquad 1\frac{1}{8}$$

Investigation 4 ▶ Estimating with Fractions

In this lesson, you will see how you can use familiar fractions to estimate the values of other fractions.

Think & Discuss

Keisha wants to make $\frac{1}{3}$ of a recipe of her grandmother's spaghetti sauce. When she divided the amount of each ingredient by 3, she found that she needed $\frac{4}{9}$ of a cup of olive oil. Keisha has only the measuring cups at right. How can she use these cups to measure *approximately* $\frac{4}{9}$ a cup of oil?

Most people are familiar with such fractions as $\frac{1}{4}$, $\frac{1}{3}$, $\frac{1}{2}$, $\frac{2}{3}$, and $\frac{3}{4}$, and have a good sense of their value. For this reason, familiar fractions like these, along with the numbers 0 and 1, are often used as benchmarks. *Benchmarks,* or reference points, can help you approximate the values of other fractions.

When you estimate with benchmark fractions, you should ask yourself questions like these:

- Is the fraction closest to 0, $\frac{1}{2}$, or 1?
- Is it greater than $\frac{1}{2}$ or less than $\frac{1}{2}$?
- Is it greater than $\frac{1}{4}$ or less than $\frac{1}{4}$?
- Is it greater than $\frac{2}{3}$ or less than $\frac{2}{3}$?

Problem Set G

Hannah and Miguel were working in their school's computer lab. They began installing the same software program onto their computers at the same time. Each computer displayed a progress bar indicating how much of the program had been installed. Here's what the progress bars looked like after 1 minute:

Miguel's Bar

Hannah's Bar

1. After 1 minute, about what fraction of the program had been installed on Miguel's computer? Explain how you made your estimate.

2. After 1 minute, about what fraction of the program had been installed on Hannah's computer? Explain how you made your estimate.

3. Hannah noticed that, after 1 minute, the shaded parts of the progress bars were about the same length. She said, "Our progress bars are the same length. Your computer has completed just as much of the installation as my computer." Is she correct? Explain.

4. Miguel said, "Your computer is only $\frac{1}{2}$ as fast as my computer." Is he right? Explain your reasoning.

Fractions involving real data can be messy. Using familiar fractions to approximate actual fractions often makes information easier to understand.

Think & Discuss

In a sixth grade gym class, 28 out of the 40 students are girls. The gym teacher said, "Girls make up about $\frac{3}{4}$ of this class." Do you agree with this statement? Explain.

Problem Set H

1. Rewrite each statement using a more familiar fraction to approximate the actual fraction. Explain how you decided which fraction to use, and tell whether your approximation is a little greater than or a little less than the actual fraction.

  a. I have been in school for $\frac{43}{180}$ of the school year.

  b. Mrs. Stratton's class is $\frac{48}{60}$ of an hour long.

  c. The air distance from Washington, DC, to Los Angeles is $\frac{2,300}{4,870}$ the air distance from Washington, DC, to Moscow.

  d. The Volga River in Europe is $\frac{2,290}{3,362}$ the length of the Ob-Irtysh River in Asia.

2. Make up your own situation involving a "messy" fraction. Tell what benchmark fraction you could use as an approximation.

Share & Summarize

In what types of situations is it useful to approximate the value of a fraction with a more familiar fraction?

On Your Own Exercises

Practice  **Apply**

1. Suppose 10 friends share 4 medium-sized pizzas so that each friend gets the same amount. What fraction of a pizza does each friend receive?

2. Suppose 6 people share 8 sticks of gum so that each person gets the same amount. What fraction of a stick does each person receive? Express your answer as a fraction and as a mixed number.

3. Suppose 10 people share 15 sticks of gum so that each person gets the same amount. What fraction of a stick does each person receive? Express your answer as a fraction and as a mixed number.

4. What number is indicated by the point? Give your answer as a mixed number and as a fraction.

5. A roll of pennies holds 50 pennies. What fraction of a roll is 237 pennies? Give your answer as a mixed number and as a fraction.

6. Althea baked a plum bar and cut it into ninths. What fair trades can she make with Rosita, whose raspberry bar is divided into sixths?

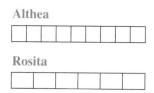

Althea

Rosita

7. Alicia's kiwifruit bar is divided into 10ths, and Rob's strawberry bar is divided into 15ths. List all the fair trades they could make.

Give two fractions that are equivalent to each given fraction.

8. $\frac{2}{3}$ 9. $\frac{5}{8}$ 10. $\frac{1}{5}$

11. Here are the four members of the $\frac{5}{9}$ fraction family. List four more.

$$\frac{5}{9} \qquad \frac{10}{18} \qquad \frac{15}{27} \qquad \frac{20}{36}$$

12. All of these fractions are in the same family:

$$\frac{33}{27} \qquad \frac{44}{36} \qquad \frac{66}{54}$$

   a. Find four more fractions in this family.

   b. What is the name of this fraction family?

impactmath.com/self_check_quiz

13. Consider the fraction $\frac{27}{36}$.

    a. To what fraction family does $\frac{27}{36}$ belong?

    b. List all the members of this fraction family with numerators less than 27.

14. Are $\frac{34}{64}$ and $\frac{18}{36}$ in the same fraction family? Explain how you know.

Rewrite each fraction or mixed number in lowest terms.

15. $\frac{12}{3}$ **16.** $\frac{9}{24}$ **17.** $5\frac{6}{9}$ **18.** $\frac{18}{45}$

Tell whether the fractions in each pair are equivalent, and explain how you know.

19. $\frac{4}{8}$ and $\frac{15}{30}$ **20.** $\frac{4}{12}$ and $\frac{8}{32}$ **21.** $\frac{50}{60}$ and $\frac{15}{18}$

Replace each ⬤ with $<$, $>$, or $=$ to make a true statement.

22. $\frac{7}{8}$ ⬤ $\frac{2}{3}$ **23.** $\frac{5}{9}$ ⬤ $\frac{3}{5}$ **24.** $\frac{5}{16}$ ⬤ $\frac{5}{17}$

25. $\frac{90}{70}$ ⬤ $\frac{45}{35}$ **26.** $\frac{1}{2}$ ⬤ $\frac{7}{11}$ **27.** $\frac{13}{8}$ ⬤ $1\frac{2}{3}$

Order each set of fractions from least to greatest.

28. $\dfrac{3}{4}, \dfrac{3}{3}, \dfrac{3}{8}, \dfrac{3}{5}, \dfrac{3}{16}, \dfrac{3}{7}, \dfrac{3}{1}$ **29.** $\dfrac{7}{7}, \dfrac{3}{4}, \dfrac{1}{2}, \dfrac{2}{5}, \dfrac{2}{3}, \dfrac{11}{8}, \dfrac{1}{3}$

30. Three computers begin installing the same program at the same time. Here are their progress bars after 1 minute:

Computer A   Computer B   Computer C

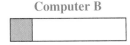

    a. About what fraction of the installation has been completed by Computer A? By Computer B? By Computer C?

    b. Order the machines from fastest to slowest. Explain how you determined the ordering.

Tell whether each fraction is closest to 0, $\frac{1}{4}$, $\frac{1}{3}$, $\frac{1}{2}$, $\frac{2}{3}$, $\frac{3}{4}$, or 1. Explain how you decided.

31. $\dfrac{5}{18}$ **32.** $\dfrac{1}{5}$ **33.** $\dfrac{9}{10}$

34. Determine whether $\frac{33}{40}$ is greater than or less than $\frac{21}{50}$ by comparing the fractions to benchmark fractions. Explain your thinking.

35. **Measurement** Five segments are shown above the yardstick. Give the length of each segment in feet. Give your answers as fractions or mixed numbers in lowest terms.

E ────────────────────────────────

D ────────────────────────

C ──────────────────────

B ──────────────

A ──────

|᠊᠊᠊᠊|᠊᠊᠊᠊|᠊᠊᠊᠊|᠊᠊᠊᠊|᠊᠊᠊᠊|᠊᠊᠊᠊|᠊᠊᠊᠊|᠊᠊᠊᠊|᠊᠊᠊᠊|᠊᠊᠊᠊|᠊᠊᠊᠊|᠊᠊᠊᠊|᠊᠊᠊᠊|᠊᠊᠊᠊|᠊᠊᠊᠊|᠊᠊᠊᠊|᠊᠊᠊᠊|
2   4   6   8   10   12   14   16   18   20   22   24   26   28   30   32   34

36. People often use mixed numbers to compare two quantities or to describe how much something has changed or grown.

a. Dion's height is about $1\frac{1}{2}$ times his younger brother Jamil's height. Jamil is about 40 inches tall. How tall is Dion?

b. Bobbi spends 40 minutes each night practicing her violin. She said, "That's $1\frac{1}{3}$ times the amount of time I spent last year." How much time did Bobbi practice each night last year?

c. The 1998 population of Seattle, Washington, was about $6\frac{3}{4}$ times the 1900 population. Seattle's 1900 population was about 80,000. Estimate Seattle's population in 1998.

37. Imagine that you have baked a delicious mango-papaya bar the same size as the bars baked by the students in Problem Set C.

Caroline   Miguel   Your Bar

Rosita   Jahmal

a. You want to trade portions of your bar with Jahmal and Rosita and still have some left for yourself. Tell how many equal-sized pieces you would cut your bar into. Describe the trade you would make with each student, and tell what fraction of your bar you would have left.

b. Give the same information for trading with Caroline and Rosita instead.

c. Give the same information for trading with Caroline and Miguel instead.

In your **own words**

Explain how factors and multiples can be used to find members of a fraction family.

38. Jing baked a coconut bar and would like to trade with all three students below.

Caroline

Conor

Rosita

a. Into how many equal-sized pieces should Jing divide her bar?

b. List the trades Jing could make, and tell how much of the bar she would have left for herself.

39. Prove It! Write a convincing argument to show that $\frac{3}{4}$ of a fruit bar is not a fair trade for $\frac{3}{5}$ of a fruit bar.

40. Biology Water makes up about $\frac{2}{3}$ of a person's body weight.

a. A student weighs 90 pounds. Determine how many pounds of the student's weight are attributed to water by finding a fraction equivalent to $\frac{2}{3}$ with a denominator of 90.

b. A student weighs 75 pounds. Determine how many pounds of the student's weight are attributed to water.

41. Preview Percent means "out of 100." You can think of a percent as the numerator of a fraction with a denominator of 100. For example, 25% means $\frac{25}{100}$. You can change a fraction to a percent by first finding an equivalent fraction with a denominator of 100.

a. Change the following fractions to percents: $\frac{3}{4}, \frac{1}{5}, \frac{20}{50}, \frac{8}{25}$.

b. In Althea's homeroom, 14 of the 20 students ride the bus to school. What percent of the students take the bus?

c. Of the 500 people in the audience at the school play, 350 bought their tickets in advance. What percent of the audience bought tickets in advance?

42. Measurement Between which two twelfths of a foot will you find each measurement? For example, $\frac{1}{8}$ is between $\frac{1}{12}$ and $\frac{2}{12}$.

a. $\frac{3}{5}$ of a foot

b. $\frac{1}{10}$ of a foot

c. $\frac{5}{8}$ of a foot

43. Of the 560 students at Roosevelt Middle School, 240 participate in after-school sports. Of the 720 students at King Middle School, 300 participate in after-school sports.

    a. In which school does the greater *number* of students participate in sports?

    b. In which school does the greater *fraction* of students participate in sports?

44. **Statistics** In 1995, about 58,300,000 people lived in the West region of the United States. About 16,200,000 of these people were under 18 years of age. At the same time, about 23,900,000 of the 91,700,000 people living in the South region were under 18 years of age. Which region had the greater fraction of children and teenagers?

45. **Statistics** A survey asked all the sixth graders at Belmont Middle School how much time they spent on homework each week. Here are the results.

Time Spent Doing Homework	Fraction of Class
0 to 1 hours	$\frac{0}{100}$
1 to 2 hours	$\frac{1}{100}$
2 to 3 hours	$\frac{12}{100}$
3 to 4 hours	$\frac{3}{100}$
4 to 5 hours	$\frac{22}{100}$
5 to 6 hours	$\frac{57}{100}$
more than 6 hours	$\frac{5}{100}$

    a. Malik said, "About half of the students in the sixth grade spend 5 to 6 hours each week doing homework." Do you agree with this statement? Explain why or why not.

    b. Tamika said, "Hardly anyone spends 4 to 5 hours each week on homework." Do you agree with this statement? Explain why or why not.

    c. About what fraction of the students spend 2 to 3 hours each week doing homework?

46. Challenge This table shows the populations of the five most heavily populated countries in 1998.

1998 Population

Country	Population
China	1,256 million
India	982 million
United States	271 million
Indonesia	206 million
Brazil	166 million

a. The world population in 1998 was about 5,926 million people. Which of these countries had about $\frac{1}{5}$ of the world population?

b. About what fraction of the total 1998 world population lived in the United States?

Mixed Review

47. Number Sense In the number 35,217, which digit is in the thousands place?

48. In the number 73.412, which digit is in the tenths place?

49. In the number 892,341.7, which digit is in the tens place?

50. Write 9,322 in words.   **51.** Write 10,010,010 in words.

Measurement In Exercises 52–55, tell which quantity is greater.

52. 0.44 meter or $\frac{1}{2}$ meter   **53.** 75% of an hour or 35 minutes

54. $6\frac{1}{4}$ feet or 76 inches   **55.** 14 nickels or $.68

56. Consider this pattern of shapes:

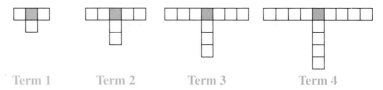

Term 1   Term 2   Term 3   Term 4

a. Copy and complete the table to show the number of squares in each term.

Term	1	2	3	4
Squares	4			

b. Predict the number of squares in Term 5 and in Term 6. Check your answers by drawing these terms.

c. Write a rule for finding the number of squares for any term number. Explain how you know your rule works.

2.3

Patterns in Decimals

You encounter decimals every day. Prices displayed in stores and advertisements, statistics in the sports section of the newspaper, and call numbers on books in the library are often given as decimals. In this lesson, you will review the meaning of decimals, and you will practice working with decimals in a variety of situations.

M A T E R I A L S

- *Spare Change cards (1 set per group)*
- *dollar charts (1 per player)*

Explore

Read the rules of the *Spare Change* game, and then play two rounds with your group.

Spare Change Game Rules

- Place four *Spare Change* cards face up on the table. Place the rest of the deck face down in a pile.

- To take a turn, a player chooses one of the four cards showing and shades that fraction of a dollar on his or her dollar chart. The player then places the card on the bottom of the deck, and replaces it with the top card from the deck.

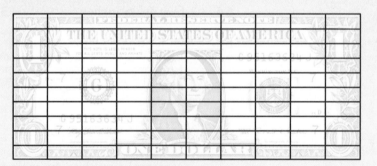

- Play continues until one player has shaded the entire card or is unable to shade any of the four amounts showing.

- The player who has shaded the amount closest to a dollar at the end of the game is the winner.

Describe some strategies you used while playing the game.

Discuss what you learned about decimals.

Investigation Understanding Decimals

Just the facts

The prefix "deci" comes from the Latin word *decem*, meaning "ten."

Decimals are equivalent to fractions whose denominators are 10, 100, 1,000, 10,000, and so on. Each decimal place has a name based on the fraction it represents.

Decimal	Equivalent Fraction	In Words
0.1	$\frac{1}{10}$	one tenth
0.01	$\frac{1}{100}$	one hundredth
0.001	$\frac{1}{1,000}$	one thousandth
0.0001	$\frac{1}{10,000}$	one ten-thousandth

EXAMPLE

How is 9.057 different from 9.57?

These decimals look similar, but they represent different numbers. You can see this by looking at the place values of the digits.

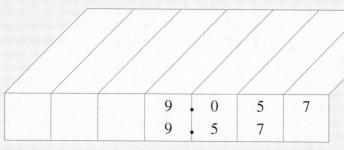

9.057 means $9 + \frac{0}{10} + \frac{5}{100} + \frac{7}{1,000}$, or $9\frac{57}{1,000}$, or $\frac{9,057}{1,000}$.

9.57 means $9 + \frac{5}{10} + \frac{7}{100}$, or $9\frac{57}{100}$, or $\frac{957}{100}$.

The number 9.057 is read "nine and fifty-seven thousandths." The number 9.57 is read "nine and fifty-seven hundredths."

In Problem Set A, you will see how the ideas discussed above relate to the *Spare Change* game.

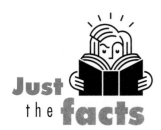

Just the facts

In many countries, a decimal comma is used instead of a decimal point, and a space is used to separate groups of three digits.

Problem Set A

1. Consider the values $0.3 and $0.03.

   a. Are these values the same?

   b. Explain your answer to Part a by writing both amounts as fractions.

   c. Illustrate your answer to Part a by shading both amounts on a dollar chart.

2. Do $0.3 and $0.30 represent the same value? Explain your answer by writing both amounts as fractions and by shading both amounts on a dollar chart.

3. In her first four turns of the *Spare Change* game, Rosita chose $0.45, $0.1, $0.33, and $0.05.

   a. Complete a dollar chart showing the amount Rosita should have shaded after the first four turns.

   b. What part of a dollar is shaded on Rosita's chart? Express your answer as a fraction and as a decimal.

   c. How much more does Rosita need to have $1.00? Express your answer as a decimal and as a fraction.

4. How could you shade a dollar chart to represent $0.125?

Now you will explore how multiplying or dividing a number by 10, 100, 1,000, and so on changes the position of the decimal point.

Problem Set B

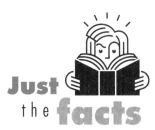
1. Copy this table:

Calculation		Result
81.07	$= 81.07 \cdot 1$	81.07
$81.07 \cdot 10$	$= 81.07 \cdot 10$	
$81.07 \cdot 10 \cdot 10$	$= 81.07 \cdot 100$	
$81.07 \cdot 10 \cdot 10 \cdot 10$	$= 81.07 \cdot 1,000$	
$81.07 \cdot 10 \cdot 10 \cdot 10 \cdot 10$	$= 81.07 \cdot 10,000$	
$81.07 \cdot 10 \cdot 10 \cdot 10 \cdot 10 \cdot 10 = 81.07 \cdot 100,000$		

a. Enter the number 81.07 on your calculator. Multiply it by 10, and record the result in the second row of the table.

b. Find $81.07 \cdot 100$ by multiplying your result from Part a by 10. Record the result in the table.

c. Continue to multiply each result by 10 to find $81.07 \cdot 1,000$; $81.07 \cdot 10,000$; and $81.07 \cdot 100,000$. Record your results.

d. Describe how the position of the decimal point changed each time you multiplied by 10.

2. In Parts a–c, predict the value of each product without doing any calculations. Check your prediction by using your calculator.

a. $7.801 \cdot 10,000$ **b.** $0.003 \cdot 100$ **c.** $9,832 \cdot 1,000$

d. When you predicted the results of Parts a–c, how did you determine where to put the decimal point?

3. Think about how the value of a number changes as you move the decimal point to the right.

a. How does the value of a number change when you move the decimal point one place to the right? Two places to the right? Three places to the right? (Hint: Look at your completed table from Problem 1, or test a few numbers to see what happens.)

b. Challenge In general, what is the relationship between the number of places a decimal is moved to the right and the change in the value of the number?

4. Tell what number you must multiply the given number by to get 240. Explain how you found your answer.

a. 2.4   **b.** 0.24   **c.** 0.00024

Problem Set C

1. Copy this table:

Calculation		Result
81.07		81.07
81.07 ÷ 10	$= \frac{1}{10}$ of 81.07	
81.07 ÷ 10 ÷ 10	$= \frac{1}{100}$ of 81.07	
81.07 ÷ 10 ÷ 10 ÷ 10	$= \frac{1}{1,000}$ of 81.07	
81.07 ÷ 10 ÷ 10 ÷ 10 ÷ 10	$= \frac{1}{10,000}$ of 81.07	
81.07 ÷ 10 ÷ 10 ÷ 10 ÷ 10 ÷ 10 $= \frac{1}{100,000}$ of 81.07		

a. Find $\frac{1}{10}$ of 81.07 by entering 81.07 on your calculator and dividing by 10. Record the result in the second row of the table.

b. Find $\frac{1}{100}$ of 81.07 by dividing your result from Part a by 10. Record the result in the table.

c. Continue to divide each result by 10 to find $\frac{1}{1,000}$ of 81.07; $\frac{1}{10,000}$ of 81.07; and $\frac{1}{100,000}$ of 81.07. Record your results.

d. Describe how the position of the decimal point changed each time you divided by 10 (that is, each time you found $\frac{1}{10}$).

2. In Parts a–c, predict each result without doing any calculations. Check your prediction by using your calculator.

a. $\frac{1}{10,000}$ of 14.14   **b.** 34,372 ÷ 100   **c.** $\frac{1}{1,000}$ of 877

d. When you predicted the results of Parts a–c, how did you determine where to put the decimal point?

3. Think about how the value of a number changes as you move the decimal point to the left.

a. How does the value of a number change when you move the decimal point one place to the left? Two places to the left? Three places to the left? (Hint: Look at your completed table from Problem 1, or test a few numbers to see what happens.)

b. Challenge In general, what is the relationship between the number of places a decimal is moved to the left and the change in the value of the number?

4. Tell what number you must divide the given number by to get 1.8. Explain how you found your answer.

a. 18   **b.** 180   **c.** 18,000

Share & Summarize

1. A shirt is on sale for $16.80. Write $16.80 as a mixed number.

2. A big-screen television costs 100 times as much as the shirt. How much does the TV cost?

3. A fresh-cooked pretzel costs $\frac{1}{10}$ as much as the shirt. How much is the pretzel?

Investigation 2 — Measuring with Decimals

In the metric system, units of measure are based on the number 10. This makes converting from one unit to another as easy as moving a decimal point!

The basic unit of length in the metric system is the meter. Each meter can be divided into 100 centimeters.

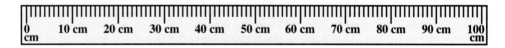

Each centimeter can be divided into 10 millimeters.

Think & Discuss

Fill in the blanks. Give your answers as both decimals and fractions.

1 cm = ___ m   1 mm = ___ cm   1 mm = ___ m

5 cm = ___ m   15 mm = ___ cm   15 mm = ___ m

Remember

The abbreviation for meter is m.
The abbreviation for centimeter is cm.
The abbreviation for millimeter is mm.

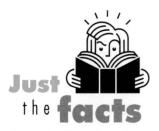

Just the facts

Since 1983, the meter has been defined as the distance light travels in a vacuum in $\frac{1}{299,792,458}$ of a second.

Problem Set D

Convert each measurement to meters. Write your answers as fractions and as decimals.

1. 35 cm **2.** 9 mm **3.** 23 mm

4. Give the lengths of Segments A and B in meters. Express your answers as fractions and as decimals.

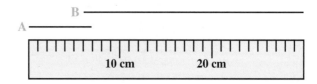

For Problems 5 and 6, tape four sheets of paper together lengthwise. Tape a meterstick on top of the paper as shown, so you can draw objects above and below the meterstick.

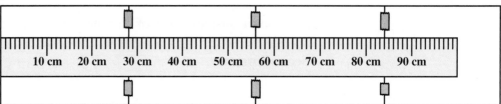

5. Collect objects whose lengths you can measure, such as pencils, books, staplers, and screwdrivers.

  a. Place the objects end to end along your meterstick until the combined length is as close to 1 meter as possible. Sketch the objects *above* the meterstick at their actual lengths.

  b. Find the length of each object to the nearest millimeter. Label the sketch of each object with its length in millimeters, centimeters, and meters.

6. In this problem, you will try to find the combination of the following objects with a length as close to 1 meter as possible.

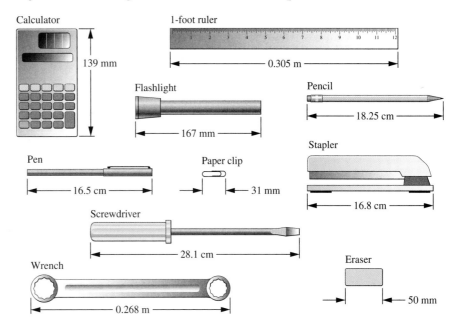

Calculator — 139 mm

1-foot ruler — 0.305 m

Flashlight — 167 mm

Pencil — 18.25 cm

Pen — 16.5 cm

Paper clip — 31 mm

Stapler — 16.8 cm

Screwdriver — 28.1 cm

Wrench — 0.268 m

Eraser — 50 mm

a. Choose one of the objects. Below your meterstick, begin at 0 and sketch the object at its actual length.

b. Choose a second object. Starting at the right end of the previous drawing, sketch the second object at its actual length.

c. Continue to choose objects and sketch them until the total length is as close to 1 meter as possible.

d. How much of a meter is left over? Express your answer as a decimal and as a fraction.

e. Which object is longest? Which object is shortest? Explain how you found your answers.

Share & Summarize

Suppose you are given a measurement in centimeters.

1. How would you move the decimal point to change the measurement to meters? Explain why this technique works.

2. How would you move the decimal point to change the measurement to millimeters? Explain why this technique works.

Investigation 3   Comparing and Ordering Decimals

You will now play a game that will give you practice finding decimals between other decimals.

Problem Set E

Read the rules for *Guess My Number,* and then play four rounds with your partner, switching roles for each round.

Guess My Number Game Rules

- Player 1 thinks of a number between 0 and 10 with no more than four decimal places and writes it down so that Player 2 cannot see it.

- Player 2 asks "yes" or "no" questions to try to figure out the number. Player 2 writes down each question and answer on a record sheet like this one:

Question	Answer	What I Know about the Number
Is the number greater than 6?	no	It is less than 6.
Is the number less than 3?	no	It is between 3 and 6.

- Play continues until Player 2 guesses the number. Player 1 receives 1 point for each question Player 2 asked.

- The winner is the player with the most points after four rounds.

1. What strategies did you find helpful when you asked questions?

2. Jahmal and Caroline are playing *Guess My Number.* Caroline is the Asker. Here is her record sheet so far:

Question	Answer
Is it more than 1?	yes
Is it between 4 and 10?	yes
Is it between 5 and 10?	yes
Is it more than 8?	no
Is it between 7 and 8?	yes

   a. What do you think of Caroline's questions? What, if anything, do you think she should have done differently?

   b. What question do you think Caroline should ask next?

3. Suppose you have asked several questions and you know the number is between 4.71 and 4.72. List at least four possibilities for the number.

4. What is the greatest decimal that can be made in this game? What is the least decimal that can be made in this game?

One reason decimals are so useful for reporting measurements is because they are easy to compare.

Problem Set F

1. This table shows the winning times for the women's 100-meter run for Summer Olympic Games from 1928 to 2000.

Year	Winner	Time (seconds)
1928	Elizabeth Robinson, United States	12.2
1932	Stella Walsh, Poland	11.9
1936	Helen Stephens, United States	11.5
1948	Francina Blankers-Koen, Netherlands	11.9
1952	Majorie Jackson, Australia	11.5
1956	Betty Cuthbert, Australia	11.5
1960	Wilma Rudolph, United States	11.0
1964	Wyomia Tyus, United States	11.4
1968	Wyomia Tyus, United States	11.0
1972	Renate Stecher, East Germany	11.07
1976	Annegret Richter, West Germany	11.08
1980	Lyudmila Kondratyeva, USSR	11.60
1984	Evelyn Ashford, United States	10.97
1988	Florence Griffith-Joyner, United States	10.54
1992	Gail Devers, United States	10.82
1996	Gail Devers, United States	10.94
2000	Marion Jones, United States	10.75

Source: *World Almanac and Book of Facts 2003.* Copyright © 2003 Primedia Reference Inc.

a. Who holds the Olympic record for the women's 100-meter run? What was her winning time? How much faster is the record-holder's time than Wilma Rudolph's time?

b. Order the times from fastest to slowest.

2. Here are the winning times (in seconds) for the men's 200-meter run for Summer Olympic Games from 1956 to 2000:

$$20.6 \quad 20.00 \quad 19.32 \quad 20.3 \quad 19.83 \quad 20.23$$

$$20.19 \quad 20.5 \quad 19.80 \quad 19.75 \quad 20.01 \quad 20.09$$

a. Order these times from fastest to slowest.

b. What is the difference between the fastest time and the slowest time?

c. List three times between 20.09 seconds and 20.19 seconds.

3. Felix says, "I think the decimal with the most digits is always the greatest number." Do you agree with him? Explain.

4. Write a rule for comparing any two decimals. Check that your rule works on each pair of numbers below.

$$23.45 \text{ and } 25.67 \qquad 3.5 \text{ and } 3.41 \qquad 16.0125 \text{ and } 16.0129$$

Share & Summarize

1. Imagine that you are asking the questions in a round of *Guess My Number.* You know that the number is between 8.344 and 8.345. What question would you ask next?

2. Which is a faster time, 25.52 seconds or 25.439 seconds? Explain how you know.

On Your Own Exercises

Practice & Apply

1. In his first two turns in the *Spare Change* game, Marcus chose $0.03 and $0.8.

   a. Complete a dollar chart showing the amount Marcus should have shaded after his first two turns.

   b. What part of a dollar is shaded on Marcus' chart? Express your answer as a fraction and as a decimal.

   c. How much more does Marcus need to have $1.00? Express your answer as a fraction and as a decimal.

2. Ms. Picó added cards with three decimal places to the *Spare Change* game deck. In her first three turns, Jing chose $0.77, $0.1, and $0.115.

   a. Complete a dollar chart showing the amount Jing should have shaded after her first three turns.

   b. What part of a dollar is shaded on Jing's chart? Express your answer as a fraction and as a decimal.

Write each decimal as a mixed number.

   3. 1.99   **4.** 7.016   **5.** 100.5

Find each product without using a calculator.

   6. 100×0.0436 **7.** $100,000 \times 754.01$ **8.** $1,000 \times 98.9$

Find each quantity without using a calculator.

   9. $\frac{1}{10}$ of 645   **10.** $7.7 \div 1,000$ **11.** $\frac{1}{10,000}$ of 55.66

Measurement Convert each measurement to meters. Write your answers as both fractions and decimals.

   12. 50 cm   **13.** 50 mm   **14.** 700 mm

15. Give the length of this baseball bat in centimeters and in meters:

10 cm   20 cm   30 cm   40 cm   50 cm   60 cm   70 cm   80 cm   90 cm

Tell the nearest tenths of a centimeter that each given measurement is between. For example, 3.66 is between 3.6 and 3.7.

16. 5.75 cm   **17.** 0.25 cm   **18.** 1.01 cm

Tell the nearest hundredths of a meter that each given measurement is between.

19. 0.555 m   **20.** 1.759 m   **21.** 0.0511 m

22. Imagine you are playing *Guess My Number* and have narrowed the possibilities to a number between 9.9 and 10. What are three possibilities for the number?

23. You are playing *Guess My Number* and have narrowed the possibilities to a number between 5.78 and 5.8. What are three possibilities for the number?

Order each set of numbers from least to greatest.

24. 7.31, 7.4, 7.110, 7.3, 7.04, 7.149

25. 21.5, 20.50, 22.500, 20.719, 21.66, 21.01, 20.99

26. **Sports** Participants in the school gymnastics meet are scored on a scale from 1 to 10, with 10 being the highest score. Here are the scores for the first event. Rob has not yet had his turn.

a. List the students from highest score to lowest score.

b. Rob is hoping to get third place in this event. List five possible scores that would put him in third place.

Student	Score
Kent	9.4
Elijah	8.9
Santiago	9.25
Matt	8.85
Emilio	9.9
Rob	
Terry	9.1
Craig	8.0
Arnon	8.7
Paul	9.2

Connect & Extend

27. **Economics** The FoodStuff market is running the following specials:

• Bananas: $0.99 per pound

• Swiss cheese: $3.00 per pound

• Rolls: $0.25 each

a. Jesse paid $9.90 for bananas. How many pounds did she buy?

b. Alex bought $\frac{1}{10}$ of a pound of Swiss cheese. How much did the cheese cost?

c. Ms. Washington is organizing the school picnic. How many rolls can she purchase with $250.00?

In your
own
words

Explain the
purpose of the
decimal point in
decimal numbers.

28. Economics A grocery store flyer advertises bananas for 0.15¢ each. Does this make sense? Explain.

29. Today is Tony's 10th birthday. His parents have decided to start giving him a monthly allowance, but they each suggest a different plan.

• Tony's mother wants to give him $0.01 each month this year, $0.10 each month next year, $1.00 each month the third year, and so on, multiplying the monthly amount by 10 each year until Tony's 16th birthday.

• Tony's father wants to give him $10 each month this year, $20 each month next year, $30 each month the next year, and so on, adding $10 to the monthly amount each year until Tony's 16th birthday.

His parents told Tony he could decide which plan to use. Which plan do you think he should choose? Explain your reasoning.

30. Science *Nanotechnology* is a branch of science that focuses on building very small objects from molecules. These tiny objects are measured with units such as microns and nanometers.

• 1 micron = 1 millionth of a meter

• 1 nanometer = 1 billionth of a meter

a. This is a nanoguitar. Although this guitar is only 10 microns long, it actually works. However, the sound it produces cannot be heard by the human ear. Express the length of the nanoguitar in meters. Give your answer as a decimal and as a fraction.

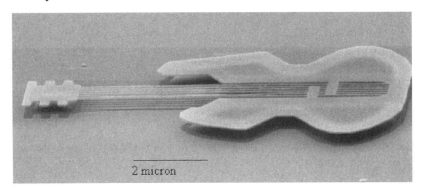

2 micron

Just *the* **facts**

A nanoguitar is about the size of a human being's white blood cell.

b. Two human hairs, side by side, would be about 0.001 meter wide. What fraction of this width is the length of the nanoguitar?

c. Microchips inside the processors of computers can have widths as small as 350 nanometers. Express this width in meters. Give your answer as a fraction and as a decimal.

d. A paper clip is about 0.035 meter long. What fraction of the length of a paper clip is the width of a microchip?

31. How much greater than 5.417 meters is 5.42 meters?

32. If a person is 2 meters 12 centimeters tall, we can say that he is 2.12 meters tall. If a person is 5 feet 5 inches tall, can we say that she is 5.5 feet tall? Why or why not?

Economics The table at right gives the value of foreign currencies in U.S. dollars on April 1, 2000.

Currency	Value in U.S. Dollars
Australian dollar	0.6075
British pound	1.5916
Canadian dollar	0.6886
Chinese renminbi	0.1208
Danish krone	0.1283
Greek drachma	0.0029
German mark	0.4884
Mexican new peso	0.1095
Singapore dollar	0.5844

33. If you exchanged 1 Canadian dollar for U.S. currency, how much money would you receive? (Assume that values are rounded to the nearest penny.)

34. Of those listed in the table, which currency is worth the most in U.S. dollars?

35. Of those listed in the table, which currency is worth the least in U.S. dollars?

36. Which currency listed in the table is worth closest to 1 U.S. dollar? How much more or less than 1 U.S. dollar is this currency worth?

37. How many Greek drachmas could you exchange for 1 penny?

EXCHANGE RATE

RATES INDICATION

	GERMANY	1.956
	GREAT BRITAIN	0.651
	CANADA	1.526
	ITALY	1934.19
	FRANCE	6.563
	JAPAN	138.47
	SWITZERLAND	1.602
	BELGIUM	4.26
	HOLLAND	2.22
	AUSTRALIA	1.61
	SPAIN	166.87
	DENMARK	7.58
	NORWAY	8.20
	SWEDEN	8.65

Find three fractions equivalent to each given fraction.

38. $\frac{7}{9}$ **39.** $\frac{12}{54}$ **40.** $\frac{6}{13}$ **41.** $\frac{14}{5}$

42. Order these fractions from least to greatest:

$$\frac{1}{3} \qquad \frac{12}{30} \qquad \frac{9}{28} \qquad \frac{11}{30} \qquad \frac{12}{29}$$

Find the prime factorization of each number.

43. 234   **44.** 1,890   **45.** 7,425

46. Statistics The pictograph shows the numbers of new dogs of seven breeds that were registered with the American Kennel Club during 2001.

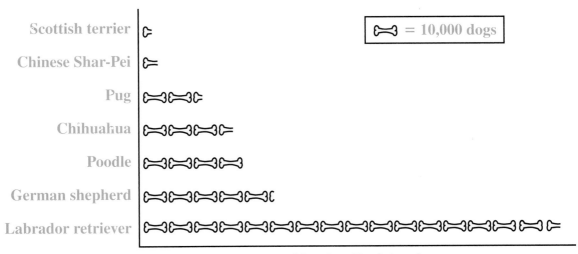

Dogs Registered During 2001

Source: *World Almanac and Book of Facts 2003.* Copyright © 2003 Primedia Reference Inc.

a. About how many Labrador retrievers were registered with the AKC during 2001?

b. About how many poodles were registered with the AKC during 2001?

c. In 2000, about 43,000 Chihuahuas were registered with the AKC. About how many fewer Chihuahuas were registered in 2001?

d. The number of German shepherds registered is about how many times the number of Chinese Shar-Peis registered?

2.4 Fractions and Decimals

Fractions and decimals are two ways of expressing quantities that are not whole numbers. You already know how to write a decimal in fraction form by thinking about the place values of its digits. In this lesson, you will find decimals that estimate the values of fractions, and you will learn to write fractions in decimal form.

Think & Discuss

Use both a fraction and a decimal to describe the approximate location of each point.

What methods did you use to make your estimates?

Investigation 1 ▶ Estimating Fraction and Decimal Equivalents

Number lines can help you understand the relationship between decimals and fractions. The diagram on the opposite page shows 10 number lines. The first is labeled with decimals, and the others are labeled with fractions. The fractions on each number line have the same denominator.

MATERIALS

copy of the number-line diagram

Problem Set A

1. Describe at least two patterns you notice in the diagram.

2. Consider decimal values greater than 0.5.

   a. Choose any decimal greater than 0.5 and less than 1.

   b. Find all the fractions in the diagram that appear to be equivalent to the decimal you chose. If there are no equivalent fractions, say so.

   c. Find two fractions in the diagram that are a little less than your decimal and two fractions that are a little greater. Try to find fractions as close to your decimal as possible.

Decimal and Fraction Number Lines

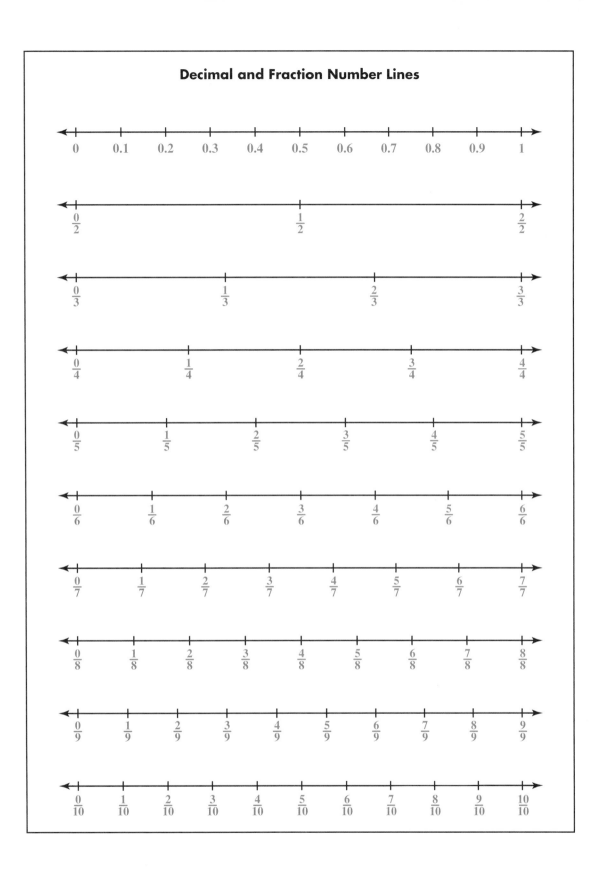

3. Consider the fractions that are not equivalent to $\frac{1}{2}$.

a. Choose any fraction in the diagram that is not equivalent to $\frac{1}{2}$.

b. Name the fraction on each of the other fraction number lines that is closest to the fraction you chose.

c. If possible, find an exact decimal value for your fraction to the hundredths place. If this is not possible, tell which two decimals, to the hundredths place, your fraction is between.

d. Repeat Parts a–c for two more fractions. Each fraction you choose should have a different denominator.

4. Decimals that are familiar make good benchmarks for estimating the values of other decimals and fractions.

a. Which decimals do you think would be useful as benchmarks? Explain why you chose those numbers.

b. For each fraction labeled in the diagram, find the benchmark decimal from Part a closest to that fraction. Organize your answers in any way you like.

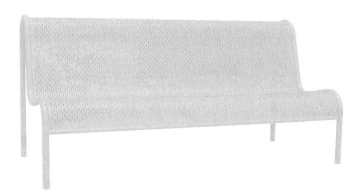

In the next problem set, you will estimate fractions and decimals greater than 1.

MATERIALS

copy of the number-line diagram

Problem Set B

Use both a mixed number and a decimal to describe the approximate location of each point.

1.

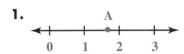

2.

3.

4. Consider the mixed number $3\frac{1}{3}$.

   a. Which of the following decimals is closest to $3\frac{1}{3}$? Explain how you decided.

   $$3.1 \qquad 3.2 \qquad 3.3 \qquad 3.5$$

   b. Find a decimal that is even closer to $3\frac{1}{3}$. Explain how you found your answer.

5. Consider the mixed number $81\frac{5}{6}$.

   a. Which of the following decimals is closest to $81\frac{5}{6}$? Explain how you decided.

   $$81.5 \qquad 81.6 \qquad 81.8 \qquad 81.9$$

   b. Find a decimal that is even closer to $81\frac{5}{6}$. Explain how you found your answer.

6. Use your estimation skills to order these numbers from least to greatest:

   $$\frac{5}{6} \qquad 0.7 \qquad \frac{2}{5} \qquad 3.1 \qquad 0.5 \qquad 3\frac{1}{3} \qquad 3.6 \qquad 5.2$$

Share & Summarize

Below is part of a student's homework assignment. Using your estimation skills, find the answers that are definitely wrong. Explain how you know they are wrong.

1. $\frac{2}{3} = 0.9$ **2.** $3\frac{7}{10} = 3.7$ **3.** $\frac{1}{5} = 2.0$ **4.** $2\frac{3}{4} = 2.75$

Investigation 2 Changing Fractions to Decimals

In Investigation 1, you found decimals approximations for given fractions. Now you will find exact decimal values for fractions.

Think & Discuss

Find a decimal equivalent to each fraction.

$$\frac{47}{1,000} \qquad\qquad \frac{59}{10} \qquad\qquad \frac{7}{20} \qquad\qquad \frac{3}{5}$$

Here's how Jing and Marcus found a decimal equivalent to $\frac{7}{20}$.

Problem Set C

Find a decimal equivalent to each given fraction.

1. $\frac{16}{25}$ **2.** $\frac{5}{8}$ **3.** $\frac{320}{200}$ **4.** $\frac{8}{125}$

5. Without using a calculator, try to find a decimal equivalent to $\frac{2}{3}$ by dividing. What happens?

VOCABULARY
repeating decimal

When you divided to find a decimal equivalent to $\frac{2}{3}$, you got 0.6666 . . ., in which the 6s repeat forever. Decimals with a pattern of digits that repeat without stopping are called **repeating decimals.** Repeating decimals are usually written with a bar over the repeating digits.

- $0.\overline{6}$ means 0.66666 . . .
- $3.1\overline{24}$ means 3.1242424 . . .

All fractions whose numerators and denominators are whole numbers have decimal equivalents that end (like 0.25) or that repeat forever (like $0.4\overline{6}$).

Calculators are useful for determining whether a fraction is equivalent to a repeating decimal. However, because the number of digits a calculator can display is limited, you sometimes cannot be certain.

Think & Discuss

Use your calculator to find a decimal approximation for each fraction below. Write your answers exactly as they appear in the calculator's display.

$$\frac{6}{9} \qquad \frac{6}{13} \qquad \frac{6}{15} \qquad \frac{6}{17} \qquad \frac{6}{21} \qquad \frac{666}{1,000}$$

Which of the fractions above definitely *are* equivalent to repeating decimals? Which of the fractions definitely *are not* equivalent to repeating decimals?

Which of the fractions above do you *think* might be repeating decimals? How could you find out for sure?

MATERIALS
copy of the fraction and decimal equivalents chart

Problem Set D

1. Fill in the decimal equivalents for each fraction in the chart. Some of the cells have been filled for you. Start by writing the decimal equivalents you know, and then use your calculator to find the others.

$\frac{0}{1}$	$\frac{0}{2}$	$\frac{0}{3}$	$\frac{0}{4}$	$\frac{0}{5}$	$\frac{0}{6}$	$\frac{0}{7}$	$\frac{0}{8}$	$\frac{0}{9}$	$\frac{0}{10}$
$\frac{1}{1}$	$\frac{1}{2}$	$\frac{1}{3}$	$\frac{1}{4}$ 0.25	$\frac{1}{5}$	$\frac{1}{6}$	$\frac{1}{7}$ $0.\overline{142857}$	$\frac{1}{8}$	$\frac{1}{9}$	$\frac{1}{10}$
$\frac{2}{1}$	$\frac{2}{2}$	$\frac{2}{3}$	$\frac{2}{4}$	$\frac{2}{5}$	$\frac{2}{6}$	$\frac{2}{7}$ $0.\overline{285714}$	$\frac{2}{8}$	$\frac{2}{9}$	$\frac{2}{10}$
$\frac{3}{1}$	$\frac{3}{2}$	$\frac{3}{3}$	$\frac{3}{4}$	$\frac{3}{5}$	$\frac{3}{6}$	$\frac{3}{7}$ $0.\overline{428571}$	$\frac{3}{8}$	$\frac{3}{9}$ $0.\overline{3}$	$\frac{3}{10}$
$\frac{4}{1}$	$\frac{4}{2}$	$\frac{4}{3}$	$\frac{4}{4}$	$\frac{4}{5}$	$\frac{4}{6}$	$\frac{4}{7}$ $0.\overline{571428}$	$\frac{4}{8}$	$\frac{4}{9}$	$\frac{4}{10}$
$\frac{5}{1}$	$\frac{5}{2}$	$\frac{5}{3}$	$\frac{5}{4}$	$\frac{5}{5}$	$\frac{5}{6}$	$\frac{5}{7}$ $0.\overline{714285}$	$\frac{5}{8}$	$\frac{5}{9}$	$\frac{5}{10}$
$\frac{6}{1}$	$\frac{6}{2}$ 3	$\frac{6}{3}$	$\frac{6}{4}$	$\frac{6}{5}$	$\frac{6}{6}$	$\frac{6}{7}$ $0.\overline{857142}$	$\frac{6}{8}$	$\frac{6}{9}$	$\frac{6}{10}$
$\frac{7}{1}$	$\frac{7}{2}$	$\frac{7}{3}$	$\frac{7}{4}$	$\frac{7}{5}$	$\frac{7}{6}$	$\frac{7}{7}$	$\frac{7}{8}$	$\frac{7}{9}$	$\frac{7}{10}$
$\frac{8}{1}$	$\frac{8}{2}$	$\frac{8}{3}$	$\frac{8}{4}$	$\frac{8}{5}$	$\frac{8}{6}$	$\frac{8}{7}$	$\frac{8}{8}$	$\frac{8}{9}$	$\frac{8}{10}$
$\frac{9}{1}$	$\frac{9}{2}$	$\frac{9}{3}$	$\frac{9}{4}$	$\frac{9}{5}$	$\frac{9}{6}$	$\frac{9}{7}$	$\frac{9}{8}$	$\frac{9}{9}$	$\frac{9}{10}$
$\frac{10}{1}$	$\frac{10}{2}$	$\frac{10}{3}$	$\frac{10}{4}$	$\frac{10}{5}$	$\frac{10}{6}$	$\frac{10}{7}$	$\frac{10}{8}$	$\frac{10}{9}$	$\frac{10}{10}$

2. Which columns contain fractions equivalent to repeating decimals?

3. Describe at least two patterns you see in the completed chart.

Save your chart for the On Your Own Exercises and for Investigation 3.

Share & Summarize

1. Describe a method for finding a decimal equivalent to a given fraction.

2. Explain what a repeating decimal is. Give an example of a fraction whose decimal equivalent is a repeating decimal.

Investigation ▶3 Patterns in Fractions and Decimals

In the next problem set, you will look for patterns in your chart of fraction and decimal equivalents from Investigation 2.

MATERIAL

completed fraction and decimal equivalents chart

Problem Set E

Find all the fractions in the chart that are equivalent to each given decimal.

1. 1.25

2. $0.\overline{6}$

3. Color all the cells with fractions equivalent to $\frac{1}{2}$. What pattern do you notice? Why does this happen?

4. Why can't the chart have a column showing fractions with a denominator of 0?

5. Look at the column containing fractions with denominator 10.

a. Describe the pattern in the decimals in this column. Explain why this pattern occurs.

b. Write $5\frac{7}{10}$ as a decimal.

c. Write 68.3 as a mixed number.

6. Look at the column containing fractions with denominator 2.

a. How do the decimal values change as you move down the column? Why?

b. Use the pattern from Part a to find the decimal equivalent of $\frac{11}{2}$, the number that would be next in the chart if a row were added.

7. Look at the column containing fractions with denominator 4.

    a. How do the decimal values change as you move down the column? Why?

    b. Use the pattern from Part a to find the decimal equivalent of $\frac{11}{4}$, the number that would be next in the chart if a row were added.

8. Look again at the fractions with denominator 2. The decimals in this column are 0, 0.5, 1, 1.5, 2, 2.5, 3, 3.5, 4, 4.5, 5. Notice that the "decimal parts" alternate between 0 (no decimal part) and 0.5.

    a. Look for a similar pattern in the column for fractions with denominator 4. Describe the pattern.

    b. Look for a similar pattern in the column for fractions with denominator 6. Describe the pattern.

    c. Look at a few other columns. Do similar patterns hold?

    d. **Challenge** Explain why these patterns occur.

Use the chart to help find the decimal equivalent for each fraction or mixed number.

    9. $10\frac{1}{2}$

    10. $32\frac{7}{9}$

    11. $62\frac{4}{5}$

    12. $23\frac{5}{6}$

Use the chart to help find a fraction equivalent to each decimal.

    13. 14.125

    14. $4.\overline{6}$

Lions and hyenas are the zebra's main predators in the wild. Zebras have a maximum speed of 40 mph, which is $\frac{4}{5}$ or 0.8, of the maximum speed of a lion. Hyenas have a maximum speed that is equal to the zebra's.

Problem Set F

In this problem set, you will use what you've learned about comparing and converting between fractions and decimals to build a fraction tower.

Building a Fraction Tower

- Choose a fraction less than 1 whose numerator and denominator are whole numbers between 1 and 9. Write both the fraction and its decimal equivalent on the bottom level of the tower.

- Choose another fraction whose numerator and denominator are between 1 and 9 and that is *less than* the fraction in the bottom level. Write the fraction and its decimal equivalent on the next level of the tower.

- Continue this process of choosing fractions and adding levels until you are unable to make a fraction with a value less than the fraction in the top level.

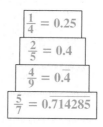

$$\frac{1}{4} = 0.25$$
$$\frac{2}{5} = 0.4$$
$$\frac{4}{9} = 0.\overline{4}$$
$$\frac{5}{7} = 0.\overline{714285}$$

1. Work with your partner to build several fraction towers. Record each tower you build. Try to build the highest tower possible.

2. What strategies did you use when building your towers?

Share & Summarize

1. Choose a pattern in the chart of fraction and decimal equivalents. It can be a pattern you discussed in class or a new pattern you have discovered. Describe the pattern, and explain why it occurs.

2. Write a letter to a student who is just learning how to build fraction towers. Explain strategies that he or she might use to build a high tower.

On Your Own Exercises

Practice & Apply

For Exercises 1–3, use the number-line diagram on page 129.

1. Name the three fractions in the diagram that are closest to 0.8.

2. Tell which two decimals, to the hundredths place, $\frac{6}{7}$ is between.

3. Tell which two decimals, to the hundredths place, $\frac{7}{8}$ is between.

4. Use a mixed number and a decimal to approximate the location of each point.

5. Consider the fraction $\frac{17}{8}$.

a. Which of these decimals is closest to $\frac{17}{8}$? Explain how you decided.

$$2.1 \qquad 2.2 \qquad 2.8 \qquad 2.9 \qquad 17.1 \qquad 17.8$$

b. Give a decimal that would be even closer to $\frac{17}{8}$. Explain how you found your answer.

6. Order these numbers from least to greatest:

$$4\frac{1}{3} \qquad 4\frac{5}{8} \qquad 4.51 \qquad 4.1 \qquad 4.491 \qquad 4\frac{8}{9}$$

7. Which of the following is equivalent to $1.2\overline{34}$?

$$1.23412341234\ldots \qquad 1.23434343434\ldots \qquad 1.23444444\ldots$$

8. Which of the following is equivalent to 2.393939…?

$$2.3\overline{9} \qquad\qquad 2.\overline{39}$$

Find a decimal equivalent for each fraction or mixed number.

9. $\frac{16}{5}$ **10.** $\frac{15}{11}$

11. $\frac{70}{250}$ **12.** $\frac{33}{24}$

13. $\frac{14}{3}$ **14.** $\frac{376}{20,000}$

15. $5\frac{1}{16}$ **16.** $\frac{9}{12}$

17. In the chart you completed in Problem Set D, color all the cells with fractions equivalent to $\frac{1}{3}$.

    a. What pattern do you notice? Why does this happen?

    b. Why doesn't every column have a colored cell?

Use the chart you completed in Problem Set D to help find a decimal equivalent to each fraction or mixed number.

18. $\frac{12}{5}$ **19.** $32\frac{7}{10}$ **20.** $65\frac{2}{3}$ **21.** $3\frac{1}{7}$

Use the chart you completed in Problem Set D to help find a fraction or mixed number equivalent to each decimal.

22. 4.125 **23.** 32.5 **24.** 4.75 **25.** $8.\overline{1}$

26. Refer to the chart you completed in Problem Set D. Look at the column containing fractions with denominator 5.

    a. How do the decimal values change as you move down the column? Why?

    b. Use the pattern in Part a to find the decimal equivalent of $\frac{11}{5}$, the number that would be next in the chart if a row were added.

In Exercises 27 and 28, the start of a fraction tower is given. If you want to build the highest tower possible, what number should you choose for the next level? Explain your answer, and give your number in both fraction and decimal form.

27.

$$\frac{6}{9} = 0.\overline{6}$$
$$\frac{7}{9} = 0.\overline{7}$$
$$\frac{8}{9} = 0.\overline{8}$$

28.

$$\frac{3}{9} = 0.\overline{3}$$
$$\frac{4}{5} = 0.8$$

29. Order these numbers from least to greatest:

        2.3 $\frac{11}{5}$ $2\frac{3}{7}$ 2.05

Devils Tower National Monument, in northeast Wyoming, rises 1,267 feet above the nearby Belle Fourche River.

30. Copy the number line. Mark and label the point corresponding to each number.

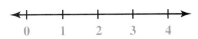

Point A: 2.4   Point B: $3\frac{8}{9}$ Point C: 0.67   Point D: 1.2

31. Look back at the number-line diagram on page 129. Imagine creating two more number lines to represent 11ths and 12ths.

a. Is $\frac{1}{11}$ less than or greater than 0.1? How do you know?

b. Is $\frac{1}{12}$ less than or greater than $\frac{1}{11}$? How do you know?

c. Is $\frac{10}{11}$ less than or greater than 0.9? How do you know?

d. Is $\frac{11}{12}$ less than or greater than $\frac{10}{11}$? How do you know?

32. Look back at the number-line diagram on page 129. Notice that 0.5 is between $\frac{1}{3}$ and $\frac{2}{3}$.

a. Between which two fifths is 0.5?

b. Between which two sevenths is 0.5?

c. Between which two ninths is 0.5?

Now think beyond the diagram.

d. Between which two 15ths is 0.5?

e. Between which two 49ths is 0.5?

In each pair, tell which fraction is closer to 0.5.

33. $\frac{2}{5}$ or $\frac{3}{7}$ **34.** $\frac{4}{9}$ or $\frac{11}{23}$ **35.** $\frac{2}{5}$ or $\frac{4}{5}$

36. $\frac{4}{9}$ or $\frac{6}{9}$ **37.** $\frac{3}{17}$ or $\frac{16}{17}$ **38.** $\frac{19}{44}$ or $\frac{27}{44}$

39. Refer to your completed chart from Problem Set D.

a. Look at the column containing fractions with denominator 9. What pattern do you see as you move down this column?

b. Find the decimal equivalents of $\frac{1}{99}$, $\frac{2}{99}$, $\frac{3}{99}$, and so on, up to $\frac{9}{99}$. What pattern do you see?

c. Find the decimal equivalents of $\frac{1}{999}$, $\frac{2}{999}$, $\frac{3}{999}$, and so on, up to $\frac{9}{999}$. What pattern do you see?

d. Predict the decimal equivalents for $\frac{1}{9,999}$, $\frac{5}{9,999}$, and $\frac{8}{9,999}$. Use your calculator to check your prediction.

In your **own words**

Explain how to find a fraction between any two given fractions. Give an example to illustrate your method.

40. Consider fractions with denominator 11.

    a. Find decimal equivalents for $\frac{1}{11}$, $\frac{2}{11}$, $\frac{3}{11}$, $\frac{4}{11}$, and $\frac{5}{11}$. What pattern do you see?

    b. Use the pattern you discovered to predict the decimal equivalents of $\frac{7}{11}$ and $\frac{9}{11}$.

41. Refer to your completed chart from Problem Set D. Compare the *row* containing fractions with *numerator* 10 to the row containing fractions with numerator 1. The decimals for fractions with numerator 10 are the same as those for fractions with numerator 1, except the decimal point is moved one place to the right. Explain why.

42. In your completed chart from Problem Set D, color the cells containing fractions equivalent to 1. Use a different color from before.

    a. Describe the pattern you see.

    b. The numbers below the line of 1s are greater than 1. The numbers above the line of 1s are less than 1. Why?

In Exercises 43–45, the rules for building a fraction tower have been changed. Tell what fraction you would choose as your starting number in order to build the tallest possible tower.

43. Use whole numbers between 1 and 20 for numerators and denominators. As before, fractions must be less than 1.

44. Use whole numbers between 1 and 9 for numerators and denominators. There is no limit on the value of the fraction.

45. Use whole numbers between 1 and 9 for numerators and denominators. The fractions must be less than $\frac{1}{2}$.

46. Suppose the rules for building a fraction tower are changed so you can choose only numbers between 1 and 4 for numerators and denominators. Fractions still must be less than 1. Draw a tower with the maximum number of levels, and explain why it is the maximum.

Find the least common multiple of the given numbers.

47. 5, 7, and 14 **48.** 20, 16, and 12 **49.** 11, 13, and 17

Economics Tell how much change you would get if you paid for each item with a $5 bill.

50. a slice of pizza for $1.74 **51.** a pack of gum for $.64

52. a magazine for $3.98 **53.** a pack of PlanetQuest cards for $2.23

Measurement Express each measurement in meters.

54. 13 mm **55.** 123 cm **56.** 0.05 cm **57.** 430 mm

58. Statistics Monica asked students in her homeroom class which cafeteria lunch was their favorite. She recorded her findings in a table. Make a bar graph to display Monica's results.

Lunch	Number of Students
Pizza	10
Veggie lasagna	4
Macaroni and cheese	5
Hamburger	7
Tuna casserole	2
Never buy lunch	2

You have had a lot of experience working with **positive numbers,** numbers that are greater than 0. You are familiar with positive whole numbers, positive decimals, and positive fractions. In this lesson, you will turn your attention to **negative numbers,** numbers that are less than 0.

Explore

Below, five students describe the current temperatures in the cities where they live. Use the thermometer to help figure out the temperature in each city.

- Bill lives in Buloxi, Mississippi. He says, "If the temperature goes up 5 degrees, it will be 62°F."

- Carlita lives in Cincinnati, Ohio. She says, "It's not very cold here. If the temperature rises 5 degrees, it will be 47.5°F."

- Kelly lives in Kennebunkport, Maine. She says, "It's freezing here, and the snow is 3 feet deep! If the temperature would just go up $30\frac{1}{4}$ degrees, it would be 32°F, and some of this snow might melt."

- Nate lives in Niagara Falls, New York. He says, "You think that's cold? If the temperature rises 30 degrees here, it will be only 30°F!"

- Jean lives in Juneau, Alaska. She says, "We can top all of you up here! If our temperature went up 30 degrees, it would be only 10°F!"

Make up a temperature puzzle, like those above, for which the answer is negative. Exchange puzzles with your partner, and solve your partner's puzzle.

Investigation Understanding Negative Numbers

In the Explore, you found that the temperature in Juneau was ⁻20°F, meaning 20 degrees *below* 0. There are many other contexts in which the number ⁻20 might be used. For example, if you were standing at a location 20 feet below sea level, your elevation would be ⁻20 feet. If you wrote checks for $20 more than you had in your bank account, your account balance would be ⁻$20.

Here are a few facts about the number ⁻20:

• ⁻20 is read as "negative twenty" or "the opposite of twenty."

• ⁻20 is located 20 units to the left of 0 on a horizontal number line.

• ⁻20 is located 20 units below 0 on a vertical number line or thermometer.

• ⁻20 is located halfway between ⁻21 and ⁻19 on a number line.

VOCABULARY
opposites

The numbers ⁻20 and 20 are *opposites*. Two numbers are **opposites** if they are the same distance from 0 on the number line but on different sides of 0.

As you know, the number 20 is positive. You will occasionally see 20 written as ⁺20. The notations 20 and ⁺20 mean the same thing, and both can be read as "positive twenty" or just "twenty."

VOCABULARY
absolute value

The **absolute value** of a number is its distance from 0 on the number line. The absolute value of a number is indicated by drawing a bar on each side of the number. For example, |⁻20| means "the absolute value of ⁻20." Since ⁻20 and 20 are each 20 units from 0 on the number line, |20| = 20 and |⁻20| = 20.

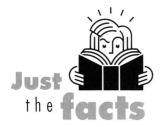

Just the facts

The lowest temperature ever recorded on Earth is ⁻128.6°F. This temperature occurred on July 21, 1983 at Vostok, a Russian station in Antarctica.

Problem Set A

1. Copy the number line below, and plot points to show the locations of the numbers listed. Label each point with the corresponding number.

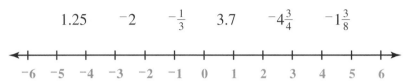

$$1.25 \qquad -2 \qquad -\frac{1}{3} \qquad 3.7 \qquad -4\frac{3}{4} \qquad -1\frac{3}{8}$$

2. On your number line from Problem 1, plot and label points for three more negative mixed numbers.

3. Give a number that describes the approximate location of each labeled point.

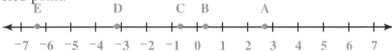

4. Find the opposite of each number.

   a. 3.2

   b. $-\frac{3}{4}$

   c. $^-2$

   d. 317

5. Find the absolute value of each number in Problem 4.

6. Is the opposite of a number always negative? Explain.

7. Is the absolute value of a number always greater than or equal to 0? Explain.

Number lines are useful for showing the order and size of numbers.

Think & Discuss

How can you tell which of two numbers is greater by looking at their locations on a horizontal number line?

Which is the warmer temperature, $^-20°F$ or $^-15°F$? How do you know?

Problem Set B

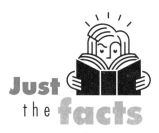

1. The table shows record low temperatures for each continent.

Continent	Location	Temperature
Africa	Ifrane, Morocco	$-11°$F
Antarctica	Vostok Station	$-129°$F
Asia	Oimekon and Verkhoyansk, Russia	$-90°$F
Australia	Charlotte Pass, New South Wales	$-9.4°$F
Europe	Ust'Shchugor, Russia	$-67°$F
North America	Snag, Yukon, Canada	$-81.4°$F
South America	Sarmiento, Argentina	$-27°$F

Source: *World Almanac and Book of Facts 2000.* Copyright © 1999 Primedia Reference Inc.

a. Order the temperatures from coldest to warmest.

b. When temperatures drop below $-30°$F, people often experience frostbite. For which locations in the table is the record low temperature below $-30°$F?

c. How many degrees below the warmest temperature in the table is the coldest temperature?

2. Ten students measured the outside temperature at different times on the same winter day. Their results are shown in the table.

Student	Temperature
Kiran	$-0.33°$F
Jill	$1.30°$F
Fabiana	$-0.80°$F
Xavier	$0.33°$F
Micheala	$-1.30°$F
Brad	$-1.80°$F
Paul	$1.08°$F
Jazmin	$-1.75°$F
Matt	$-1.00°$F
Jessica	$0.80°$F

a. List the temperatures in order from coldest to warmest.

b. On the night the students recorded the temperatures, a weather reporter said, "The average temperature today was a chilly 0°F." Which students recorded temperatures that were closest to the average temperature?

Share & Summarize

1. Explain what the opposite of a number is.

2. Name three negative numbers between -3 and -2.

On Your Own Exercises

1. If the temperature rises 10°F, it will be ⁻25°F. What is the temperature now?

2. If the temperature goes down 33°F, it will be ⁻10°F. What is the temperature now?

3. If the temperature goes up $1\frac{1}{4}$°F, it will be 0°F. What is the temperature now?

4. Copy the number line, and plot points to show the locations of the numbers listed. Label each point with the corresponding number.

$$-2.5 \qquad ^{-}1.75 \qquad 2\frac{1}{3} \qquad ^{-}3 \qquad 0.4 \qquad 4\frac{1}{2} \qquad ^{-}0.8$$

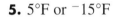

In Exercises 5–8, tell which temperature is warmer.

5. 5°F or ⁻15°F

6. ⁻35°F or ⁻25°F

7. $^{-}5\frac{1}{2}$°F or $^{-}5\frac{3}{4}$°F

8. ⁻100.9°F or ⁻100.5°F

In your
own
words

Explain how a number line can help you order a set of positive and negative numbers.

impactmath.com/self_check_quiz

Earth Science Elevations are measured from sea level, which is considered to have an elevation of 0 feet. Elevations above sea level are positive, and elevations below sea level are negative. In Exercises 9–13, use this table, which shows the elevation of the lowest point on each continent.

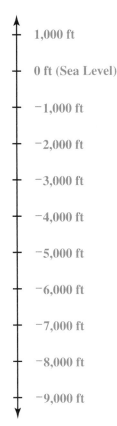

Continent	Location of Lowest Point	Elevation
North America	Death Valley	$^-282$ ft
South America	Valdes Peninsula	$^-131$ ft
Europe	Caspian Sea	$^-92$ ft
Asia	Dead Sea	$^-1,312$ ft
Africa	Lake Assal	$^-512$ ft
Australia	Lake Eyre	$^-52$ ft
Antarctica	Bentley Subglacial Trench	$^-8,327$ ft

Source: *World Almanac and Book of Facts 2000.* Copyright © 1999 Primedia Reference Inc.

9. Order the elevations in the table from lowest to highest.

10. Copy the number line at left. Plot each elevation given in the table, and label the point with the name of the continent.

11. How much lower is the Dead Sea than the Caspian Sea?

12. One of the elevations in the table is significantly lower than the others.

    a. On which continent is this low point located?

    b. How much lower is this point than the next lowest point?

13. **Challenge** The highest point in North America is the summit of Denali, a mountain in Alaska with an elevation of 20,320 feet. How many feet higher than Death Valley is the Denali summit?

Write each fraction or mixed number as a decimal.

14. $\frac{3}{5}$ **15.** $\frac{132}{10,000}$ **16.** $\frac{7}{9}$

17. $3\frac{17}{20}$ **18.** $\frac{72}{2,500}$ **19.** $\frac{173}{12}$

Geometry The measure of two angles of a triangle are given. Find the measure of the third angle.

20. $23°$ and $47°$ **21.** $60°$ and $60°$ **22.** $10°$ and $120°$

Tell whether segments of the given lengths could be joined to form a triangle.

23. 3, 5, and 7   **24.** 10, 10, and 21   **25.** 17, 5, and 9

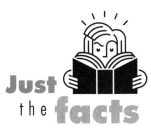

Just the facts

In the winter, 40,000 square miles of seawater around Antarctica freezes every day. This winter ice almost doubles the size of the continent!

Chapter Summary

In this chapter, you used the idea of stretching machines to help think about *factors, multiples,* and *prime and composite numbers.* You learned to find the *prime factorization* of a number and to find the *greatest common factor* or *least common multiple* of a group of numbers.

You used what you learned about factors and multiples to find *equivalent fractions* and to compare fractions. You also saw how you could use *benchmark fractions* like $\frac{1}{4}$, $\frac{1}{2}$, and $\frac{2}{3}$ to estimate the values of more complicated fractions.

You reviewed the meaning of decimals, and you investigated how multiplying or dividing by 10, 100, 1,000, 10,000, and so on affects the position of the decimal point. Then you compared and ordered decimals and found a number between two given decimals.

You saw how you could write a fraction as a decimal. You discovered that sometimes the decimal representation of a fraction is a *repeating decimal,* with a pattern of digits that repeats forever.

Finally, you studied negative numbers. You saw how negative numbers are represented on a number line, and you compared and ordered sets of negative numbers.

Strategies and Applications

The questions in this section will help you review and apply the important ideas and strategies developed in this chapter.

Understanding and applying concepts related to factors and multiples

1. Describe the difference between a prime number and a composite number. Give two examples of each.

2. Explain how to use a factor tree to find the prime factorization of a number. Illustrate your explanation with an example.

3. Joni set her computer's clock to ding every 20 minutes, quack every 30 minutes, and chime every 45 minutes. At noon, her computer dinged, quacked, and chimed at the same time.

a. In how many minutes will Joni's computer ding and quack at the same time?

b. In how many minutes will her computer quack and chime at the same time?

c. In how many minutes will her computer ding, chime, and quack at the same time?

d. What sounds will the computer make at 4:30 P.M.?

Finding equivalent fractions and comparing fractions

4. Find two fractions equivalent to $\frac{4}{6}$. Use diagrams or another method to explain why the fractions are equivalent to $\frac{4}{6}$.

5. Explain what it means for a fraction to be in *lowest terms*. Then describe a method for writing a given fraction in lowest terms.

6. Describe two methods for comparing fractions. Then use one of the methods to determine whether $\frac{7}{10}$ is greater than, less than, or equal to $\frac{8}{11}$.

Understanding and comparing decimals

7. Describe a rule for comparing two decimals. Demonstrate your rule by comparing 307.63 with 308.63 and by comparing 3.786 with 3.779.

8. How does the value of a number change when you move the decimal point three places to the right? Illustrate your answer with an example.

9. How does the value change when you move the decimal point two places to the left? Illustrate your answer with an example.

Converting decimals to fractions and fractions to decimals

10. Explain how you would write a decimal in fraction form. Illustrate by writing 0.97 and 0.003 as fractions.

11. Explain how you would find a decimal equivalent to a given fraction. Give an example.

12. Explain what a *repeating decimal* is. Give an example of a fraction that is equivalent to a repeating decimal.

Understanding negative numbers and opposites

13. What is the *opposite* of a number? Give an example.

14. The temperature at 6 A.M. was $^{-}13°F$. By noon, the temperature had risen 24 degrees. What was the temperature at noon?

15. Describe how you would compare two negative numbers.

Demonstrating Skills

List the factors of each number.

16. 18   **17.** 64   **18.** 43

Give the prime factorization of each number. Use exponents when appropriate.

19. 246   **20.** 56   **21.** 81

Find the greatest common factor and the least common multiple of each pair of numbers.

22. 9 and 12   **23.** 11 and 23   **24.** 14 and 28

List three fractions equivalent to each given fraction.

25. $\frac{6}{7}$ **26.** $\frac{13}{39}$ **27.** $\frac{32}{720}$

Replace each ● with $<$, $>$, or $=$ to make a true statement.

28. $\frac{5}{16}$ ● $\frac{7}{24}$ **29.** $\frac{14}{22}$ ● $\frac{35}{55}$ **30.** $\frac{9}{49}$ ● $\frac{3}{14}$

Order each set of decimals from least to greatest.

31. 0.7541, 1.754, 0.754, 0.75411, 0.7641

32. 251.889, 249.9, 251.9, 251.8888, 252.000001

Compute each result mentally.

33. $0.00012 \cdot 1,000$

34. $\frac{1}{10,000}$ of 344

35. 100×77.5

Fill in each blank.

36. 13 cm = _____ m

37. 557 mm = _____ m

38. 14 m = _____ cm

Find a number between each given pair of numbers.

39. 11.66 and 11.67

40. 0.0001 and 0.001

41. 3.04676 and 3.04677

Write each fraction or mixed number in decimal form.

42. $\frac{7}{8}$ **43.** $\frac{8}{6}$

44. $\frac{11}{15}$ **45.** $5\frac{17}{20}$

Write each decimal as a fraction or mixed number.

46. 2.003   **47.** 0.62

48. 0.90   **49.** 31.031031

Find the opposite of each number.

50. $^-12$ **51.** 99

52. $^-0.001$ **53.** 17.26

Order each set of numbers from least to greatest.

54. $5, \ ^-12, \ ^-5, \ ^-11.999, \ 4.75, \ 1$

55. $\frac{3}{4}, \ ^-\frac{1}{3}, \ ^-\frac{2}{5}, \ \frac{7}{8}, \ ^-0.1, \ \frac{1}{10}$

CHAPTER 3

Working with Fractions and Decimals

Real-Life Math

The House that Fractions Built Did you know that planning and building a house requires lots of calculations with fractions and decimals? For example, the architects who make blueprints, and the contractors who read them, need to know how to add, subtract, multiply, and divide fractions and mixed numbers.

Think About It Blueprints are detailed drawings of the floorplan, front, back, and side views of a building. They are drawn to a particular scale—for example, $\frac{1}{4}$ inch on a blueprint might represent 1 foot on the actual house. Suppose an architect wants to represent a floor with dimensions 12 feet by 14 feet. What dimensions should she use on the drawing?

Family Letter

Dear Student and Family Members,

We live in a digital age! So, it is becoming more and more likely that a number is shown as a decimal instead of as a fraction. For example, a weather forecaster may say that we've had 2.5 inches more rain than normal this month, but you know that means $2\frac{1}{2}$ inches. However, it is still important to be able to compute with fractions because many real-life measurement problems, especially in cooking and the building trades, use fractions. Here are two examples:

- A chocolate chip cookie recipe calls for $\frac{3}{4}$ cup brown sugar. You want to double the recipe. How much brown sugar should you use?
- Suppose you have a wooden board $36\frac{1}{2}$ inches long and cut off a piece that is $6\frac{3}{8}$ inches long. How much do you have left?

In the next few weeks, you'll be learning to add, subtract, multiply, and divide with fractions and how to multiply and divide with decimals. Working with fractions and decimals can be tricky, and there are many ways to make errors. That is why it's so important to decide whether your answers seem reasonable.

Vocabulary There is only one new term in this chapter—*reciprocal*. You'll use reciprocals when you divide fractions.

What can you do at home?

You might help your student think about or work with fractions and decimals as they come up day to day:

- In the grocery store, find the unit cost (the cost of 1 ounce, 1 liter, or 1 piece) of different brands or different packaging for the same item, and decide which is the better value.
- Figure out the correct quantities when doubling or halving a recipe.
- Work with measurements, customary or metric, when you are measuring, sewing, or doing carpentry work.

3.1 Adding and Subtracting Fractions

You know how to compare fractions and how to find equivalent fractions. Now you will begin to think about adding and subtracting fractions. You probably already know how to add and subtract fractions with the same denominator.

Think & Discuss

Solve each problem in your head.

$\frac{1}{5} + \frac{2}{5} =$ _____ $\qquad$ $\frac{5}{7} - \frac{2}{7} =$ _____ $\qquad$ $\frac{1}{2} + \frac{1}{2} =$ _____

$\frac{1}{3} +$ _____ $= 1$ $\qquad$ _____ $+ \frac{2}{6} = 1$ $\qquad$ $\frac{5}{8} +$ _____ $= 1$

$1 - \frac{1}{3} =$ _____ $\qquad$ $1 - \frac{3}{8} =$ _____ $\qquad$ $1 - \frac{1}{6} =$ _____

Explain how to add or subtract fractions with the same denominator.

Explain how to find a fraction that adds to a given fraction to give 1.

Explain how to subtract a fraction from 1.

Investigation Adding and Subtracting with Fraction Pieces

You can use *fraction pieces* and a *fraction mat* to add and subtract fractions.

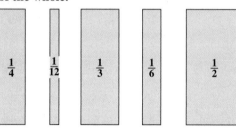

The square on the fraction mat is the whole. It represents 1.

1

The fraction pieces represent fractions of the whole.

$\frac{1}{4}$ $\qquad$ $\frac{1}{12}$ $\qquad$ $\frac{1}{3}$ $\qquad$ $\frac{1}{6}$ $\qquad$ $\frac{1}{2}$

One $\frac{1}{2}$ piece and three $\frac{1}{6}$ pieces cover the square. You can represent this with the addition equation

Removing three $\frac{1}{6}$ pieces leaves a $\frac{1}{2}$ piece. You can represent this with the subtraction equation

$$\frac{1}{2} + \frac{3}{6} = 1$$

$$1 - \frac{3}{6} = \frac{1}{2}$$

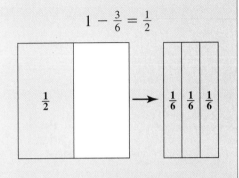

MATERIALS

set of fraction pieces and fraction mat

Problem Set A

1. Choose fraction pieces with two different denominators.

    a. Find as many ways as you can to cover the square on your fraction mat with those two types of fraction pieces. (Rearranging the same pieces in different positions on the mat does *not* count as a new way to cover the square.)

    For each combination you find, make a sketch and write an equation. Each equation should be a sum of two fractions equal to 1. For example, the equation in the Example above is $\frac{1}{2} + \frac{3}{6} = 1$.

    b. Choose another pair of denominators, and look for ways to cover the square with those two types of fraction pieces. Record a sketch and an equation for each combination.

    c. Continue this process until you think you have found all the ways to cover the square with two types of fraction pieces.

2. For each addition equation, you can write two related subtraction equations. In the Example above, the sixths pieces were removed to get $1 - \frac{3}{6} = \frac{1}{2}$. You could instead remove the $\frac{1}{2}$ piece to get $1 - \frac{1}{2} = \frac{3}{6}$.

    Choose three of the addition equations you wrote in Problem 1, and write two related subtraction equations for each.

3. Now find as many ways as you can to cover the square with *three* different types of fraction pieces. Record a sketch and an equation for each combination.

All the sums in Problem Set A equal 1. You can also use your fraction pieces to form sums greater than or less than 1.

EXAMPLE

Use fraction pieces to find $\frac{2}{4} + \frac{1}{6}$.

Choose the appropriate pieces, and arrange them on your fraction mat.

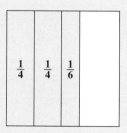

Look for a set of identical fraction pieces that cover the same area. You could use two $\frac{1}{3}$ pieces. So,

$$\frac{2}{4} + \frac{1}{6} = \frac{2}{3}$$

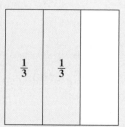

The addition equation above has two related subtraction equations:

$$\frac{2}{3} - \frac{2}{4} = \frac{1}{6} \qquad \qquad \frac{2}{3} - \frac{1}{6} = \frac{2}{4}$$

MATERIALS

set of fraction pieces and fraction mat

Problem Set B

1. Form sums less than 1 by combining two different types of fraction pieces. Find at least six different combinations.

   a. Record each combination by drawing a sketch and writing an addition equation.

   b. For each addition equation you wrote in Part a, write two related subtraction equations.

2. Now combine two different types of fraction pieces to form sums greater than 1. Find at least six different combinations.

   a. Record a sketch and an addition equation for each combination.

   b. For each addition equation you wrote in Part a, write two related subtraction equations.

3. Now you will look for a pair of fractions with sum $\frac{7}{12}$.

    a. Use the $\frac{1}{12}$ pieces to cover $\frac{7}{12}$ of the square on your fraction mat.

    b. Look for a combination of two types of fraction pieces that cover $\frac{7}{12}$ of the square. Record the result with a sketch and an addition equation.

    c. Write two related subtraction equations for the addition equation.

MATERIALS

set of fraction pieces and fraction mat

Share & Summarize

Consider the equation $\frac{2}{3} + \frac{1}{4} = \frac{10}{12}$.

1. Explain how you could use fraction pieces to determine whether the equation is true.

2. Is the equation true? If it isn't, change it to a true equation by replacing one of the fractions in it.

Investigation 2 — Adding and Subtracting Using Common Denominators

By now you are probably pretty good at using fraction pieces to add and subtract fractions with unlike denominators. But what do you do if you don't have your fraction pieces or if you want to add fractions with denominators other than 2, 3, 4, 6, or 12?

MATERIALS

set of fraction pieces and fraction mat

Explore

Try to find this sum without using fraction pieces or writing anything:

$$\frac{1}{6} + \frac{3}{4}$$

Were you able to calculate the sum in your head? If so, what strategy did you use? If not, why did you find the problem difficult?

The sum would be much easier to find if the fractions had the same denominator.

Now use your fraction pieces or another method to find two fractions with the same denominator, one equivalent to $\frac{1}{6}$ and another equivalent to $\frac{3}{4}$. Then add the two fractions in your head.

Problem Set C

In this problem set, use your fraction pieces or any other method you
know to help rewrite the fractions.

1. Choose two of the addition equations you wrote for Problem 1 of
   Problem Set A. Rewrite each equation so the fractions being added
   have a common denominator. Check to be sure each sum is equal to 1.
   For example:

$$\frac{1}{3} + \frac{8}{12} = 1 \quad \rightarrow \quad \frac{4}{12} + \frac{8}{12} = \frac{12}{12} = 1$$

2. Choose two of the addition equations you wrote for Problem 3 of
   Problem Set A. Rewrite each equation so the fractions being added
   have a common denominator. Check to be sure each sum is equal
   to 1.

3. Choose two of the addition equations you wrote for Problem 1 of
   Problem Set B. Rewrite each equation, and the two related subtraction equations, using common denominators. Check each equation to
   make sure it is correct.

4. Choose two of the addition equations you wrote for Problem 2 of
   Problem Set B. Rewrite each equation, and the two related subtraction equations, using common denominators. Check that each equation is correct.

Many of the problems in the next problem set would be difficult or impossible to solve using fraction pieces. Use what you know about common
denominators and equivalent fractions to rewrite the sums and differences.

Problem Set D

Rewrite each problem using a common denominator, and then find the
sum or difference. Give your answers in lowest terms. If an answer is
greater than 1, write it as a mixed number.

1. $\frac{1}{2} + \frac{7}{8}$

2. $\frac{2}{5} + \frac{2}{3}$

3. $\frac{1}{9} + \frac{5}{6}$

4. $\frac{4}{5} - \frac{1}{4}$

5. $\frac{7}{8} - \frac{2}{3}$

6. $\frac{5}{9} - \frac{1}{3}$

7. Daniela has $\frac{1}{2}$ of
   a bag of potting
   soil. She uses $\frac{1}{3}$
   of a bag to plant
   some flower seeds.
   How much of the
   bag remains?

Remember

A fraction is in *lowest
terms* if the only
common factor of its
numerator and denomi-
nator is 1.

Jahmal, Marcus, and Caroline discuss some strategies they use to find common denominators.

As you work on the next problem set, you might try some of the methods described in the Example above.

Problem Set E

Find each sum or difference, showing each step of your work. Give your answers in lowest terms. If an answer is greater than 1, write it as a mixed number.

1. $\frac{1}{3} + \frac{3}{7}$

2. $\frac{1}{4} + \frac{3}{9}$

3. $\frac{4}{5} - \frac{3}{8}$

4. $\frac{13}{27} - \frac{2}{9}$

5. $\frac{7}{15} + \frac{1}{3} + \frac{1}{5}$

6. $\frac{11}{12} + \frac{3}{8}$

7. $1 - \frac{3}{4} - \frac{1}{5}$

8. $1\frac{31}{42} + \frac{17}{21}$

9. $1\frac{3}{4} - \frac{5}{8}$

10. $\frac{32}{75} + \frac{32}{50}$

A *magic square* is a square grid of numbers in which every row, column, and diagonal has the same sum. This magic square has a sum of 15:

8	1	6
3	5	7
4	9	2

Problem Set F

You can create your own magic squares using a set of fraction cards and a grid.

1. Arrange the following numbers into a magic square with a sum of 1. Record your grid.

$$\frac{1}{15} \qquad \frac{2}{15} \qquad \frac{4}{15} \qquad \frac{7}{15} \qquad \frac{8}{15} \qquad \frac{1}{5} \qquad \frac{2}{5} \qquad \frac{3}{5} \qquad \frac{1}{3}$$

2. Arrange the following numbers into a magic square with a sum of $\frac{1}{2}$. Record your grid.

$$\frac{5}{36} \qquad \frac{7}{36} \qquad \frac{1}{18} \qquad \frac{5}{18} \qquad \frac{1}{12} \qquad \frac{1}{9} \qquad \frac{2}{9} \qquad \frac{1}{6} \qquad \frac{1}{4}$$

Share & Summarize

Use an example to help explain the steps you follow to add or subtract two fractions with different denominators.

Investigation 3 Adding and Subtracting Mixed Numbers

In this investigation, you will apply what you've learned about adding and subtracting fractions to solve problems involving mixed numbers.

Explore

Rosita is making a box to hold her pencils. She needs five pieces of wood: one piece for the bottom and four pieces for the sides. The top of the box will be open.

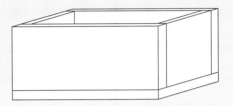

Rosita plans to cut the five pieces from a long wooden board that is 4 inches wide and $\frac{5}{8}$ inch thick. Her pencils are about $7\frac{1}{2}$ inches long when they are new. Rosita wants to make the inside of the box $\frac{3}{4}$ inch longer, so it is easy to take the pencils out. She made this sketch of the top view of the box. All measurements are in inches.

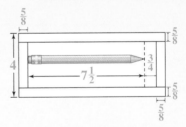

• What lengths will Rosita need to cut for the bottom, ends, and sides of her box? Give your answers as fractions or mixed numbers. Show how you found your answers.

• What will the height of the box be?

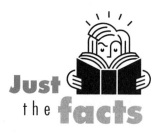

1. Suppose a piece of wood $6\frac{7}{8}$ inches long is cut from a $36\frac{1}{2}$-inch board. How much of the board is left? Explain how you found your answer.

2. Juana needs two red ribbons for a costume she is making. One ribbon must be $2\frac{1}{3}$ feet long, and the other must be $3\frac{1}{2}$ feet long. What total length of red ribbon does she need?

3. Juana has a piece of blue ribbon $6\frac{1}{2}$ feet long. If she cuts off a piece $3\frac{3}{4}$ feet long, how much blue ribbon will be left?

4. Juana has a length of green ribbon that is $3\frac{2}{3}$ yards long. If she cuts off a piece $1\frac{3}{4}$ yards long, how many yards of green ribbon will remain?

Althea and Jing are comparing how they thought about Problem 4 in Problem Set G.

Think & Discuss

Althea changed both mixed numbers to fractions. Finish her solution.

Jing changed $3\frac{2}{3}$ to $2\frac{5}{3}$. Explain why these two numbers are equal. How does this step help Jing find the solution?

Finish Jing's solution.

Just the facts

For most people around the world, decimals are easier to understand and work with than fractions. In 1997, the New York Stock Exchange broke with 200 years of tradition by voting to phase out its use of fractions in quoting stock prices.

Problem Set H

Use any method you like to solve these problems.

1. Jahmal is following the progress of Amresco stock. On Monday, the starting value of the stock was $16\frac{1}{8}$. During the week, it changed by the following amounts:

Monday	Tuesday	Wednesday	Thursday	Friday
$-\frac{3}{4}$	$+1\frac{1}{8}$	$+\frac{1}{4}$	$-\frac{1}{2}$	$-\frac{7}{8}$

   a. What was the value of Amresco stock at the end of Friday?

   b. How much did the stock gain or lose during the week?

   c. What were the stock's highest and lowest values during the week? On what days did each of these values occur?

2. Miguel's brother Carlos is $69\frac{1}{2}$ inches tall. Last year Carlos was $63\frac{3}{4}$ inches tall. How much did he grow in the year?

Find each sum or difference. Give your answers in lowest terms.

3. $3\frac{7}{8} + 2\frac{1}{4}$

4. $3\frac{7}{8} - 2\frac{1}{4}$

5. $6\frac{1}{3} - 5\frac{3}{4}$

6. $13\frac{3}{4} + 8\frac{19}{20}$

7. $22\frac{7}{10} - 13\frac{3}{4}$

8. $9\frac{1}{2} + 3\frac{7}{8}$

Share & Summarize

Look back at Althea's and Jing's methods for subtracting mixed numbers.

1. For which types of problems do you prefer Althea's method? Give an example.

2. For which types of problems do you prefer Jing's method? Give an example.

Lab
Investigation ▶ Using a Fraction Calculator

Many calculators allow you to perform operations with fractions and mixed numbers. In this lab, you will use a calculator to add and subtract fractions.

MATERIALS

calculator with fraction capabilities

TI-34 II

Learning the Basics

To enter a fraction on your calculator:

- Enter the numerator.
- Press ⬚.
- Enter the denominator.

To enter a mixed number:

- Enter the whole-number part.
- Press (UNIT).
- Enter the numerator of the fraction part.
- Press ⬚.
- Enter the denominator of the fraction part.

1. Use your calculator to find $\frac{1}{4} + 1\frac{2}{3}$.

If you enter a fraction that is not in lowest terms, or if a calculation results in a fraction that is not in lowest terms, the calculator may display something like N/D → n/d. To put the fraction in lowest terms:

- Press (SIMP).
- Enter a common factor of the numerator and denominator.
- Press (ENTER).

The calculator will divide the numerator and denominator by the factor you specify and display the result. If the fraction is still not in lowest terms, the calculator will continue to display N/D → n/d. In that case, repeat the steps above to divide by another common factor.

2. Estimate the value of $\frac{180}{210}$. Then, use your calculator to help you write $\frac{180}{210}$ in lowest terms. Compare your answer to your estimate.

To use your calculator to change a fraction to a mixed number or to change a mixed number to a fraction:

- Press ⟨2nd⟩ [A^b/c ◀▶ ^{d}/e].

- Press ⟨ENTER =⟩.

3. Use your calculator to change $3\frac{5}{7}$ to a fraction and to change $\frac{43}{21}$ to a mixed number.

MATERIALS

2 decks of Fraction Match cards

Playing *Fraction Match*

Fraction Match is a memory game for two players. Here are the rules:

- Choose Deck 1 or Deck 2.

- Shuffle the cards. Place them face down in five rows of six cards each.

- The first player turns over two cards. He or she uses a calculator, if needed, to determine whether the values on the cards are the same.

- If the cards have the same value, the player keeps them and takes another turn. If they have different values, the player turns them back over and his or her turn ends.

- Play continues until all the cards have been taken. The player with the most cards at the end of the game wins.

When you finish the game, play again with the other deck.

What Did You Learn?

4. Describe step by step how to use a calculator to find $2\frac{11}{14} - \frac{2}{7}$.

5. Design your own deck of *Fraction Match* cards. Your deck should have at least 16 cards. Test your deck by playing *Fraction Match* with a friend or classmate.

On Your Own Exercises

Practice &
Apply

Solve each problem in your head.

1. $\frac{3}{4} - \frac{2}{4} =$ _____

2. $\frac{12}{17} - \frac{4}{17} =$ _____

3. _____ $- \frac{3}{8} = \frac{1}{8}$

4. $\frac{5}{12} -$ _____ $= \frac{1}{12}$

5. _____ $- \frac{8}{9} = \frac{1}{9}$

6. _____ $- \frac{7}{11} = \frac{5}{11}$

Use your fraction pieces or another method to help fill in each blank.

7. $\frac{1}{4} +$ _____ $= 1$

8. $\frac{1}{2} + \frac{1}{3} +$ _____ $= 1$

9. $\frac{1}{12} + \frac{2}{6} +$ _____ $+ \frac{1}{4} = 1$

10. $\frac{1}{6} + \frac{5}{12} +$ _____ $+ \frac{1}{3} = 1$

11. $\frac{1}{2} + \frac{1}{3} +$ _____ $= 1\frac{1}{6}$

12. $\frac{2}{3} + \frac{11}{12} -$ _____ $= \frac{5}{6}$

13. $\frac{5}{6} + \frac{1}{3} + \frac{1}{2} +$ _____ $= 2$

14. $\frac{1}{6} + \frac{1}{2} +$ _____ $= \frac{3}{4}$

Use your fraction pieces or another method to write each sum or difference with a common denominator. Then find the sum or difference. Give your answers in lowest terms. If an answer is greater than 1, write it as a mixed number.

15. $\frac{5}{6} - \frac{1}{2}$

16. $\frac{11}{12} - \frac{3}{4}$

17. $\frac{7}{6} - \frac{1}{3}$

18. $\frac{13}{12} + \frac{3}{4}$

Find each sum or difference. Give your answers in lowest terms. If an answer is greater than 1, write it as a mixed number.

19. $\frac{3}{8} + \frac{2}{3}$

20. $\frac{9}{22} - \frac{3}{10}$

21. $\frac{25}{32} + \frac{7}{24}$

22. $\frac{8}{26} + \frac{9}{39}$

23. On a sheet of paper, create a magic square with a sum of 2 using the numbers $\frac{11}{12}, \frac{3}{4}, \frac{5}{6}, \frac{1}{2}, \frac{2}{3}, \frac{5}{12}, \frac{1}{3}$, 1, and $\frac{7}{12}$.

24. On a sheet of paper, create a magic square with a sum of $1\frac{1}{2}$ using the numbers $\frac{1}{2}, \frac{1}{4}, \frac{2}{3}, \frac{1}{6}, \frac{5}{12}, \frac{5}{6}, \frac{1}{3}, \frac{3}{4}$, and $\frac{7}{12}$.

Find each sum or difference, showing each step of your work. Give your answers in lowest terms. If an answer is greater than 1, write it as a mixed number.

25. $2\frac{1}{2} - \frac{7}{9}$

26. $1\frac{8}{15} - \frac{3}{5}$

27. $10\frac{2}{5} - 4\frac{1}{3}$

28. $3\frac{5}{6} + \frac{6}{7}$

29. $3\frac{1}{4} + 1\frac{1}{3}$

30. $4\frac{1}{3} - 2\frac{3}{8}$

impactmath.com/self_check_quiz

31. Economics On Monday, the starting value of EdCorp stock was $9\frac{1}{4}$. During the week, the value changed by the following amounts:

Monday	Tuesday	Wednesday	Thursday	Friday
$+1\frac{1}{2}$	$-\frac{7}{8}$	$-1\frac{3}{4}$	$+\frac{5}{8}$	$+1\frac{1}{4}$

a. What was the value of EdCorp stock at the end of Friday?

b. How much did the stock gain or lose during the week?

c. What were the stock's highest and lowest values during the week? On what days did each of these values occur?

Connect & Extend

32. Cover your fraction mat in as many different ways as you can using *four* different types of fraction pieces.

a. Write an equation for each combination you find.

b. Describe the strategy you used to find all the possible equations.

Give the rule for finding each term in the sequence from the previous term. Then use your rule to find the missing terms. Use the last term to help check your answers.

33. $0, \frac{1}{4}, \frac{2}{4}, \frac{3}{4},$ ___, ___, ___, ___, $\frac{8}{4}$

34. $0, \frac{2}{3}, \frac{4}{3}, \frac{6}{3},$ ___, ___, ___, ___, $\frac{16}{3}$

35. $\frac{24}{4}, \frac{21}{4}, \frac{18}{4}, \frac{15}{4},$ ___, ___, ___, ___, $\frac{0}{4}$

In Exercises 36–41, use this information:

Rolling Fractions is played with a number cube with faces labeled $\frac{1}{2}$, $\frac{1}{3}$, $\frac{1}{4}$, $\frac{1}{6}$, $\frac{1}{8}$, and $\frac{1}{12}$. Each player has a game card divided into 24 equal rectangles. Players take turns rolling the cube and shading that fraction of the card. For example, if a player rolls $\frac{1}{2}$, he would shade 12 of the 24 rectangles.

If the fraction rolled is greater than the unshaded fraction of the card, the player shades no rectangles for that turn. The first player to shade the card completely is the winner.

Rolling Fractions
Game Card

36. In her first two turns, Caroline rolled $\frac{1}{2}$ and $\frac{1}{6}$. To win the game on her next turn, which fraction would she need to roll?

37. In his first two turns, Miguel rolled $\frac{1}{3}$ and $\frac{1}{12}$. To win the game in *two* more turns, which two fractions would he need to roll?

38. In their first three turns, Conor rolled $\frac{1}{8}$, $\frac{1}{6}$, and $\frac{1}{3}$, and Jahmal rolled $\frac{1}{12}$, $\frac{1}{2}$, and $\frac{1}{3}$. Who is more likely to win on his next roll? Explain why.

39. In her first two turns, Rosita rolled $\frac{1}{2}$ and $\frac{1}{3}$.

   a. To win on her next turn, which fraction would she need to roll?

   b. To win in *two* more turns, which fractions would she need to roll?

   c. Which fractions would give Rosita a sum greater than 1 on her third turn?

40. What is the fewest number of turns it could take to win this game? Which fractions would a player have to roll in that number of turns?

41. Luke rolled $\frac{1}{2}$, $\frac{1}{3}$, and $\frac{1}{8}$. He says he might as well quit because there is no way for him to win. Do you agree? Explain.

42. Arrange the numbers $\frac{3}{4}$, $\frac{2}{3}$, $\frac{1}{4}$, $\frac{11}{12}$, $\frac{5}{12}$, $\frac{1}{3}$, $\frac{7}{12}$, $\frac{1}{2}$, and $\frac{5}{6}$ into a magic square. What is the sum for your magic square?

43. Arrange the numbers 1, $\frac{2}{3}$, $\frac{5}{6}$, $\frac{1}{3}$, $\frac{1}{2}$, $\frac{7}{6}$, $\frac{3}{2}$, $\frac{4}{3}$, and $\frac{5}{3}$ into a magic square. What is the sum for your magic square?

44. **Challenge** Create a magic square with a sum of 1 in which the denominators of the fractions are factors of 24.

Remember

In a magic square, each row, column, and diagonal has the same sum.

45. Arrange 1, 2, 3, 4 in the boxes to create the least possible sum. Use each number exactly once.

$$\frac{\square}{\square} + \frac{\square}{\square}$$

46. Arrange 1, 2, 3, 4 in the boxes to create the least possible positive difference. Use each number exactly once.

$$\frac{\square}{\square} - \frac{\square}{\square}$$

47. Arrange 2, 3, 4 and 12 in the boxes to create the least possible sum. Use each number exactly once.

$$\frac{\square}{\square} + \frac{\square}{\square}$$

48. Arrange 2, 3, 4, and 12 in the boxes to create the least possible positive difference. Use each number exactly once.

$$\frac{\square}{\square} - \frac{\square}{\square}$$

Give the rule for finding each term in the sequence from the previous term. Then use your rule to find the missing terms. Use the last term to check your answers.

49. $1\frac{2}{5}, 2\frac{4}{5}, 4\frac{1}{5}, 5\frac{3}{5},$ ____ , ____ , ____ , ____ , $12\frac{3}{5}$

50. $3\frac{1}{2}, 6\frac{3}{4}, 10, 13\frac{1}{4},$ ____ , ____ , ____ , ____ , $29\frac{1}{2}$

51. $15, 14\frac{5}{8}, 14\frac{1}{4}, 13\frac{7}{8},$ ____ , ____ , ____ , ____ , 12

52. Measurement Below is part of a ruler.

a. What fraction of an inch does the smallest division of this ruler represent? How do you know?

b. How could you use this ruler to find $1\frac{1}{8} + \frac{5}{16}$? Find the sum.

c. How could you use this ruler to find $2\frac{1}{4} - \frac{5}{16}$? Find the difference.

d. Could you use this ruler to find $2\frac{1}{2} - \frac{5}{12}$? Explain.

e. Could you use this ruler to find $\frac{3}{8} + 1\frac{7}{16}$? Explain.

Mixed Review

Find each quantity.

53. $\frac{1}{5}$ of 300 **54.** $\frac{2}{6}$ of 180 **55.** $\frac{3}{4}$ of 32 **56.** $\frac{1}{4}$ of 1.6

57. $\frac{1}{2}$ of 3 **58.** $\frac{1}{3}$ of $\frac{1}{3}$ **59.** $\frac{1}{3}$ of $\frac{1}{6}$ **60.** $\frac{1}{3}$ of $\frac{1}{9}$

Number Sense Name a fraction between the given fractions.

61. $\frac{1}{3}$ and $\frac{1}{2}$ **62.** $\frac{1}{4}$ and $\frac{4}{15}$ **63.** $\frac{13}{16}$ and $\frac{11}{12}$

64. Geometry Choose which of these terms describe each polygon (list all that apply): quadrilateral, pentagon, hexagon, concave, symmetric, regular.

a. **b.** **c.**

65. Measurement The bar graph shows the air distance from several cities to New York City.

Air Distances from NYC

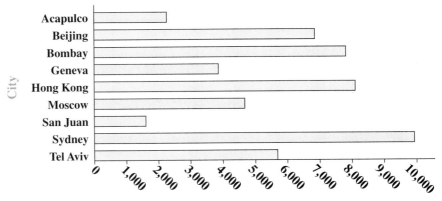

Source: *New York Public Library Desk Reference.* New York: Macmillan, 1998.

a. Which city on the graph is closest to New York City? What is the approximate air distance between this city and New York?

b. Which city on the graph is farthest from New York City? What is the approximate air distance between this city and New York?

c. Last week, Ms. Frankel flew from New York City to Geneva for a business meeting, back to New York City, and then to Hong Kong to visit a friend. About how many miles did she fly last week?

d. About how many times farther from New York City is Moscow than San Juan?

Pagoda in
Hong Kong

3.2 Multiplying and Dividing with Fractions

In this lesson, you will learn how to multiply and divide with fractions. As you work on the problems, it may be helpful to think about what multiplication and division mean and about how these operations work with whole numbers.

Explore

Work with your group to solve these problems. Try to find more than one way to solve each problem.

- You want to serve lemonade to 20 people. Each glass holds $\frac{3}{4}$ cup. How many cups of lemonade do you need?

- You grew 12 pounds of peas. You give some away, keeping $\frac{2}{3}$ of the peas for yourself. How many pounds do you have left?

The problems you just solved can be represented with multiplication equations. Write a multiplication equation to represent each problem, and explain why the equation fits the situation.

Investigation 1 ▶ Multiplying Fractions and Whole Numbers

In this investigation, you will explore more problems involving multiplication with fractions. As you work on the problems, you might try some of the strategies you and your classmates used in the Explore.

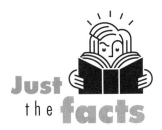

Just the facts

The vanilla extract used in recipes is derived from the vanilla planifolia orchid. This plant is native to Mexico, where it is pollinated by bees and tiny hummingbirds. In other parts of the world, the plant must be pollinated by hand.

Problem Set A

1. Suppose you want to bake a fraction cake.

a. If you want enough cake to serve 12 people, how much of each ingredient do you need?

b. For each ingredient, write a multiplication equation to represent the work you did in Part a.

> **Fraction Cake**
> $\frac{3}{4}$ cup sugar
> $2\frac{1}{2}$ cups flour
> $\frac{1}{2}$ teaspoon salt
> $1\frac{1}{2}$ teaspoons vanilla
> 2 eggs
>
> Serves four people.

2. Find each product using any method you like.

a. $\frac{1}{4} \times 20$ **b.** $20 \cdot \frac{1}{2}$ **c.** $20 \times \frac{3}{4}$ **d.** $\frac{3}{2} \cdot 20$

3. Describe the strategies you used to find the products in Problem 2.

Think & Discuss

Hannah and Jahmal have different strategies for multiplying a whole number by a fraction.

Hannah says:

I multiply the whole number by the numerator of the fraction, and then divide the result by the denominator.

Jahmal says:

I do just the opposite. I divide the whole number by the denominator of the fraction, and then multiply the result by the numerator.

Try both methods on Parts a–d of Problem 2 above. Do they both work?

You will probably find that some multiplication problems are easier to do with Hannah's method and others are easier with Jahmal's method. For each problem below, tell whose method you think would be easier to use.

$$7 \cdot \frac{5}{14} \qquad \frac{3}{4} \cdot 8 \qquad 10 \cdot \frac{7}{36} \qquad \frac{9}{10} \cdot 5$$

The methods described above also work for multiplying a fraction by a decimal.

Problem Set B

Use Hannah's or Jahmal's method, or one of your own, to solve these problems. Show how you find your answers.

1. A cocoa recipe for one person calls for $\frac{3}{4}$ cup milk. Tell how much milk is needed to make the recipe for the following numbers of people.

   a. 3 people   **b.** 5 people   **c.** 6 people   **d.** 8 people

2. Fudge costs $5.80 per pound. Use the fact that 16 ounces = 1 pound to help find the cost of each amount of fudge.

   a. $\frac{3}{4}$ pound   **b.** 8 ounces   **c.** 4 ounces   **d.** $1\frac{1}{2}$ pounds

3. At Fiona's Fabrics, plaid ribbon costs $0.72 per yard. Give the cost of each length of ribbon.

   a. $\frac{2}{3}$ yard   **b.** $1\frac{1}{2}$ feet   **c.** 8 inches   **d.** $1\frac{3}{4}$ yards

Problem Set C

Complete each multiplication table.

1.

×	24	120	60	72
$\frac{1}{2}$	12			
$\frac{2}{3}$				
$\frac{1}{4}$			15	
$\frac{3}{4}$				

2.

×	180		60	120
$\frac{1}{5}$	36			
$\frac{1}{2}$			30	
$\frac{2}{3}$		200		
$\frac{1}{4}$				30

3.

×				120
$\frac{3}{2}$			36	
	64	96		
$\frac{3}{8}$		54		
			20	100

Share & Summarize

1. Make up a word problem that can be solved by multiplying a whole number and a fraction. Explain how to solve your problem.

2. What calculation do you need to do to find $\frac{3}{4}$ of 6 inches? What is $\frac{3}{4}$ of 6 inches?

Investigation 2 ▶ A Model for Multiplying Fractions

Remember

The area of a rectangle is its length times its width.

You can visualize the product of two whole numbers by drawing a rectangle with side lengths equal to the numbers. The area of the rectangle represents the product. This rectangle represents 4×6:

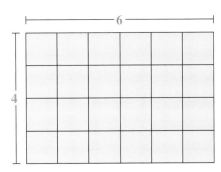

$4 \times 6 = 24$

You can represent the product of a whole number and a fraction in the same way. The shaded portion of this diagram represents the product $\frac{1}{2} \times 6$:

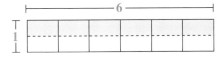

Think & Discuss

What is the area of each small rectangle in the diagram above? ☐

How many small rectangles are shaded?

Use your answers to the previous questions to find the total shaded area. Explain why this area is equal to $\frac{1}{2} \times 6$.

Problem Set D

1. Consider this diagram.

   a. What two numbers does the shaded region show the product of?

   b. What is the area of each small rectangle? How many small rectangles are shaded?

   c. Use your answers from Part b to find the total shaded area.

   d. Write a multiplication equation to represent the product of the numbers from Part a.

2. You can use similar diagrams to represent the product of two fractions. The shaded portion of this diagram represents the product $\frac{2}{3} \times \frac{3}{4}$.

   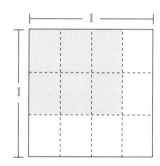

   a. What are the dimensions of the entire figure?

   b. Look at the entire shaded region. What is the height of this region? What is the width of this region?

   c. Use your answers from Part b to explain why the area of the shaded region is $\frac{2}{3} \times \frac{3}{4}$.

   d. What is the area of each small rectangle? How many small rectangles are shaded?

   e. Use your answers from Part d to find $\frac{2}{3} \times \frac{3}{4}$.

Draw a diagram like the one in Problem 2 to represent each product. Then use your diagram to find the product. Give your answer as a multiplication equation (for example, $\frac{2}{3} \times \frac{3}{4} = \frac{6}{12}$).

3. $\dfrac{1}{2} \times \dfrac{1}{3}$ 4. $\dfrac{3}{5} \cdot \dfrac{3}{4}$ 5. $\dfrac{1}{2} * \dfrac{5}{6}$

6. Look back at your diagrams and equations from Problems 3–5. Can you see a shortcut for multiplying two fractions *without* making a diagram? If so, use your shortcut to find $\frac{4}{7} \times \frac{2}{3}$. Then draw a diagram to see if your shortcut worked.

Think & Discuss

You may have noticed the following shortcut for multiplying two fractions:

The product of two fractions is the product of the numerators over the product of the denominators.

Use one of your rectangle diagrams from Problem Set D to explain why this shortcut works.

Problem Set E

Use the shortcut described above to find each product. Express your answers in both the original form and in lowest terms.

1. $\frac{3}{5} \times \frac{1}{4}$ **2.** $\frac{1}{6} \cdot \frac{2}{3}$ **3.** $\frac{4}{5} \times \frac{5}{8}$

4. $\frac{2}{3} \cdot \frac{3}{7}$ **5.** $\frac{2}{3} * \frac{1}{8}$ **6.** $\frac{2}{3} \cdot \frac{3}{8}$

7. Rob wants to create a small herb garden in his back yard. The space he marked off is a square, $\frac{7}{8}$ of a meter on each side. What will the area of his herb garden be?

Share & Summarize

Draw a diagram to represent the product of two fractions you have not yet multiplied together in this investigation. Explain how the diagram shows the product.

Investigation 3 More Multiplying with Fractions

In the last investigation, you found a shortcut for multiplying fractions. Now you will look at products involving mixed numbers.

EXAMPLE

This diagram illustrates $1\frac{1}{2} \cdot 2$.
Each shaded section is labeled with its area. The total area is
$1 + 1 + \frac{1}{2} + \frac{1}{2} = 3$, so

$$1\frac{1}{2} \cdot 2 = 3$$

Problem Set F

Draw a diagram to illustrate each product. Then use your diagram to find the product. Give your answer as a multiplication equation.

1. $1\frac{1}{2} \times \frac{1}{3}$

2. $1\frac{1}{2} \cdot 2\frac{1}{2}$

3. Marcus suggested this shortcut for multiplying mixed numbers:

Multiply the whole number parts, multiply the fraction parts, and add the two results.

Try Marcus' method to find the products in Problems 1 and 2. Does it work?

4. Miguel suggested this shortcut for multiplying mixed numbers:

Change the mixed numbers to fractions, and multiply.

Try Miguel's method on the products in Problems 1 and 2. Does it work?

Since there are several calculations involved in multiplying two mixed numbers, it is a good idea to estimate the product before you multiply.

Think & Discuss

Consider this problem:

$$1\frac{1}{3} \times 5\frac{2}{3}$$

Before multiplying, make an estimate of the product. Explain how you found your answer.

Now change both mixed numbers to fractions, and multiply.

How does your result compare to your estimate?

Problem Set G

In Problems 1–6, complete Parts a and b.

a. Estimate the product.

b. Find the product, showing all your steps, and give your result as a mixed number. If your answer is far from your estimate, check your calculations.

1. $1\frac{3}{8} \cdot 2\frac{1}{2}$

2. $3\frac{1}{3} \times \frac{8}{5}$

3. $\frac{1}{4} \times 8\frac{3}{5}$

4. $3\frac{1}{2} * 1\frac{2}{3}$

5. $9\frac{2}{3} \times 1\frac{1}{2}$

6. $2\frac{1}{4} \cdot \frac{7}{8}$

7. Wei-Ling wants to hang wallpaper on two walls in her kitchen. One wall measures $11\frac{1}{2}$ feet by $8\frac{1}{2}$ feet, and the other measures $15\frac{2}{3}$ feet by $8\frac{1}{2}$ feet. About how many square feet of wallpaper will she need? Estimate first, and then calculate.

After you multiply fractions or mixed numbers, you often have to put the product in lowest terms. Sometimes you can save yourself work by simplifying *before* you multiply. To do this, first check whether the numerator of each fraction shares a common factor with the denominator of the other fraction.

Find $\frac{1}{3} \times \frac{3}{4}$.

Notice that **3** is a factor of the denominator of $\frac{1}{3}$ and the numerator of $\frac{3}{4}$.

$$\frac{1}{3} \times \frac{3}{4} = \frac{1 \times 3}{3 \times 4}$$
Rewrite as the product of numerators over the product of denominators.

$$= \frac{3}{3} \times \frac{1}{4}$$
Group the common factors to form a fraction equal to 1.

$$= 1 \times \frac{1}{4}$$
Simplify.

$$= \frac{1}{4}$$

Find $\frac{2}{3} \times \frac{9}{16}$.

Notice that **2** is a factor of 2 and 16, and **3** is a factor of 3 and 9. As in the previous example, group these common factors to form a fraction equal to 1.

$$\frac{2}{3} \times \frac{9}{16} = \frac{2 \times 9}{3 \times 16}$$
Rewrite as the product of numerators over the product of denominators.

$$= \frac{2 \times 3 \times 3}{3 \times 2 \times 8}$$
Rewrite 9 as 3×3 and 16 as 2×8.

$$= \frac{2 \times 3}{2 \times 3} \times \frac{3}{8}$$
Group the common factors to form a fraction equal to 1.

$$= 1 \times \frac{3}{8}$$
Simplify.

$$= \frac{3}{8}$$

In Problems 1–6, find the product in two ways:

- Multiply the fractions, and write the product in lowest terms.
- Simplify before finding the product.

Show all your steps.

1. $\frac{3}{5} \times \frac{15}{6}$ **2.** $\frac{5}{6} \cdot \frac{3}{10}$ **3.** $\frac{1}{8} \times \frac{2}{3}$

4. $\frac{7}{12} \cdot \frac{3}{5}$ **5.** $\frac{3}{10} \times \frac{2}{3}$ **6.** $\frac{1}{2} \cdot 2\frac{4}{5}$

Find each product.

7. $\frac{3}{8} \cdot \frac{16}{7}$ **8.** $\frac{2}{3} \times \frac{7}{8}$ **9.** $\frac{2}{5} \cdot \frac{5}{6}$

10. $\frac{4}{5} \times \frac{3}{4}$ **11.** $\frac{1}{5} \cdot 1\frac{1}{2}$ **12.** $\frac{4}{5} \times 5\frac{3}{16}$

13. Mali wants to make enough fraction pasta for six servings. Rewrite the recipe for her.

> **Fraction Pasta**
>
> $1\frac{1}{2}$ cups vegetable stock
>
> $\frac{2}{3}$ cup diced carrots
>
> $\frac{3}{4}$ cup asparagus tips
>
> $\frac{5}{6}$ cup peas   Makes five servings.
>
> $1\frac{1}{3}$ pounds pasta
>
> 2 tablespoons olive oil
>
> $1\frac{1}{6}$ cups Parmesan cheese

Share & Summarize

1. Explain how to multiply two mixed numbers. Tell how you can use estimation to determine whether your answer is reasonable.

2. Explain how you could find the product $\frac{3}{4} \times \frac{8}{9}$ by simplifying before you multiply.

Investigation Dividing Whole Numbers by Fractions

You know how to add, subtract, and multiply fractions. Now you will learn how to divide a whole number by a fraction.

Just the facts

Apples belong to the rose family. More than 7,500 varieties of apple are grown worldwide—including 2,500 varieties grown in the United States.

Explore

Work with your group to solve the following problems. Try to find more than one way to solve each problem.

- Suppose you have five apples to share with your friends. If you divide each apple in half, how many halves will you have to share?

- There are 10 cups of punch left in the punchbowl. Each glass holds $\frac{2}{3}$ of a cup. How many glasses can you fill?

The problems above can be represented by division equations. Write an equation to represent each problem, and explain why the equation fits the situation.

Caroline and Marcus solved the second Explore problem in different ways.

Problem Set ▌

Use Caroline's or Marcus' method to find each quotient. Try each method at least once.

1. $5 \div \frac{1}{3}$ **2.** $6 \div \frac{1}{6}$

3. $4 \div \frac{2}{3}$ **4.** $8 \div \frac{4}{5}$

5. $3 \div \frac{3}{6}$ **6.** $5 \div \frac{5}{6}$

Every multiplication problem has two related division problems. Here are two examples:

Multiplication Problem	Related Division Problems	
$2 \times 10 = 20$	$20 \div 10 = 2$	$20 \div 2 = 10$
$\frac{1}{2} \times 40 = 20$	$20 \div 40 = \frac{1}{2}$	$20 \div \frac{1}{2} = 40$

You can use this idea to solve division problems involving fractions.

EXAMPLE

Find $20 \div \frac{1}{4}$.

The *quotient* is the number that goes in the blank in this division equation:

$$20 \div \frac{1}{4} = \underline{\quad}$$

You can find the quotient by thinking about the related multiplication equation:

$$\frac{1}{4} \times \underline{\quad} = 20$$

Now just think, "One fourth of what number equals 20?" The answer is 80. So, $20 \div \frac{1}{4} = 80$.

Problem Set J

Fill in the blanks in each pair of related equations.

1. $15 \div \frac{1}{2} =$ _____ $\frac{1}{2} \cdot$ _____ $= 15$

2. $20 \div \frac{2}{3} =$ _____ $\frac{2}{3} \cdot$ _____ $= 20$

Find each quotient by writing and solving a related multiplication equation.

3. $20 \div \frac{1}{5} =$ _____   **4.** $14 \div \frac{2}{3} =$ _____

5. $15 \div \frac{3}{5} =$ _____   **6.** $12 \div \frac{3}{4} =$ _____

All the quotients you have found so far are whole numbers. Of course, this is not always the case. Find each quotient below using any method you like. Explain how you found your answer.

7. $3 \div \frac{2}{5} =$ _____   **8.** $7 \div \frac{3}{4} =$ _____

Problem Set K

Use any methods you like to complete each division table. Each entry is the result of dividing the first number in that row by the top number in that column. As you work, look for patterns that might help you complete the table without computing every quotient.

1.

÷	$\frac{1}{3}$	$\frac{2}{3}$	$\frac{3}{3}$
6			
4			
2			

2.

÷	$\frac{1}{4}$	$\frac{2}{4}$	$\frac{3}{4}$	$\frac{4}{4}$
12				
18				
24				

3.

÷	$\frac{1}{5}$	$\frac{2}{5}$	$\frac{3}{5}$	$\frac{4}{5}$	$\frac{5}{5}$
24					
18					
9					

4. What patterns did you notice as you completed each table?

5. Suppose you know that $16 \div \frac{1}{3} = 48$. Describe a quick way to find $16 \div \frac{2}{3}$. What is $16 \div \frac{2}{3}$?

6. Suppose you know that $15 \div \frac{1}{4} = 60$. Describe a quick way to find $15 \div \frac{2}{4}$ and $15 \div \frac{3}{4}$. What are the results?

7. Use the fact that $15 \div \frac{1}{5} = 75$ to find each quotient.

    a. $15 \div \frac{2}{5}$ **b.** $15 \div \frac{3}{5}$ **c.** $15 \div \frac{4}{5}$

Share & Summarize

1. Describe how to find $6 \div \frac{2}{3}$ by using a related multiplication equation.

2. Describe one other method for computing $6 \div \frac{2}{3}$.

Investigation 5 ▶ Dividing Fractions by Fractions

In the last investigation, you saw how to divide a whole number by a fraction. You can use the same methods to divide a fraction by a fraction, but it can be much more difficult.

Explore

Find $\frac{5}{8} \div \frac{1}{4}$ by using a diagram, a number line, or another model to figure out how many $\frac{1}{4}$s are in $\frac{5}{8}$.

Find $\frac{5}{8} \div \frac{1}{4}$ by writing a related multiplication equation.

Now try to use one of the methods above to find $\frac{2}{3} \div \frac{3}{5}$. (Warning: It isn't easy!) If you find the answer, explain how you found it.

You probably found it difficult to use the methods you know to compute $\frac{2}{3} \div \frac{3}{5}$ in the Explore. Luckily, there is an easier method for dividing fractions. To understand how it works, you will need to use two facts you learned earlier.

Fact 1: A division problem can be written in the form of a fraction. For example, $2 \div 3$ can be written $\frac{2}{3}$.

Fact 2: The value of a fraction does not change when its numerator and denominator are multiplied by the same number.

Just the facts

The $\div$ symbol for division is called an *obelus*. The word comes from the Greek word *obelos*, meaning "spike." The word *obelisk*, a four-sided tapering pillar, has a similar origin.

EXAMPLE

Find $\frac{3}{4} \div \frac{2}{3}$.

Start by using the first fact above to rewrite the division problem as a fraction.

$$\frac{\frac{3}{4}}{\frac{2}{3}}$$

This fraction looks complicated, but you can make it simpler by using the second fact above.

You want to multiply the numerator and denominator by a number that will change *both* the numerator and the denominator to whole numbers. Any common multiple of 4 and 3 will do, such as 12.

$$\frac{\frac{3}{4} \cdot \frac{12}{1}}{\frac{2}{3} \cdot \frac{12}{1}} = \frac{\frac{36}{4}}{\frac{24}{3}} = \frac{9}{8}$$

So, $\frac{3}{4} \div \frac{2}{3} = \frac{9}{8}$, or $1\frac{1}{8}$.

It's always a good idea to check the answer by multiplying:

$$\frac{2}{3} \times \frac{9}{8} = \frac{18}{24} = \frac{3}{4}$$

The answer is correct.

Problem Set L

Use the method described in the Example to find each quotient. Show all your steps.

1. $\frac{7}{8} \div \frac{2}{3}$

2. $\frac{5}{6} \div \frac{1}{4}$

3. $5 \div \frac{3}{8}$

4. $\frac{1}{4} \div \frac{3}{4}$

5. $\frac{2}{5} \div 4$

6. $1\frac{1}{3} \div \frac{4}{5}$

Fill in each blank with one of the following: *greater than, less than,* or *equal to.* Give an example to illustrate each completed sentence.

7. When you divide a fraction by a greater fraction, the quotient is _____ 1.

8. When you divide a fraction by a lesser fraction, the quotient is _____ 1.

Many people use a shortcut when dividing with fractions. The patterns you find in the Think & Discuss below will help you understand the shortcut and why it works.

Think & Discuss

Find each product in your head. Describe any patterns you see in the problems and the answers.

$$\frac{3}{4} \times \frac{4}{3} \qquad \frac{2}{15} \times \frac{15}{2}$$

$$\frac{5}{8} \times \frac{8}{5} \qquad \frac{25}{100} \times \frac{100}{25}$$

Now find these products. How are these problems similar to those above?

$$\frac{1}{4} \times 4 \qquad \frac{1}{6} \times 6$$

$$\frac{1}{20} \times 20 \qquad \frac{1}{100} \times 100$$

Two numbers with a product of 1 are **reciprocals** of one another. Every number except 0 has a reciprocal. You can find the reciprocal of a fraction by switching its numerator and denominator. Here is the shortcut for dividing fractions:

To divide a fraction by a fraction, multiply the first fraction by the reciprocal of the second fraction.

EXAMPLE

Find $\frac{5}{7} \div \frac{10}{12}$.

To find $\frac{5}{7} \div \frac{10}{12}$, multiply $\frac{5}{7}$ by the reciprocal of $\frac{10}{12}$:

$$\frac{5}{7} \div \frac{10}{12} = \frac{5}{7} \times \frac{12}{10} = \frac{60}{70} = \frac{6}{7}$$

To see why the shortcut works, rewrite the division problem as a fraction, and multiply *both* the numerator and denominator by the reciprocal of the denominator. The denominator becomes 1.

$$\frac{5}{7} \div \frac{10}{12} = \frac{\frac{5}{7}}{\frac{10}{12}} = \frac{\frac{5}{7} \cdot \frac{12}{10}}{\frac{10}{12} \cdot \frac{12}{10}} = \frac{\frac{5}{7} \cdot \frac{12}{10}}{1} = \frac{5}{7} \cdot \frac{12}{10}$$

So, $\frac{5}{7} \div \frac{10}{12} = \frac{5}{7} \cdot \frac{12}{10}$.

Problem Set M

Find each quotient using any method you like.

1. $\frac{3}{2} \div \frac{9}{6}$ **2.** $\frac{2}{5} \div \frac{5}{2}$ **3.** $\frac{1}{8} \div \frac{1}{9}$

Estimate whether each quotient will be greater than, less than, or equal to 1. Then find the quotient.

4. $\frac{3}{5} \div \frac{3}{4}$ **5.** $2 \div \frac{3}{5}$ **6.** $\frac{2}{3} \div \frac{5}{6}$

7. $1 \div 3\frac{1}{2}$ **8.** $3\frac{1}{2} \div \frac{2}{7}$ **9.** $4\frac{1}{2} \div 2\frac{1}{4}$

Share & Summarize

1. Describe two methods for dividing a fraction by a fraction.

2. Write two fraction division problems—one with a quotient greater than 1 and the other with a quotient less than 1.

On Your Own Exercises

Practice & Apply

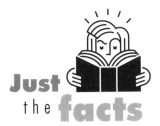

Just the **facts**

The first practical sewing machines were built in the mid 1800s and could sew only straight seams. Modern sewing machines are often equipped with computer technology—for instance, scanners that can take an image and reproduce it in an embroidered version on cloth.

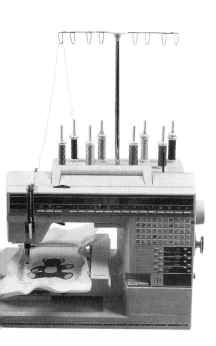

1. Caroline is sewing a costume with 42 small ribbons on it. Each ribbon is $\frac{1}{3}$ of a yard long. How many yards of ribbon does she need?

2. Caroline is sewing a costume from a pattern that calls for 4 yards of material. Because the costume is for a small child, she is reducing all the lengths to $\frac{2}{3}$ of the lengths given by the pattern. How many yards of material should she buy?

3. Consider this table of products.

    a. Copy the table, and write the result of each multiplication problem in the second column.

    b. What patterns do you see in the problems and the results?

    c. What would be the next two problems and results?

Problem	Result
$\frac{1}{4} \cdot 10$	
$\frac{2}{4} \cdot 10$	
$\frac{3}{4} \cdot 10$	
$\frac{4}{4} \cdot 10$	
$\frac{5}{4} \cdot 10$	
$\frac{6}{4} \cdot 10$	
$\frac{7}{4} \cdot 10$	
$\frac{8}{4} \cdot 10$	

4. Consider this table of products.

    a. Copy the table, and write the result of each multiplication problem in the second column.

    b. What relationships do you see between the fraction in each problem and the result?

    c. If you changed the problems so the numerator of each fraction were 2 instead of 1, how would the products change?

Problem	Result
$\frac{1}{30} \times 60$	
$\frac{1}{20} \times 60$	
$\frac{1}{15} \times 60$	
$\frac{1}{12} \times 60$	
$\frac{1}{6} \times 60$	
$\frac{1}{5} \times 60$	
$\frac{1}{4} \times 60$	
$\frac{1}{3} \times 60$	
$\frac{1}{2} \times 60$	

5. Consider the product $\frac{3}{4} \cdot 8$.

    a. Draw a rectangle diagram to represent this product.

    b. What is the area of each small rectangle in your diagram? How many small rectangles are shaded?

    c. What is $\frac{3}{4} \cdot 8$ equal to?

6. Consider the product $\frac{2}{3} \times \frac{4}{5}$.

    a. Draw a rectangle diagram to represent this product.

    b. What is the area of each small rectangle in your diagram? How many small rectangles are shaded?

    c. What is $\frac{2}{3} \times \frac{4}{5}$ equal to?

Find each product. Give your answers in lowest terms.

7. $\frac{7}{8} \times \frac{2}{5}$ **8.** $\frac{2}{3} \cdot \frac{7}{12}$ **9.** $\frac{6}{11} \times \frac{2}{3}$

10. $\frac{30}{50} \times \frac{15}{20}$ **11.** $\frac{4}{7} \cdot \frac{7}{9}$ **12.** $\frac{13}{15} * \frac{5}{6}$

13. Copy and complete this multiplication table. Give each answer as both a fraction and a mixed number. One answer is given for you.

$\times$	$1\frac{1}{2}$	$2\frac{2}{3}$	$3\frac{3}{4}$
$1\frac{1}{2}$	$\frac{9}{4} = 2\frac{1}{4}$		
$2\frac{1}{2}$			
$3\frac{1}{2}$			

Find each product by simplifying before you multiply.

14. $\frac{3}{4} \times \frac{8}{9}$ **15.** $\frac{2}{3} \cdot \frac{3}{8}$ **16.** $\frac{3}{5} \times \frac{4}{9}$ **17.** $\frac{2}{5} \cdot \frac{5}{9}$

18. Consider this multiplication table:

    a. Copy and complete the table. Give your answers in lowest terms. Express answers greater than 1 as mixed numbers.

$\times$	$\frac{2}{3}$	$1\frac{1}{3}$	2	$2\frac{2}{3}$	$3\frac{1}{3}$
$\frac{1}{2}$					
$\frac{1}{4}$					
$\frac{1}{8}$					

    b. What patterns and relationships do you see in the table?

Fill in each blank.

19. $\frac{2}{3} \cdot$ ___ $= 20$ **20.** $\frac{2}{3} \times$ ___ $= 15$ **21.** $\frac{2}{3} \cdot$ ___ $= 10$

22. In Parts a–f, use any method you like to find the answer.

a. How many $\frac{1}{4}$s are in 16? **b.** $\frac{1}{4}$ of what number is 16?

c. How many $\frac{2}{3}$s are in 16? **d.** $\frac{2}{3}$ of what number is 16?

e. How many $\frac{4}{3}$s are in 16? **f.** $\frac{4}{3}$ of what number is 16?

g. How are the answers to Parts a and b related? Parts c and d? Parts e and f? Explain why this makes sense.

23. Find the first quotient, and use your result to predict the second quotient. Check your answers by multiplying.

a. $14 \div \frac{1}{3}$ $\qquad$ $14 \div \frac{2}{3}$

b. $5 \div \frac{5}{8}$ $\qquad$ $15 \div \frac{5}{8}$

c. $6 \div \frac{1}{5}$ $\qquad$ $6 \div \frac{3}{5}$

d. $6 \div \frac{3}{7}$ $\qquad$ $24 \div \frac{3}{7}$

Fill in each blank.

24. $\frac{2}{3} \cdot$ ____ $= 1$ **25.** ____ $\times \frac{1}{8} = 1$ **26.** $1\frac{1}{2} \cdot \frac{2}{3} =$ ____

27. ___ $\times 1\frac{1}{8} = 1$ **28.** $\frac{5}{3} \times$ ____ $= 1$ **29.** $2\frac{1}{2} \times \frac{4}{5} =$ ____

30. In Parts a–f, estimate whether the quotient will be greater than, less than, or equal to 1. Then find the quotient. Leave answers greater than 1 in fraction form.

a. $\dfrac{1}{3} \div \dfrac{2}{5}$ $\qquad\qquad$ **b.** $\dfrac{2}{5} \div \dfrac{1}{3}$

c. $\dfrac{7}{8} \div \dfrac{3}{4}$ $\qquad\qquad$ **d.** $\dfrac{3}{4} \div \dfrac{7}{8}$

e. $\dfrac{1}{5} \div \dfrac{1}{7}$ $\qquad\qquad$ **f.** $\dfrac{1}{7} \div \dfrac{1}{5}$

g. Look at the problems and answers for Parts a–f. How are the answers to Parts a and b related? Parts c and d? Parts e and f? Explain why this makes sense.

Find each quotient. Give all answers in lowest terms. If an answer is greater than 1, write it as a mixed number.

31. $7\frac{1}{2} \div 1\frac{1}{2}$

32. $3\frac{1}{3} \div \frac{3}{4}$

33. $2\frac{2}{3} \div \frac{8}{5}$

34. $\frac{9}{8} \div \frac{2}{3}$

35. $\frac{8}{10} \div \frac{1}{100}$

36. $5\frac{1}{3} \div 2\frac{2}{3}$

37. $2\frac{1}{3} \div 3\frac{1}{3}$

38. $18 \div 4\frac{1}{2}$

39. $7\frac{1}{2} \div 4\frac{1}{2}$

Preview Find each missing factor.

40. $\frac{1}{3} \times$ _____ $= 60$

41. $\frac{2}{3} \cdot$ _____ $= 60$

42. $\frac{1}{5} \times$ _____ $= 60$

43. $\frac{2}{5} \cdot$ _____ $= 60$

44. $\frac{3}{5} \times$ _____ $= 60$

45. $\frac{4}{5} \cdot$ _____ $= 60$

46. Economics Rosita paid $8 for a CD marked "$\frac{1}{3}$ off." How much did she save?

47. Life Science The St. Louis Zoo has about 700 species of animals.

   a. The Detroit Zoological Park has $\frac{2}{5}$ as many species as the St. Louis Zoo. About how many species does the Detroit park have?

   b. The Toronto Zoo has $\frac{23}{70}$ *fewer* species than the St. Louis Zoo. About how many species does the Toronto Zoo have?

The tapir, an inhabitant of Central and South America and Southeast Asia

Tell whether each statement is true or false.

48. $\frac{2}{3}$ of $300 = \frac{3}{4}$ of 200

49. $\frac{1}{3}$ of $150 = \frac{1}{2}$ of 100

50. $\frac{2}{3}$ of $300 = \frac{1}{2}$ of 400

51. $\frac{2}{3}$ of $100 = \frac{1}{3}$ of 200

52. Economics Last week, Sal's Shoe Emporium held its semiannual clearance sale. All winter boots were marked down to $\frac{4}{5}$ of the original price. This week, the sale prices of the remaining boots were cut in half. What fraction of the original price is the new sale price?

53. Use what you have learned about multiplying fractions to explain why multiplying the numerator and denominator of a fraction by the same number does not change the fraction's value. Give examples if they help you to explain.

54. Measurement The Danson's horse ranch is a rectangular shape measuring $\frac{3}{5}$ mile by $\frac{4}{7}$ mile.

a. What is the area of the ranch?

b. There are 640 acres in 1 square mile. What is the area of the Danson's ranch in acres?

55. Measurement The left column of the table shows the ingredient list for a spice cake that serves 12 people. Complete the table to show the amount of each ingredient needed for the given numbers of people.

	12 People	10 People	8 People	6 People	4 People	2 People
Flour	$2\frac{1}{4}$ c					
Sugar	$1\frac{1}{3}$ c					
Salt	$\frac{3}{4}$ tsp					
Butter	$1\frac{1}{2}$ sticks					
Ginger	$\frac{1}{2}$ tsp					
Raisins	$\frac{2}{3}$ c					

56. In Parts a and b, use the numbers 1, 2, 3, 4, 5, and 6.

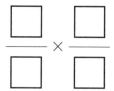

$$\frac{\square}{\square} \times \frac{\square}{\square}$$

   a. Fill in each square with one of the given numbers to create the greatest possible product. Use each number only once.

   b. Fill in each square with one of the given numbers to create the least possible product. Use each number only once.

   c. Choose four different whole numbers. Repeat Parts a and b using your four numbers.

57. Consider this multiplication table:

×	$4\frac{1}{2}$	$2\frac{1}{4}$	$1\frac{1}{8}$	$\frac{9}{16}$	$\frac{9}{32}$
$\frac{1}{2}$	$2\frac{1}{4}$				
$1\frac{1}{2}$					
$2\frac{1}{2}$					
$3\frac{1}{2}$					

   a. Copy the table, and use your knowledge of fraction multiplication and number patterns to help complete it. Try to do as few paper-and-pencil calculations as possible.

   b. Describe the pattern in the table as you read across the rows from left to right.

   c. Describe the pattern in the table as you read down the columns.

58. Rosita bought 6 yards of ribbon for a sewing project.

   a. She wants to cut the ribbon into pieces $\frac{1}{2}$ foot long. How many pieces can she cut from her 6-yard ribbon?

   b. What calculations did you do to solve Part a?

   c. Rosita decides instead to cut pieces $\frac{2}{3}$ foot long. How many pieces can she cut?

   d. What calculations did you do to solve Part c?

59. Hannah made up riddles about the ages of some people in her family. See if you can solve them.

a. My brother Tim is $\frac{3}{4}$ my cousin Janice's age. Tim is 15 years old. How old is Janice?

b. To find my grandpa Henry's age, divide my aunt Carol's age by $\frac{4}{5}$ and then add 10. Aunt Carol is 40. How old is Grandpa Henry?

c. Today is my uncle Mike's 42nd birthday. He is now $\frac{2}{3}$ of his father's age and $\frac{7}{3}$ of his daughter's age. How old are Uncle Mike's father and daughter?

60. Measurement In this problem, use the following facts:

1 cup = 8 ounces   1 quart = 4 cups   1 gallon = 4 quarts

a. Conor has 3 gallons of lemonade to serve at his party. If he pours $\frac{3}{4}$-cup servings, how many servings can he pour?

b. What calculations did you do to solve Part a?

c. If Conor pours 4-ounce servings instead, how many servings can he pour?

d. What calculations did you do to solve Part c?

61. Consider this division table:

$\div$	$\frac{1}{2}$	$\frac{1}{4}$	$\frac{1}{8}$	$\frac{1}{16}$	$\frac{1}{32}$	$\frac{1}{64}$
$\frac{1}{2}$						
$\frac{1}{4}$						
$\frac{1}{8}$						
$\frac{3}{2}$						
$\frac{3}{4}$						
$\frac{3}{8}$						

a. Copy the table, and use your knowledge of fraction division and number patterns to help complete it. Try to do as few paper-and-pencil calculations as possible.

b. Describe the number pattern in the table as you read across the rows from left to right. Explain why this pattern occurs.

c. Describe the number pattern in the table as you read down the columns. Explain why this pattern occurs.

Tell whether each statement is true or false. If a statement is false, give the correct quotient.

62. $\frac{9}{12} \div \frac{1}{4} = 3$

63. $\frac{1}{4} \div \frac{9}{12} = \frac{2}{3}$

64. $3\frac{1}{2} \div \frac{7}{4} = 2$

65. $\frac{7}{4} \div 3\frac{1}{2} = \frac{1}{2}$

66. $\frac{10}{4} \div \frac{4}{10} = 1$

67. $\frac{4}{10} \div \frac{10}{4} = 1$

Evaluate each expression.

68. $\frac{1}{2} \cdot \frac{3}{4} + \frac{1}{2} \div \frac{5}{6}$

69. $\left(\frac{3}{5} + \frac{7}{8}\right) * \frac{1}{5} \div \frac{2}{15}$

70. In Parts a and b, use the numbers 2, 4, 6, 8, 10, and 12.

$$\frac{\square}{\square} \div \frac{\square}{\square}$$

a. Fill in each blank with one of the given numbers to create the greatest possible quotient. Use each number only once.

b. Fill in each blank with one of the given numbers to create the least possible quotient. Use each number only once.

c. Choose four different whole numbers. Repeat Parts a and b using your four numbers.

71. Explain how you could solve the problem below without doing any calculations. Give the result, and check your answer by performing the calculations.

$$\left(\frac{5}{6} \div \frac{4}{7}\right) \times \left(\frac{4}{7} \div \frac{5}{6}\right)$$

Find each sum or difference.

Mixed Review

72. $5\frac{6}{7} + 1\frac{5}{14}$

73. $\frac{11}{12} - \frac{5}{18}$

74. $\frac{5}{9} + \frac{13}{7}$

75. $3 - 1\frac{31}{72}$

76. $2\frac{5}{8} - 1\frac{3}{4}$

77. $\frac{7}{8} + \frac{7}{12} + \frac{7}{16}$

78. How many times 0.024 is 24,000?

79. What fraction of 564 is 0.564?

80. **Geometry** Give the coordinates of each point on the grid.

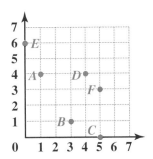

Find the opposite of each number.

81. 4.2

82. $^-0.32$

83. $^-184$

84. $\frac{14}{17}$

85. $12\frac{1}{3}$

86. $^-0.0041$

Express the shaded portion of each figure as both a fraction in lowest terms and a decimal.

87.

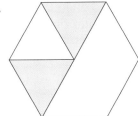

88.

3.3 Multiplying and Dividing with Decimals

Figuring out the tip on a restaurant bill, converting measurements from one unit to another, finding lengths for a scale model, and exchanging money for different currencies are just a few activities that involve multiplying and dividing with decimals. In this lesson, you will learn how to multiply and divide decimals and to use estimation to determine whether your results are reasonable.

Think & Discuss

Fill in the blank with a whole number or a decimal so the product is

$16 \times \underline{\qquad}$

• greater than 100.

• greater than 32 but less than 100.

• at least 17 but less than 32.

• equal to 16.

• greater than 8 but less than 16.

• less than 8 but greater than 0.

Investigation ▶ 1 Multiplying Whole Numbers and Decimals

Although you may not be able to calculate 172×97 in your head, you know the product is greater than both 172 and 97. In fact, when you multiply any two numbers greater than 1, the product must be greater than either of the numbers.

In the next problem set, you will explore what happens when one of the numbers is less than 1.

Problem Set A

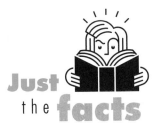

Just the facts

Cats purr at the same frequency as an idling diesel engine, about 26 cycles per second.

1. Luke considered two brands of food for his cats. Kitty Kans cost $0.32 per can, and Purrfectly Delicious cost $0.37 per can. Find the cost of each of the following.

a. 10 cans of Kitty Kans

b. 10 cans of Purrfectly Delicious

c. 12 cans of Kitty Kans

d. 12 cans of Purrfectly Delicious

e. Look at your answers to Parts a–d. In each case, was the cost more or less than the number of cans? Explain why this makes sense.

2. After buying cat food, Luke went to CD-Rama and bought a CD for $14.00, plus 4% sales tax. To calculate a 4% sales tax, multiply the cost of the purchase by 0.04.

a. The cashier tried to charge Luke $19.60—$14.00 for the CD and $5.60 for sales tax. Without actually calculating the tax, how do you know that the cashier made a mistake?

b. What is the correct amount of sales tax on Luke's $14 purchase?

3. Multiply each number by 0.01. If you see a shortcut for finding the answer without doing any calculations, use it.

a. 1,776   **b.** 28,000   **c.** 29,520   **d.** 365.1

4. If you used a shortcut for Problem 3, explain what you did. If you didn't use a shortcut, compare the four original numbers with the answers. What pattern do you see?

You can multiply with decimals by first ignoring the decimal points and multiplying whole numbers. Then you can use your estimation skills to help determine where to put the decimal point in the answer.

Problem Set B

Jahmal's calculator is broken. It calculates correctly but no longer displays decimal points! The result 452.07 is displayed as 45207. Unfortunately, 4.5207 is displayed the same way.

1. List all other numbers that would be displayed as 45207.

2. Jahmal's calculator displays 45207 when he enters 5.023×9. Without using a calculator, estimate the product. Use your estimate to figure out where to place the decimal point in the result.

3. Jahmal's calculator also displays 45207 when 5×904.14 is entered. Use estimation to figure out the correct result of this calculation. Explain how you found your answer.

4. When Jahmal enters 279×0.41, his calculator displays 11439.

   a. Is the correct result greater than or less than 279? Explain.

   b. Jahmal estimated the result by multiplying 300 by $\frac{40}{100}$. Will this calculation give a good estimate of the actual answer? Explain.

   c. What estimate will Jahmal get for the product?

   d. What is the exact answer to 279×0.41? Explain how you know.

5. Below are some calculations Jahmal entered and the results his calculator displayed. Estimate each product, and explain how you made your estimate. Then give the actual product by placing the decimal point in the correct place in the display number.

   a. 203×1.8 (3654)   **b.** $0.5 \times 28,714$ (14357)

   c. 42×0.11 (462)   **d.** $8,975 \times 70.07$ (62887825)

   e. 0.22×216 (4752)   **f.** 0.09×71 (639)

6. Look back at your results for Problem 5. For each part, tell whether the product is greater than or less than the whole number in the problem, and explain why.

7. Jahmal's calculator displayed a result as 007. What might the actual result be? Give all the possibilities.

1. Suppose you are multiplying a whole number by a decimal.

   a. When do you get a result less than the whole number? Give some examples.

   b. When do you get a result greater than the whole number? Give some examples.

2. Explain how using estimation is helpful when multiplying with decimals.

Investigation 2 Multiplying Decimals as Fractions

One strategy for multiplying decimals is to write them as fractions, multiply the fractions, and then write the answer as a decimal. For example, here's how you could calculate 0.7×0.02:

$$0.7 \times 0.02 = \frac{7}{10} \times \frac{2}{100} = \frac{14}{1,000} = 0.014$$

Think & Discuss

Althea said the $\frac{2}{100}$ in the calculation above should have been written in lowest terms before multiplying. Rosita disagreed. She said it is easier to change the answer to a decimal if the fractions are not written in lowest terms before multiplying.

What do you think? Defend your answer.

Remember

A decimal greater than 1 can be represented by a fraction or a mixed number. For example, 5.12 can be written as $\frac{512}{100}$ or $5\frac{12}{100}$.

Problem Set C

Write the decimals as fractions and then multiply. Give each product as a decimal. Show all your steps.

1. $0.6 \cdot 0.7$

2. $1.06 \cdot 0.07$

3. $5.12 \cdot 0.2$

4. $0.002 \cdot 0.003$

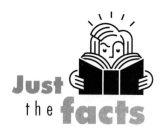

5. Conor calculated 0.625×0.016 on his calculator and got 0.01. Jing told him, "You must have made a mistake, because when you write the problem with fractions, you get $\frac{625}{1,000} \times \frac{16}{1,000}$. When you multiply these fractions, you get a denominator of $1,000,000$. Your answer is equal to $\frac{1}{100}$, which has a denominator of 100."

Find 0.625×0.016 using your calculator. Did Conor make a mistake? If so, explain it. If not, explain the mistake in Jing's reasoning.

6. In a particular state, the sales tax is 5%. To calculate the sales tax on an item, multiply the cost by 0.05. Assume the sales tax is always rounded up to the next whole cent.

a. Calculate the sales tax on an $0.89 item. Show your steps.

b. Find a price less than $1 so that the sales-tax calculation results in a whole number of cents without rounding.

Multiplying decimals is just like multiplying whole numbers, except you need to determine where to place the decimal point in the product. In the last investigation, you used estimation to locate the decimal point. The idea of writing decimals as fractions before multiplying leads to another method for multiplying decimals.

Think & Discuss

Write each number as a fraction.

    10.7   2.43   0.073   13.0601

How does the number of digits to the right of the decimal point compare to the number of 0s in the denominator of the fraction?

Find each product.

    $10 \times 10,000$ 100×100 $1,000,000 \times 10,000$

How is the number of 0s in each product related to the numbers of 0s in the numbers being multiplied?

Now consider the product 0.76×0.041.

If you wrote 0.76 and 0.041 as fractions, how many 0s would be in each denominator? How many 0s would be in the denominator of the product?

Use the fact that $76 \times 41 = 3,116$ and your answers to the above questions to find the product 0.76×0.041. Explain how you found your answer.

Here is the rule you used to multiply decimals in the Think & Discuss:

- Ignore the decimal points, and multiply the numbers as if they were whole numbers.

- Place the decimal point so that the number of digits to its right is equal to the total number of digits to the right of the decimal points in the numbers being multiplied.

Problem Set D

1. Consider the product 0.1123×0.92.

    a. Use the rule above to find the product. Explain each step.

    b. Check your answer to Part a by changing the numbers to fractions and multiplying.

    c. Without multiplying, how could you know that the correct answer couldn't be 1.03316 or 0.0103316?

2. Use the rule above to recalculate the answers to Problems 1–4 of Problem Set C. Do you get the same results you found by writing the decimals as fractions and then multiplying?

3. Jing said, "The rule doesn't work for the product 0.625×0.016. When I find the product on my calculator, I get 0.01, which has only two digits to the right of the decimal point. According to the rule, there should be six digits to the right of the decimal point. What's going on?" Explain why this is happening.

Find each product without using a calculator.

4. $43.32 \cdot 3.07$ **5.** 0.0005×12.6 **6.** $102.667 * 0.11$

Share & Summarize

1. Explain how you can use fractions to determine where to place the decimal point in the product of two decimals. Give an example to illustrate your answer.

2. The total number of digits to the right of the decimal points in 0.25 and 0.012 is five. Explain why the product $0.25 \cdot 0.012$ has only three decimal places.

Decimals are easy to understand and compare, and they are easier to use than fractions when doing computations with calculators and computers. For these reasons, numerical information is often given in decimal form. In this lesson, you will explore some real situations involving calculations with decimals.

Think & Discuss

You can predict approximately how tall a four-year-old will be at age 12 by multiplying his or her height by 1.5.

Hannah's four-year-old brother, Jeremy, likes to play with her calculator. She let him press the buttons as she predicted his height at age 12. He is now 101.5 cm tall, so she told him to enter 1.5 × 101.5. The calculator's display read 1522.5.

Hannah realized right away that Jeremy had pressed the wrong keys. How did she know?

Problem Set E

1. On her fourth birthday, Shanise was exactly as tall as a meterstick. Predict her height at age 12.

2. Ynez is a four-year-old who likes to use very precise numbers. She claims she is exactly 98.75 cm tall.

   a. Use this value to predict Ynez's height at age 12.

   b. Rosita says it doesn't make sense to use the exact decimal answer as a prediction for Ynez's height. Do you agree? Explain.

   c. What is a reasonable prediction for Ynez's height at age 12?

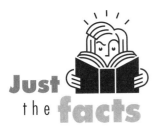

Just the facts

The practice of tipping began in British inns in the 18th century. Patrons gave waiters money before a meal with a note indicating the money was "to insure promptness." The word *tip* is an abbreviation of this message.

The most common decimal calculations are probably those that involve money. You will now explore some situations involving money.

Problem Set F

1. To figure out the cost of dinner at a restaurant, Mr. Rivera multiplies the total of the prices for the items by 1.25. The result includes the cost of the items ordered, the 5% sales tax, and a 20% tip.

   a. Mr. Rivera and his friend went to dinner. The cost of the items they ordered was $42. When Mr. Rivera tried to calculate the cost with tax and tip, he got $85. Without doing any calculations, explain why this result cannot be correct.

   b. Estimate the cost of their dinner, including tax and tip, in your head. Explain how you made your estimate.

   c. Find the exact cost of the dinner.

Since you can't pay for something in units smaller than a penny, most prices are given in dollars and whole numbers of cents. Gasoline prices, however, are often given to tenths of a cent (thousandths of a dollar) and displayed in a form that combines decimals and fractions. For example, a price of $1.579 is given as 1.57\frac{9}{10}$. The fraction $\frac{9}{10}$ represents $\frac{9}{10}$ of a cent.

2. Imagine you are buying exactly 10 gallons of gas priced at 1.57\frac{9}{10}$ per gallon.

   a. Without using a calculator, figure out exactly how much the gas will cost.

   b. If the price were rounded to $1.58 per gallon, how much would you pay?

   c. How much difference is there in your answers to Parts a and b? Why do you think gas prices aren't just rounded up to, for example, $1.58 per gallon?

3. Ms. Kenichi filled her sports utility vehicle with 38.4 gallons of gas. The gas cost 1.57\frac{9}{10}$ per gallon.

   a. Estimate the total cost for the gas.

   b. Now calculate the exact price, rounding off to the nearest penny at the end of your calculation.

When visiting foreign countries, travelers exchange the currency of their home country for the currency used in the country they are visiting. Converting from one unit of currency to another involves operations with decimals.

Problem Set G

The unit of currency in Japan is the *yen*. On April 23, 2000, 1 yen was worth $0.009463, slightly less than 1 U.S. penny. In the following problems, round your answers to the nearest cent.

1. On April 23, 2000, what was the value of 100 yen in U.S. dollars?

2. Dr. Kuno was traveling in Japan on April 23, 2000. On that day, she purchased a video-game system priced at 14,100 yen for her nephew. What was the equivalent price in U.S. dollars?

3. Copy and complete the table to show the dollar equivalents for the given numbers of yen.

Yen	Dollars
10	
50	
100	
150	
200	
300	
1,000	
2,000	
3,000	
10,000	
1,000,000	

4. Use your table to help estimate the dollar equivalent of 6,000 yen. Explain how you made your estimate.

5. Use your table to help estimate the yen equivalent of $100. Explain how you made your estimate.

Just the facts

The dollar is the unit of currency in the United States, but coins allow you to pay amounts less than a dollar. In Japan, there are no coins worth less than a yen, so you can pay only whole numbers of yen.

Share & Summarize

On her return from Japan, Dr. Kuno traded her yen for dollars. She handed the teller 23,500 yen, and the teller gave her a $20 bill, a $1 bill, and some change. Use estimation to explain how you know the teller gave Dr. Kuno the wrong amount.

Investigation 4 ▶ Dividing Decimals

In this investigation, you will divide decimals. First you will examine patterns relating multiplication and division.

Problem Set H

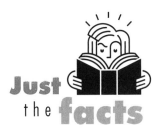

Just the facts

The ÷ symbol was used by editors of early manuscripts to indicate text to be cut. The symbol was used to indicate division as early as 1650.

1. Copy and complete this table.

3.912 × 0.1 =	3.912 ÷ 0.1 =
3.912 × 0.01 =	3.912 ÷ 0.01 =
3.912 × 0.001 =	3.912 ÷ 0.001 =
4,125.9 × 0.1 =	4,125.9 ÷ 0.1 =
4,125.9 × 0.01 =	4,125.9 ÷ 0.01 =
4,125.9 × 0.001 =	4,125.9 ÷ 0.001 =

2. Look for patterns in your completed table.

   a. What happens when a number is multiplied by 0.1, 0.01, and 0.001?

   b. What happens when a number is divided by 0.1, 0.01, and 0.001?

3. Copy and complete this table.

3.912 ÷ 10 =	3.912 × 10 =
3.912 ÷ 100 =	3.912 × 100 =
3.912 ÷ 1,000 =	3.912 × 1,000 =
4,125.9 ÷ 10 =	4,125.9 × 10 =
4,125.9 ÷ 100 =	4,125.9 × 100 =
4,125.9 ÷ 1,000 =	4,125.9 × 1,000 =

4. Compare your results for Problems 1 and 3. Then complete these statements.

   a. Multiplying a number by 0.1 is the same as dividing it by ____.

   b. Multiplying a number by 0.01 is the same as dividing it by ____.

   c. Multiplying a number by 0.001 is the same as dividing it by ____.

   d. Dividing a number by 0.1 is the same as multiplying it by ____.

   e. Dividing a number by 0.01 is the same as multiplying it by ____.

   f. Dividing a number by 0.001 is the same as multiplying it by ____.

Now you can probably divide by such decimals as 0.1, 0.01, and 0.001 mentally. For other decimal-division problems, it is not easy to find an exact answer in your head. However, the next Example shows a method for estimating the quotient of two decimals.

EXAMPLE

Estimate $0.0351 \div 0.074$.

Think of the division problem as a fraction.

$$\frac{0.0351}{0.074}$$

Multiply both the numerator and denominator by 10,000 to get an equivalent fraction involving whole numbers.

$$\frac{0.0351 \times 10,000}{0.074 \times 10,000} = \frac{351}{740}$$

$\frac{351}{740}$ is close to $\frac{1}{2}$, or 0.5. So, $0.0351 \div 0.074$ is about 0.5.

Problem Set ▌

In Problems 1–3, estimate each quotient. Then use your calculator to find the quotient to the nearest thousandth.

1. $25.27 \div 0.59$ **2.** $32.47 \div 81.5$ **3.** $0.4205 \div 0.07$

4. Find $10 \div 0.01$ without using a calculator.

5. Suppose 1 Japanese yen is worth 0.009155 U.S. dollar, and you have $10 to exchange.

   a. Which of the following calculations will determine how many yen you will receive? Explain your answer.

   0.009155×10 $0.009155 \div 10$ $10 \div 0.009155$

   b. How many yen will you receive for $10?

   c. Your answer to Part b should be fairly close to your answer for Problem 4. Explain why.

   d. Caroline calculated that she could get about 22 yen in exchange for $20. Is Caroline's answer reasonable? Explain.

When you don't have a calculator to divide decimals, you can use the method shown in the next Example to change a decimal-division problem into a division problem involving whole numbers.

EXAMPLE

Calculate $5.472 \div 1.44$.

Write the division problem as a fraction. Then multiply the numerator and denominator by 1,000 to get a fraction with a whole-number numerator and denominator.

$$\frac{5.472}{1.44} = \frac{5.472 \times 1,000}{1.44 \times 1,000} = \frac{5,472}{1,440}$$

Now just divide the whole numbers.

$$
\begin{array}{r}
3.8 \\
1440\overline{)5472} \\
4320 \\
\hline
1152\,0 \\
1152\,0 \\
\hline
0
\end{array}
$$

So, $5.472 \div 1.44 = 3.8$.

Problem Set J

Solve these problems without using a calculator.

Find each quotient.

1. $99.33 \div 0.473$ **2.** $6.1452 \div 170.7$ **3.** $0.0752 \div 3.2$

4. There are 2.54 centimeters in 1 inch.

   a. Estimate the number of inches in 25 cm.

   b. Use your answer to Part a to estimate the number of inches in 100 cm.

   c. Find the actual number of inches in 25 cm and in 100 cm.

   d. The average height of a 12-year-old girl in the United States is 153.5 cm. Convert this height to inches. Round to the nearest tenth of an inch.

Share & Summarize

1. Describe a quick way to divide a number by 0.01. Use your knowledge of fraction division to explain why your method works.

2. Luke and Rosita are arguing about whose cat is heavier. Luke says his cat Tom is huge—almost 23 pounds! Rosita says her cat Spike is even heavier—close to 11 kilograms! There are about 2.2 pounds in 1 kilogram. Describe a calculation you could do to figure out whose cat is heavier.

Investigation 5 Multiplying or Dividing?

Your calculator multiplies and divides decimals accurately (as long as you don't make any mistakes when pressing the keys). However, it can't tell you *whether* to multiply or divide. In this investigation, you will decide whether to multiply or divide in specific situations.

Think & Discuss

For each question, choose the correct calculation and explain how you decided. Although you don't have to do the calculation, you may want to make an estimate to check that the answer is reasonable.

- A package of eight blank videotapes costs $15.95. Which calculation could you do to find the cost per tape: 15.95×8 or $15.95 \div 8$?

- There are 2.54 centimeters in 1 inch. A sheet of paper is $8\frac{1}{2}$ inches wide. Which calculation could you do to find the paper's width in centimeters: 2.54×8.5 or $2.54 \div 8.5$?

- Marcus is building a model-railroad layout in HO scale, in which 0.138 inch in the model represents 1 foot in the real world. He wants to include a model of the Sears Tower in Chicago, which is about 1,450 feet tall. Which calculation could he do to find the height of the model tower in inches: $1,450 \times 0.138$ or $1,450 \div 0.138$?

In the next two problem sets, you will need to think carefully about whether to multiply or divide to find each answer.

Problem Set K

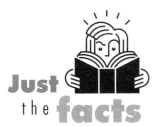
1. On January 1, 1999, the *euro* was introduced as the common currency of 11 European nations. Currency markets opened on January 4. On that day, 1 euro was worth 1.1874 U.S. dollars.

a. On January 4, 1999, what was the value of 32 euros in U.S. dollars? Explain how you decided whether to multiply or divide.

b. On January 4, 1999, what was the value of $32.64 in euros? Explain how you decided whether to multiply or divide.

2. A kilometer is equal to approximately 0.62 mile.

a. Molly ran a 42-km race. Find how far she ran in miles, and explain how you decided whether to multiply or divide.

b. The speed limit on Duncan Road is 55 miles per hour. Convert this speed to kilometers per hour (rounded to the nearest whole number), and explain how you decided whether to multiply or divide.

Problem Set L

1. Miguel's father's car can travel an average of 18.4 miles on 1 gallon of gasoline. Gas at the local station costs $1.53\frac{9}{10}$ per gallon.

a. Miguel's father filled the car with 12.8 gallons of gas. How much did he pay?

b. Miguel's brother took the car to the gas station and handed the cashier a $10 bill. How much gas could he buy with $10? Round your answer to the nearest hundredth of a gallon.

c. How far could Miguel's brother drive on the amount of gas he bought? Round your answer to the nearest mile.

d. Over spring break, Miguel's family drove the car on a 500-mile trip. About how much gas did they use? Round your answer to the nearest gallon.

2. Althea is building a model-railroad layout using the Z scale, the smallest scale for model railroads. In the Z scale, 0.055 inch on a model represents 1 foot in the real world.

a. The caboose in Althea's model is 1 inch long. How long is the real caboose?

b. Althea is making model people for her layout. She wants to make a model of her favorite basketball player, who is 6 feet 9 inches tall. If she could make her model exactly to scale, how tall would it be?

c. Althea is using an architect's ruler, which measures to the nearest tenth of an inch. Using this ruler, it is impossible for her to build her basketball-player model to the exact scale height you calculated in Part b. How tall do you think she should make the model?

Share & Summarize

In the United States, people measure land area in acres. People in many other countries use hectares, the metric unit for land area. There is approximately 0.405 hectare in 1 acre.

1. To convert 3.5 acres to hectares, do you multiply or divide 3.5 by 0.405? Explain how you know.

2. To convert 3.5 hectares to acres, do you multiply or divide 3.5 by 0.405? Explain how you know.

On Your Own Exercises

Practice & Apply

1. Dae Ho made a spreadsheet showing products of decimals and whole numbers. The toner in his printer is running low, and when he printed his spreadsheet, the decimal points were nearly invisible. Copy the spreadsheet, and insert decimal points in the first and third columns to make the products correct.

Decimal	Whole Number	Product
48398	306	14809788
364	967	351988
1706	698	1190788
167935	534	8967729
75072	976	73270272
93	160	14880

2. **Economics** Gloria, Wilton, and Alex have a band that plays for dances and parties. Gloria does most of the song writing, and Wilton acts as the manager. The band divides their earnings as follows:

   • Gloria gets 0.5 times the band's profit.

   • Wilton gets 0.3 times the band's profit.

   • Alex gets 0.2 times the band's profit.

   a. This month the band earned $210. How much money should each member get?

   b. A few months later, the band members changed how they share their profit. Now Gloria gets 0.42 times the profit, Wilton gets 0.3 times the profit, and Alex gets 0.28 times the profit. Alex said his share of $210 would now be $66. Explain why this estimate could not be correct, and calculate the correct amount.

   c. If the band makes $2,000 profit over the next several months, how much more will Wilton earn than Alex? (Try finding the answer without calculating how much money Wilton earns!)

Measurement Use the fact that 1 m = 0.001 km to convert each distance to kilometers without using a calculator.

    3. 283 m   **4.** 314,159 m

    5. 2,000,000 m   **6.** 1,776 m

    7. 7 m   **8.** 0.12 m

Write the decimals as fractions, and then multiply. Give the product as a decimal.

9. 0.17×0.003

10. $0.0005 \cdot 0.8$

11. 0.00012×12.34

12. $0.001 \times 0.2 \times 0.3$

Find each product without using a calculator.

13. 0.023×17.51

14. $0.15 \cdot 1.75$

15. $0.34 * 0.0072$

16. $3.02 \cdot 100.25$

17. 0.079×0.970

18. $0.0354 * 97.3$

Calculate each product mentally.

19. $0.0002 \cdot 2.5$

20. 7×0.006

21. 0.03×0.05

22. $0.4 \cdot 0.0105$

23. Economics The unit of currency in Guatemala is the *quetzal*. On April 23, 2000, 1 quetzal was worth 0.1295 U.S. dollar.

    a. How much was a 100-quetzal note worth in U.S. dollars?

    b. On this same day, a small rug in a Guatemalan market was priced at 52 quetzals. Convert this amount to U.S. dollars and cents.

24. Economics At Sakai's Sweet Shop, gummy worms cost $.28 per ounce and chocolate-covered raisins cost $.37 per ounce.

    a. How much do 6 ounces of gummy worms cost?

    b. Use the fact that 16 ounces = 1 pound to calculate the cost of a pound of chocolate-covered raisins.

    c. In the state where Sakai's is located, sales tax on candy is computed by multiplying the price by 0.1 and rounding up to the next penny. Without using a calculator, find the tax on 6 ounces of gummy worms and on 1 pound of chocolate-covered raisins.

Just the facts

The *quetzal* is a bird that is found in the rain forests of Central America. Today the male quetzal appears on Guatemalan currency and on the Guatemalan flag.

25. Economics When she eats at a restaurant, Viviana likes to leave a 15% tip, multiplying the price of the meal by 0.15. Franklin usually leaves a 20% tip, multiplying the price by 0.20. They both round up to the nearest 5¢.

   a. How much tip would Viviana leave for a $24.85 meal?

   b. How much tip would Franklin leave for a $24.85 meal?

   c. The price $24.85 does not include tax. Viviana and Franklin live in a state where the meal tax is 6% and fractions of a cent are rounded to the nearest penny. Figure out the tax on the $24.85 meal by multiplying by 0.06.

   d. Calculate the tips Viviana and Franklin would leave if they tipped based on the cost of the meal plus tax.

26. Measurement In Investigation 3, you learned that you can multiply a four-year-old's height by 1.5 to predict his or her height at age 12. You can work backward to estimate what a 12-year-old's height might have been at age 4.

   a. Nicky is 127.5 cm tall at age 12. Estimate her height at age 4 to the nearest centimeter.

   b. Jacabo is 152.4 cm tall at age 12. Estimate his height at age 4 to the nearest centimeter.

27. Measurement To determine a length in inches on an HO-scale model of an object, divide the actual length in feet by 7.25. For example, a 15-foot-high building would have a height of about 2 inches in the model.

   a. Without using a calculator, estimate how long a model train would be if the real train is 700 feet long. Explain how you made your estimate.

   b. Calculate the length of the model train to the nearest quarter of an inch.

   c. Estimate the length of the model train to the nearest foot. Is the length of the model shorter or longer than your estimate?

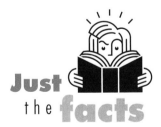

Just the facts

The HO scale is the most popular size for model railroads. Using this scale, models are $\frac{1}{87}$ the size of real trains.

28. Economics Last summer, Rosita, Miguel, Luke, and Marcus shared a paper route. At the end of the summer, they divided their $482.50 profit according to how much each had worked.

- To get Rosita's share, the profit was divided by 2.5.
- To get Miguel's share, the profit was divided by 4.0.
- To get Luke's share, the profit was divided by 6.25.
- To get Marcus' share, the profit was divided by 10.

a. Without using a calculator, estimate how much each friend earned, and explain how you made your estimates.

b. Now calculate the exact amount each friend received.

c. Do their shares add to $482.50? If not, change one person's share so they *do* add to $482.50.

Evaluate each expression without using a calculator.

29. $0.1 \cdot 17 + 15 \cdot 0.001$

30. $8.82 \div 0.63 \div 0.7$

31. $2.75 - 0.05 \cdot 10$

32. Economics Sasha wants to buy a gallon of orange juice. He is considering two brands. Sunny Skies cost $.77 per quart, and Granger's Grove cost $2.99 per gallon. Sasha likes the taste of both brands. Which brand is a better deal? Explain.

33. Economics The unit of currency in Vanuatu is the *vatu*. In March 2003, there were 125.37 vatus to 1 U.S. dollar.

a. What is the value of 853.25 vatus in dollars?

b. What is the value of $853.25 in vatus? Round your answer to the nearest hundredth of a vatu.

c. In January 1988, there were 124.56 vatus to 1 U.S. dollar. Did the value of a vatu go up or down between January 1988 and March 2003? Justify your answer.

Just the facts

Vanuatu is a Y-shaped group of 83 islands in the Southwest Pacific.

34. Measurement In this exercise, use these facts:

$$1 \text{ quart} = 0.947 \text{ liter} \qquad 1 \text{ quart} = 32 \text{ ounces}$$

   a. VineFresh grape juice is $1.25 a quart; Groovy Grape is $1.35 a liter. If you like both brands, which is the better buy? Explain.

   b. How many quarts are in 0.5 liter? In 1 liter? In 1.5 liters? In 2.0 liters? Give your answers to the nearest hundredth.

   c. How many liters are in 10 ounces? In 12 ounces? In 20 ounces? In 2 quarts? Express your answers to the nearest thousandth.

35. In this problem, you will try to get as close to 262 as you can, without going over, by multiplying 210 by a number. You can use a calculator, but the only operation you may use is multiplication.

   a. Should you multiply 210 by a number greater than 1 or less than 1?

   b. Get as close as you can to 262 by multiplying 210 by a number with only one decimal place. What number did you multiply 210 by? What is the product?

   c. Get as close as you can to 262 by multiplying 210 by a number with two decimal places. What number did you multiply 210 by? What is the product?

   d. Now get as close as you can using a number with three decimal places. Give the number you multiplied by and the product.

   e. Try multiplying 210 by numbers with up to nine decimal places to get as close to 262 as possible. Give each number you multiplied by and the product. Describe any strategies you develop for choosing numbers to try.

36. Measurement In this problem, you will figure out how much water it would take to fill a room in your home.

   a. Choose a room in your home that has a rectangular floor. Find the length, width, and height of the room to the nearest foot. (If you don't have a yardstick or tape measure, just estimate.) Multiply the three measurements to find the volume, or number of cubic feet, in the room.

   b. A cubic foot holds 7.48 gallons of water. How many gallons will it take to fill the room?

   c. If 748 gallons of water cost $1.64, how much would it cost to fill the room? Explain how you found your answer.

37. Physical Science A light bulb's wattage indicates how much energy the bulb uses in 1 hour. For example, a 75-watt bulb uses 75 watt-hours of energy per hour. Electric companies charge by the kilowatt-hour.

a. One watt-hour = 0.001 kilowatt-hour. How many kilowatt-hours does a 75-watt bulb use in an hour? In 24 hours?

b. Suppose your electric company charges $0.12 per kilowatt-hour. How much would it cost to leave a 75-watt bulb on for 24 hours?

c. Figure out how much it would cost to leave all the light bulbs in your home on for 24 hours. You will need to count the bulbs and note the wattage of each bulb. Do not look directly at a light bulb. If it is not possible to find the wattage of some bulbs, assume each is a 75-watt bulb. (Count only incandescent bulbs, not fluorescent or halogen bulbs.)

38. Economics The imaginary country of Glock uses *utils* for its currency. One util has the same value as 5 U.S. dollars. In other words, there are $5 per util.

a. What is the value of $1 in utils? That is, how many utils are there per dollar? Express your answer as both a fraction and a decimal.

b. In Part a, you used the number of dollars per util to find the number of utils per dollar. What mathematical operation did you use to find your answer?

c. Use the same process to find the number of yen per dollar if 1 yen is worth $0.009155. Round your answer to the nearest whole yen.

d. If you are given the value of 1 unit of a foreign currency in dollars, describe a rule you could use to find the value of $1 in that foreign currency.

e. If a unit of foreign currency is worth more than a dollar, what can you say about how much $1 is worth in that currency?

f. If a unit of foreign currency is worth less than a dollar, what can you say about how much $1 is worth in that currency?

39. Architecture Miguel built a scale model of the Great Pyramid of Giza as part of his history project. He used a scale factor of 0.009 for his model. This means he multiplied each length on the actual pyramid by 0.009 to find the length for his model.

On the next page are some of the measurements on Miguel's model. Find the measurements of the actual pyramid. Round your answers to the nearest meter.

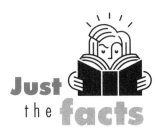
a. Height of pyramid: 1.23 m

b. Length of each side of pyramid's base: 2.07 m

c. Height of king's chamber: 0.05 m

d. Length of king's chamber: 0.04 m

e. Width of king's chamber: 0.09 m

40. Imagine you are playing a game involving multiplying and dividing decimals. The goal is to score as close to 100 as possible without going over. Each player starts with 10 points. On each turn, you do the following:

• Draw two cards with decimals on them.

• Using estimation, choose to multiply or divide your current score by one of the decimals.

• Once you have made your decision, compute your new score by doing the calculation and rounding to the nearest whole number. If the result is over 100, you lose.

• You may decide to stop at the end of any turn. If you do, your opponent gets one more turn to try to score closer to 100 than your score.

a. On one turn, you start with a score of 50 and draw 0.2 and 1.75. Tell what your new score will be if you do each of the following.

  i. divide by 0.2   **ii.** multiply by 0.2

  iii. divide by 1.75   **iv.** multiply by 1.75

b. Your score is now 88, and you draw 1.3 and 0.6. Use estimation to figure out your best move. Then calculate your new score.

c. Jahmal and Hannah are playing against each other. On his last turn, Jahmal's score was 57, and he drew 0.8 and 1.8. On her last turn, Hannah's score was 89, and she drew 0.7 and 1.2. If each player made the best move, who has the greater score now? Explain.

Find each product or quotient.

41. $\frac{4}{5} \cdot \frac{5}{7}$ **42.** $\frac{10}{13} \div \frac{5}{26}$ **43.** $\frac{345}{479} \cdot \frac{479}{345}$

44. $4\frac{2}{3} \div 1\frac{5}{6}$ **45.** $6\frac{7}{8} \cdot \frac{16}{49}$ **46.** $1\frac{1}{2} \div \frac{6}{4}$

Write each fraction as a decimal.

47. $\frac{17}{20}$ **48.** $\frac{19}{250}$ **49.** $\frac{16}{9}$

Measurement Convert each measurement to meters.

50. 32 cm **51.** 32 mm **52.** 32,000 cm

53. Statistics Toni asked students in her class what their favorite season is. She put her results in this pie chart.

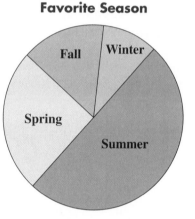

Favorite Season

 a. About what fraction of the students like spring best?

 b. About what percent of the students prefer summer?

 c. There are 20 students in Toni's class. About how many chose fall or winter as their favorite season?

54. Economics Jahmal earns money by working as a pet-sitter. He uses the following rule to charge customers:

Charge $1 per pet plus $5 per day.

 a. How much will Jahmal earn if he watches Ms. Nasca's parakeets, Bud and Lou, for four days?

 b. Mr. Ruiz has one guinea pig, three gerbils, and two ferrets. How much would Jahmal charge to pet-sit Mr. Ruiz's pets for seven days?

 c. Jahmal took care of Ms. Chou's cats for five days and charged a total of $28. How many cats does Ms. Chou have?

Chapter Summary

V O C A B U L A R Y
reciprocal

In this chapter, you learned how to do calculations with fractions and decimals. You used fraction pieces to add and subtract fractions with different denominators. You found that by rewriting the fractions with a common denominator, you could add and subtract without fraction pieces.

You then used what you know about multiplying whole numbers to figure out how to multiply a whole number by a fraction. You used rectangle diagrams to discover a method for multiplying two fractions.

You learned how to divide a whole number by a fraction by using a model and by writing a related multiplication problem. Then you learned two methods for dividing two fractions.

Finally, you turned your attention to operations with decimals. You learned that you could use estimation—along with what you already know about multiplying and dividing whole numbers—to multiply and divide decimals.

Strategies and Applications

The questions in this section will help you review and apply the important ideas and strategies developed in this chapter.

Adding and subtracting fractions and mixed numbers

1. Explain the steps you would follow to add two fractions with different denominators. Give an example to illustrate your steps.

2. Describe two methods for subtracting one mixed number from another. Use one of the methods to find $7\frac{1}{3} - 4\frac{5}{6}$. Show your work.

Multiplying fractions and mixed numbers

3. Find $\frac{2}{3} \times \frac{4}{5}$ by making a rectangle diagram. Show how this method of finding the product is related to finding the product of the numerators over the product of the denominators.

4. Describe how you would multiply two mixed numbers. Give an example to illustrate your method.

Dividing fractions and mixed numbers

5. Describe two ways to divide a whole number by a fraction. Illustrate both methods by finding $4 \div \frac{2}{3}$.

6. Use the problem $\frac{5}{6} \div \frac{4}{9}$ to illustrate a method for dividing fractions.

7. Without dividing, how can you tell whether a quotient will be greater than 1? Less than 1?

Multiplying and dividing decimals

8. Describe how to multiply 9.475×0.0012 without using a calculator. Be sure to explain how to decide where to put the decimal point.

9. Describe how you can use what you know about dividing whole numbers to divide two decimals without using a calculator. Illustrate your method by finding $15.665 \div 0.65$.

10. Suppose 1 U.S. dollar is equivalent to 10.3678 Moroccan dirham.

    a. To convert $100 to dirham, what calculation would you do? Explain how you know this calculation is correct.

    b. To convert 100 dirham to dollars, what calculation would you do? Explain how you know this calculation is correct.

    c. Caroline said that $750 is equal to about 7.5 dirham. Is Caroline's estimate reasonable? Explain.

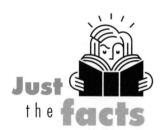

Just the facts

Morocco is located on the westernmost tip of north Africa. Although the climate in most of the country is quite warm, parts of the mountains enjoy snow for most of the year.

Demonstrating Skills

Find each sum or difference. Give your answers in lowest terms. If an answer is greater than 1, write it as a mixed number.

11. $\frac{4}{5} + \frac{1}{10}$ **12.** $\frac{3}{7} + \frac{5}{9}$ **13.** $\frac{13}{15} + \frac{1}{3}$

14. $\frac{5}{8} - \frac{1}{12}$ **15.** $\frac{11}{24} - \frac{3}{9}$ **16.** $\frac{4}{5} - \frac{9}{25}$

17. $3\frac{1}{3} + 2\frac{3}{5}$ **18.** $1\frac{3}{8} - \frac{3}{4}$ **19.** $5\frac{1}{7} - \frac{11}{3}$

Find each product or quotient. Give your answers in lowest terms. If an answer is greater than 1, write it as a mixed number.

20. $5 \cdot \frac{3}{10}$ **21.** $4 \div \frac{1}{8}$ **22.** $3\frac{1}{2} \cdot 6$

23. $\frac{2}{3} \times 12$ **24.** $\frac{5}{8} \cdot 14$ **25.** $1\frac{2}{3} \cdot \frac{7}{8}$

26. $\frac{6}{11} \times \frac{5}{3}$ **27.** $45 \div \frac{3}{5}$ **28.** $\frac{123}{12} \cdot \frac{12}{123}$

29. $\frac{1}{3} \cdot \frac{75}{100}$ **30.** $\frac{9}{14} \times \frac{16}{21}$ **31.** $\frac{15}{21} \div \frac{5}{7}$

32. $3\frac{2}{7} \div 2\frac{3}{7}$ **33.** $\frac{343}{425} \div \frac{343}{425}$ **34.** $\frac{18}{35} \cdot \frac{14}{27}$

Use the fact that $652 \times 25 = 16{,}300$ to find each product without using a calculator.

35. $65.2 \cdot 2.5$ **36.** 0.652×25 **37.** $6.52 \cdot 0.00025$

38. $6.52 \times 2{,}500$ **39.** $0.00652 \cdot 0.25$ **40.** 65.2×0.25

Find each product or quotient.

41. 0.25×400 **42.** $64 \div 0.8$ **43.** $32.07 \cdot 0.001$

44. $32.07 \div 0.001$ **45.** 7.75×12.4 **46.** 0.009×1.2

47. $0.144 \div 0.6$ **48.** $87.003 \cdot 5.5$ **49.** $19 \div 0.00038$

CHAPTER 4

Making Sense of Percents

Real-Life Math

Survey Says! Results of surveys are often reported as percents. For example, a retail association recently reported that 72% (a little less than $\frac{3}{4}$) of Americans give Mother's Day gifts. Of course, the association didn't survey every American. People who conduct surveys often use a method called *sampling.* This method involves surveying a part of a population, called a *sample,* and using the results to make predictions about the entire population.

Think About It For survey predictions to be reliable, the sample should include all the different types of people in the population. How would you design a survey to find out the favorite lunch in your school?

Family Letter

Dear Student and Family Members,

Look in any magazine or newspaper and you're likely to see numbers written as percents. Listen to any sporting event, and you'll probably hear statistics reported using percents. Percents are everywhere. But what are percents? The word *percent* means *for each 100*, so a percent like 50% is the same amount as the fraction $\frac{50}{100}$ (or $\frac{1}{2}$), or the decimal 0.50 (or 0.5). Fractions, decimals, and percents can be used interchangeably to represent parts of a whole quantity.

Often, the word *percent* is used in connection with the *percent of* some quantity. For example, you might hear, "Only 73% of registered voters voted in the last election." No matter what the quantity, 100% of a quantity always means all of it, and 50% always means half of it. The amount indicated by a certain percent changes as the size of the quantity changes. For example, 50% of 10 dogs is 5 dogs, but 50% of 100 dogs is 50 dogs.

What can you do at home?

There are many real-life occurrences of percents that you might want to explore with your student while we are studying this chapter. For example, you could ask your child to help you

- calculate the tip when you eat at a restaurant,
- calculate the price of an item that is on sale for 25% off,
- comparison shop between items that have different prices with different percents off of the regular prices, and
- compare interest rates on credit cards.

Using Percents

You see and hear percents used all the time.

You may not understand exactly what percents are, but you are probably familiar with them just from your everyday experiences. For example, you know that 95% is a good test score while 45% is not.

Think & Discuss

Use what you know about percents to answer these questions.

• In their last game, the Kane High School basketball team made 10% of their shots. Do you think the team played well? Explain.

• Marcus is going camping this weekend. The weather report for the area he's traveling to claims a 90% chance of rain. Do you think Marcus should bring his rain gear? Explain.

• The latest Digit Heads CD normally costs $16, but this week it is on sale for 25% off. Rosita has $15. Do you think she has enough money to buy the CD? Explain.

In this lesson, you will explore what percents are and how they can be used to make comparisons.

Investigation Understanding Percents

VOCABULARY

percent

Like a fraction or a decimal, a percent can be used to represent a part of a whole. The word **percent** means "out of 100." For example, 28% means 28 out of 100, or $\frac{28}{100}$, or 0.28.

Think & Discuss

Each *100-grid* below contains 100 squares. Express the part of each grid that is shaded as a fraction, a decimal, and a percent.

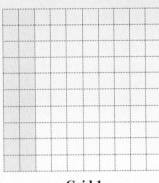

Grid 1

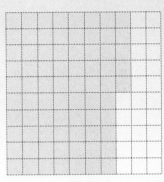

Grid 2

MATERIALS

- sheet of 100-grids
- transparent 100-grid

Of the 10,000 to 15,000 cheetahs alive today, about 10% live in captivity. In the wild, a cheetah lives about 7 years; the average life span in captivity is about 70% longer.

Problem Set A

Shade the given percent of a 100-grid. Then express the part of the area that is shaded as a fraction and as a decimal.

1. 10%   **2.** 25%   **3.** 1%

4. 15%   **5.** 50%   **6.** 110%

For each square, estimate the percent of the area that is shaded.

7.

8.

9.

10.

11.

12.

13. Describe the strategies you used to estimate the percents of the areas that were shaded.

14. Now place a 100-grid over each square in Problems 7–12. Express the exact portion that is shaded as a percent, a fraction, and a decimal.

You have seen that a percent is a way of writing a fraction with a denominator of 100. You can change a fraction to a percent by first finding an equivalent fraction with a denominator of 100. However, in many cases, it is easier to find a decimal first.

Think & Discuss

Write each fraction or decimal as a percent, and explain how you found your answers.

$$\frac{13}{20} \qquad \frac{3}{5} \qquad \frac{73}{50} \qquad 0.13 \qquad 0.9 \qquad 0.072$$

Write each fraction as a percent, and explain how you found your answers.

$$\frac{5}{8} \qquad \frac{11}{15} \qquad \frac{87}{150}$$

In Problem Set A, you used percents to represent part of an area. You can also use a percent to represent part of a collection or a group. When the group is made up of 100 items, finding a percent is easy. To find a percent in other cases, you can apply what you know about fractions and decimals.

Problem Set B

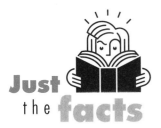

Just the facts

From 1959 through 1962, the All-Star Game was played twice a year. No game was played in 1945 because of severe travel restrictions during World War II.

1. Of the 25 students in Ms. Sunseri's homeroom, 11 are in band or choir. Express the part of the class in band or choir as a fraction, a decimal, and a percent. Explain how you found your answers.

2. Of the 70 All-Star Baseball Games played between 1933 and 2000, 40 were won by the National League. Express the portion of games won by the National League as a fraction, a decimal, and a percent. Round the decimal to the nearest hundredth and the percent to the nearest whole percent. Explain how you found your answers.

3. Last winter, Miguel worked shoveling driveways. He hoped to earn $200—the cost of a new bike he wanted. At the end of the winter, he had earned $280. Express the portion of the bike's cost Miguel earned as a fraction, a decimal, and a percent. Explain how you found your answers.

Just the facts

The first price scanner was introduced at a supermarket convention in 1974. The first product ever purchased using a checkout scanner was a pack of chewing gum.

ISBN 1-57039-941

Problem Set C

In October 1989, *Harper's* magazine printed this fact:

Percent of supermarket prices that end in 9 or 5: 80%

More than 10 years have passed since this statistic was printed. In this problem set, you will analyze some data to see whether it is still true.

1. With your group, devise a plan for testing whether the statistic is true today. Describe your plan.

2. Carry out your plan, and describe what you discovered.

3. Compare your results with those of other groups in your class. Describe how the findings of other groups are similar to or different from your findings.

4. Do you think the statistic is still true? If not, what percent do you think better describes the portion of today's supermarket prices that end in 9 or 5?

Share & Summarize

Tell what the word *percent* means, and explain how to use a percent to represent part of a whole. Give an example to illustrate your explanation.

Investigation 2 Making a Circle Graph

Data that represent parts of a whole are sometimes displayed in a circle graph, or pie chart. In this investigation, you will gather data from your class and use a circle graph to summarize your results.

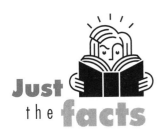

Think & Discuss

Work as a class to gather student responses to this question:

Which is your favorite sport to watch? Choose one.

❏ *football*   ❏ *soccer*   ❏ *basketball*   ❏ *baseball*

❏ *ice hockey*   ❏ *other*   ❏ *none*

Copy this table. In the second column, record the number of students who voted for each choice. In the third column, record the fraction of the total votes each choice received. You will fill in the remaining columns later.

Sport	Number of Votes	Fraction of Total Votes	Estimated Percent of Total Votes	Calculated Percent of Total Votes
Football				
Soccer				
Basketball				
Baseball				
Ice hockey				
Other				
None				

In the next problem set, you will create a circle graph to display the results of your survey.

MATERIALS
- 1-meter strip of heavy paper
- 4 copies of the quarter-circle template
- scissors
- tape
- colored pencils
- straightedge

Problem Set D

Your teacher will give you a strip of heavy paper exactly 1 meter long. The strip has been divided into equal-sized rectangles—one for each member of your class. The entire strip represents your whole class, or 100% of the votes.

1. On your strip, color groups of rectangles according to the number of votes each choice received. Use a different color for each choice. For example, if five students voted for football, color the first five rectangles blue. If seven chose soccer, color the next seven rectangles green. When you are finished, all the rectangles should be colored.

Football				Soccer															

Now you are ready to create a circle graph.

2. Tape the ends of your strip together, with no overlap, to form a loop with the colored rectangles inside.

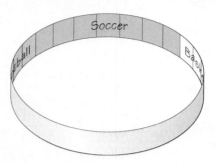

3. Tape four copies of the quarter-circle template together to form a circle.

4. Place your loop around the circle. On the edge of the circle, mark where each color begins and ends.

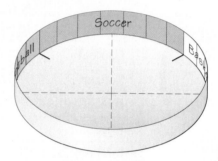

5. Remove the loop, and use a straightedge to connect each mark you made to the center of the circle.

6. Color the sections of your graph. Label each section with the sport name and the fraction of votes that sport received. For example, your circle graph might look like this:

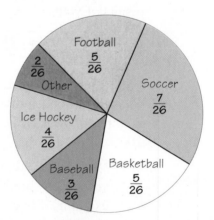

Just the **facts**

Professional football is the most-watched sport in the United States, followed by figure skating.

Circle graphs in books, magazines, and newspapers are often labeled with percents. You will now add percent labels to your circle graph.

- table from the Think & Discuss
- paper strip and circle graph from Problem Set D
- scissors
- centimeter ruler

Problem Set E

Since the strip you used in Problem Set D is exactly 100 centimeters long, you can estimate the percent of votes each choice received by finding the lengths of the sections in centimeters.

1. What percent of the whole strip does 1 centimeter represent? What percent of the votes does 1 centimeter represent?

2. Cut the tape that is holding the ends of your strip together, and lay the strip out flat. Find the length of one rectangle in centimeters. About what percent of the class does one vote, or one rectangle, represent?

3. What percent of the class do five votes represent? Explain how you found your answer.

4. By measuring the length of each section to the nearest centimeter, estimate the percent of students who voted for each choice. Record your estimates in your table.

5. Now *calculate* the percent of students who voted for each choice to the nearest whole percent. Record the results in your table, and then add percent labels to your circle graph.

6. Compare your estimates to the calculated percents. If any estimate is far from the actual percent, try to explain why.

7. Add the percents for all seven choices. If your result is not 100%, explain why.

Share & Summarize

In Problem Set E, you estimated the percent of the votes each choice received by measuring the lengths of the sections in centimeters. Why did this work?

Investigation 3 ▶ Parts of Different Wholes

In Investigation 2, you conducted a class survey. The same survey was conducted at Pioneer Middle School, but data were gathered from the entire sixth grade—160 students! Here are their results:

Sport	Number of Votes	Fraction of Total Votes
Football	12	$\frac{12}{160}$
Soccer	22	$\frac{22}{160}$
Basketball	40	$\frac{40}{160}$
Baseball	28	$\frac{28}{160}$
Ice hockey	14	$\frac{14}{160}$
Other	28	$\frac{28}{160}$
None	16	$\frac{16}{160}$

Just the facts

Field hockey is the second-most-played team sport in the world. Only soccer is more popular.

Think & Discuss

Suppose you want to compare the popularity of a particular sport among sixth graders at Pioneer Middle School with its popularity among students in your class.

• Would comparing the numbers of votes the sport received tell you whether it was more popular at Pioneer or in your class? Explain.

• Would comparing the fraction of votes each sport received tell you whether it was more popular at Pioneer or in your class? What about comparing percents? Explain.

• Which of the three types of comparisons mentioned above—comparing numbers of votes, fractions, or percents—do you think would be best for comparing popularity? Explain.

Problem Set F

1. Calculate the percent of the votes each sport received at Pioneer Middle School. Round your answers to the nearest whole percent.

2. Do the percents add to 100%? If so, why? If not, why not?

3. Write a short newspaper article comparing the Pioneer Middle School data with the data from your class.

In Problem Set F, you found that percents allow you to compare parts of different groups, even if the groups are of very different sizes. The next two problem sets will give you more practice with this idea.

Problem Set G

Mrs. Torres asked her first- and second-period classes this question:

Which continent would you most like to visit?

The results for the two classes are listed below.

Continent	Period 1 Votes	Period 2 Votes
Europe	2	3
Antarctica	0	1
Asia	3	8
Australia	10	5
South America	1	2
Africa	4	11

1. Akili is in the Period 2 class. He said Europe was a more popular choice in his class than in the Period 1 class. Is he correct? Explain.

2. Luisa is in the Period 1 class. She said Australia was twice as popular in her class as in the Period 2 class. Is she correct? Explain.

3. Lee is in the Period 2 class. He said that in his class, Asia was four times as popular as South America. Is he correct? Explain.

4. Write two true statements similar to those made by Akili, Luisa, and Lee comparing the data in the table.

Problem Set H

Marathon City held a walkathon to raise money for charity. Of the 500 students at East Middle School, 200 participated in the walkathon. At West Middle School, 150 of the 325 students participated. The sponsors of the walkathon plan to give an award to the middle school with the greater participation.

1. What argument might the principal at East present to the sponsors to convince them to give the award to her school?

2. What argument might the principal at West make to convince the sponsors to give the award to his school?

3. Which school do you think deserves the award? Defend your choice.

Share & Summarize

Suppose Ms. Washington's class has fewer students than your class. They would like to compare their results for the sports survey with the results from your class.

- Conor suggests comparing the number of votes each sport received.

- Jahmal thinks it would be better to compare the fraction of the votes each sport received.

- Rosita says it is best to compare percents.

Which type of comparison do you think is best? Defend your answer.

Just the facts

Hockey pucks are constructed of rubber and measure 3 inches in diameter.

Investigation Linking Percents, Fractions, and Decimals

Percents, fractions, and decimals can all be used to represent parts of a whole. However, in some situations, one form may be easier or more convenient.

For example, you have seen that it is often easier to compare percents than to compare fractions. To be a good problem-solver, you need to become comfortable changing numbers from one form to another.

Problem Set I

Write each given fraction or mixed number as a decimal and a percent.
Write each given percent as a decimal and a fraction or mixed number in
lowest terms.

1. The head of the cafeteria staff took a survey to find out what
   students wanted for lunch.

   a. He found that 77% of the students wanted pizza every day.

   b. He was surprised to find that $\frac{2}{5}$ of students would like to have
   a salad bar available, in case they didn't want what was served
   for lunch.

   c. He found that $\frac{11}{20}$ of students favored French fries while 45% pre-
   ferred mashed potatoes.

2. Mt. Everest, with an estimated elevation of 29,028 feet, is the highest mountain in Asia and the world.

   a. Mt. Aconcagua is the highest mountain in South America. It is
   approximately 78% of the height of Mt. Everest.

   b. Mt. McKinley is the highest mountain in North America. It is
   approximately 70% of the height of Mt. Everest.

   c. Mt. Kilimanjaro is the highest peak in Africa. It is approximately
   $\frac{2}{3}$ of the height of Mt. Everest.

3. The Ob-Irtysh, with a length of approximately 3,460 miles, is the
   longest river in Asia and the fourth-longest river in the world.

   a. The Mississippi–Missouri–Red Rock River is the longest river in
   North America and the third longest in the world. It is about $1\frac{2}{25}$
   as long as the Ob-Irtysh.

   b. The Amazon is the longest river in South America and the
   second longest in the world. It is about 113% as long as the
   Ob-Irtysh.

Just the facts

One of the greatest
challenges for moun-
tain climbers is the
"seven summits"—
climbing the highest
mountain on each of
the seven continents.

Mt. Kilimanjaro

You have been thinking about percents as parts of wholes. Like fractions and decimals, you can also think of percents simply as numbers.

In the next problem set, you will label points on a number line with fractions, decimals, and percents. As you work, you will become familiar with some common percents that are often used as benchmarks.

Remember

Benchmarks are familiar values that you can use to approximate other values.

Problem Set J

Copy each number line. Fill in the blanks so that each tick mark is labeled with a percent, a fraction, and a decimal. Write all fractions in lowest terms.

1.

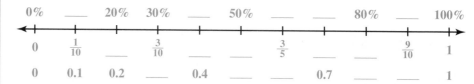

2.

3.

4.

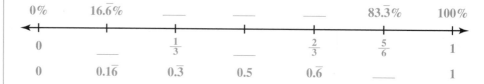

5. The percent equivalent for $\frac{1}{3}$ is often written as $33\frac{1}{3}\%$. Explain why this makes sense.

6. What fraction is equivalent to $66\frac{2}{3}\%$?

To solve the problems in Problem Set K, you will need to apply what you know about converting between fractions and percents.

Problem Set K

Sunscreens block harmful ultraviolet (UV) rays produced by the sun. Each sunscreen has a Sun Protection Factor (SPF) that tells you how many minutes you can stay in the sun before you receive 1 minute of burning UV rays. For example, if you apply sunscreen with SPF 15, you get 1 minute of UV rays for every 15 minutes you stay in the sun.

1. A sunscreen with SPF 15 blocks $\frac{14}{15}$ of the sun's UV rays. What percent of UV rays does the sunscreen block?

2. Suppose a sunscreen blocks 75% of the sun's UV rays.

   a. What fraction of UV rays does this sunscreen block? Give your answer in lowest terms.

   b. Use your answer from Part a to calculate this sunscreen's SPF. Explain how you found your answer.

3. A label on a sunscreen with SPF 30 claims the sunscreen blocks about 97% of harmful UV rays. Assuming the SPF factor is accurate, is this claim true? Explain.

Share & Summarize

1. How do you change a percent to a fraction?

2. How do you change a percent to a decimal?

3. How do you change a decimal to a percent?

4. How do you change a fraction to a percent?

On Your Own Exercises

Practice & Apply

Shade the given percent of a 100-grid. Then express the shaded portion as a fraction and a decimal.

1. 37% **2.** 72% **3.** 4% **4.** 125%

Estimate the percent of each square that is shaded, and describe how you made your estimate.

5.

6.

7.

8.

Write each fraction or decimal as a percent. Round to the nearest tenth of a percent.

9. $\frac{4}{5}$ **10.** 0.32 **11.** 0.036 **12.** $\frac{3}{71}$

13. 1 **14.** $\frac{19}{20}$ **15.** 2.7 **16.** 0.004

17. Seven of the 20 people at Rosita's birthday party are in her math class. Express the portion of the guests that are in her math class as a fraction, a decimal, and a percent. Explain how you found your answers.

18. Social Studies Of the 232 million tons of garbage generated in the United States every year, about 87 million tons are paper products. Express the portion of U.S. garbage that is paper as a fraction, a decimal, and a percent. Explain how you found your answers.

19. Last season, Jing set a goal of hitting 9 home runs. When the season was over, she had hit 12 home runs. Express the portion of her goal Jing reached as a fraction, a decimal, and a percent. Round to the nearest whole percent. Explain how you found your answers.

20. Here are the results of the sports survey for Mr. Shu's class:

Sport	Number of Votes	Fraction of Votes	Percent of Votes
Football	8		
Soccer	2		
Basketball	7		
Baseball	4		
Ice hockey	4		
Other	1		
None	2		

a. Complete the table to show the fraction of votes and the percent of votes each sport received. Round to the nearest whole percent.

b. Starting with a 100-centimeter strip divided into 28 sections to represent the 28 votes, the class created a circle graph. What is the length of each section? What percent of the votes does each section represent?

c. How long should the basketball section of the strip be?

d. Which circle graph correctly represents the data for Mr. Shu's class? Explain how you know.

Graph A

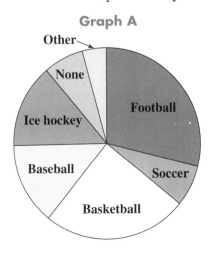

Graph B

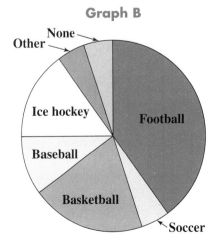

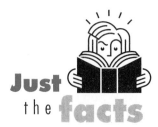

21. A sports magazine asked 3,600 of its subscribers this question:

Which of these sports do you think is most dangerous? Choose one.

❏ *football*   ❏ *ice hockey*   ❏ *skiing*

❏ *sky diving*   ❏ *rock climbing*   ❏ *other*

Ms. Kelsey's math class decided to conduct the same survey in their class. The table shows the results of the magazine's survey and Ms. Kelsey's class survey.

Sport	Fraction of Votes in Magazine Survey	Fraction of Votes in Ms. Kelsey's Class
Football	$\frac{524}{3,600}$	$\frac{3}{25}$
Ice hockey	$\frac{320}{3,600}$	$\frac{0}{25}$
Skiing	$\frac{870}{3,600}$	$\frac{8}{25}$
Skydiving	$\frac{607}{3,600}$	$\frac{11}{25}$
Rock climbing	$\frac{959}{3,600}$	$\frac{2}{25}$
Other	$\frac{320}{3,600}$	$\frac{1}{25}$

a. Find the percent of votes each sport received in the magazine survey. Round to the nearest percent.

b. Find the percent of votes each sport received in Ms. Kelsey's class. Round to the nearest percent.

c. Write a short newspaper article comparing the results of the magazine survey with the results of the survey in Ms. Kelsey's class.

22. Mr. Diaz asked his first- and fifth-period classes this question:

Which type of movie do you most like to watch? Choose one.

❏ *action*   ❏ *suspense*   ❏ *drama*

❏ *comedy*   ❏ *animation*   ❏ *other*

Movie Type	Period 1 Votes	Period 5 Votes
Action	3	6
Suspense	0	3
Drama	8	9
Comedy	11	7
Animation	2	2
Other	1	6

a. Chloe is in Period 5. She said drama was a more popular choice in her class than in the Period 1 class. Is she correct? Explain.

b. Ricki is in Period 5. She said, in her class, comedy was more than twice as popular as suspense. Is she correct? Explain.

c. Don is in Period 1. He said animation was just as popular in Period 5 as it was in his class. Is he correct? Explain.

d. Write two true statements comparing the data in the table.

23. Nutrition Harvest granola has 6.5 grams of fat per 38-gram serving. Healthy Crunch granola has 8 grams of fat per 50-gram serving.

a. The makers of Harvest granola claim it has less fat than Healthy Crunch granola. What argument could they use to defend their claim?

b. The makers of Healthy Crunch claim their granola has less fat than Harvest. What argument could they use to defend their claim?

Write each given percent as a decimal and a fraction in lowest terms. Write each given fraction as a decimal and a percent.

24. Social Studies About 7% of Americans are under age 5, and about 13% are over age 65.

25. About $\frac{2}{3}$ of U.S. households with televisions subscribe to a cable-television service.

26. In 1820, almost 72% of U.S. workers were employed in farm occupations. By 1994, only $\frac{1}{40}$ of U.S. workers were employed in farming.

27. The number of students in band is about 115% of the number in orchestra and about $\frac{5}{4}$ of the number in choir.

28. Copy and complete the table so the numbers in each row are equivalent.

Fraction	Decimal	Percent
$\frac{1}{2}$	0.5	50%
		7.8%
	5.2	
$\frac{7}{16}$		
	0.37	

29. Science What percent of the sun's harmful UV rays does a sunscreen with SPF 25 block?

30. A sunscreen blocks 95% of the sun's harmful UV rays. What is the sunscreen's SPF?

Just the facts

Cable-television signals are received from antennas and satellites by cable companies. The signals are then sent out to customers along coaxial and fiber-optic cables.

31. A public television station held a five-day fund-raiser. At the end of each day, a staff member recorded the part of the fund-raising goal reached on a large thermometer. Here are the thermometers for each of the five days:

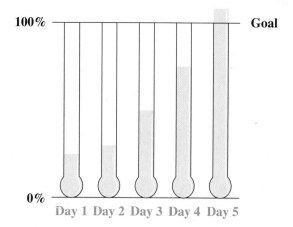

a. Estimate the percent of the goal reached each day.

b. The station set a goal of $20,000 for the fund-raiser. About how much money had they raised by the end of Day 3? By the end of Day 5?

32. Every January, Framingham Middle School holds its annual Winter Event. An article in the school paper reported that 45% of seventh graders voted that this year's event should be an ice-skating party. The president of the seventh grade class said that $\frac{9}{20}$ of seventh graders voted for ice skating. Could both reports be correct? Explain.

33. Imagine that you are in charge of planning a town park. The park will be shaped like a square. The community council has given you these guidelines:

• At least 12% of the park must be a picnic area.

• Between 15% and 30% of the park should be a play area with a sandbox and playground equipment.

• A goldfish pond should occupy no more than 10% of the park.

On a 100-grid, sketch a plan for your park. You may include any features you want as long as the park satisfies the council's guidelines. Label the features of your park, including the picnic area, play area, and goldfish pond, and tell what percent of the park each feature will occupy.

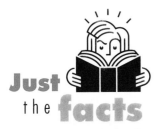

34. Statistics Conor asked the 24 students in his math class what kind of pets they have. Here are his results:

Pet	Students
Dog	12
Cat	8
Ferret	2
Bird	3
Fish	6
Other	3
None	5

a. Calculate the percent of students that have each type of pet. Round to the nearest percent.

b. The total of the numbers in the second column is greater than the number of students in the class. Explain why this makes sense.

c. Would it make sense to make a circle graph for these data? Explain why or why not.

35. Preview You have seen that circle graphs are useful for displaying data that are parts of a whole. A *stacked bar graph* is also useful for this purpose. The bar graph at right shows how the total number of new cars sold in 2001 is divided among various vehicle sizes.

a. The bar graph is exactly 10 cm tall. Describe how you could use a ruler to estimate the percent of cars in each size category.

b. Estimate the percent of cars in each size category.

2001 Car Sales by Size

36. Economics Caroline needs a new winter coat. The Winter Warehouse advertises that everything in the store is 75% of the retail price. Coats Galore advertises that all its coats are on sale for $\frac{7}{10}$ of the retail price. If the stores carry the same brands at the same prices, where will Caroline find better prices? Explain.

37. At Valley Middle School, the sixth grade has 160 students, the seventh grade has 320 students, and the eighth grade has 240 students. The student congress is traditionally made up of eight representatives from each grade.

a. For each grade, find the percent of students in the congress. Give your answers to the nearest tenth of a percent.

b. Tom is in the seventh grade class. He says that, since his class has more students, it should have more representatives. He suggests that each grade be represented by the same *percent* of its students. Devise a plan for setting up the student congress this way. Tell how many representatives each grade should elect, and the percent of each grade that is represented.

c. Which plan do you think is fairer: the original plan or the plan you devised in Part b? Defend your answer.

38. It is often useful to use benchmark percents to estimate the values of actual percents. Useful benchmarks include multiples of 10% (10%, 20%, 30%, and so on), as well as 25% and 75%.

a. Of the 910 students attending Mill Middle School, 179 walk to school. Use a benchmark to estimate the percent of students who walk to school, and explain how you decided which benchmark to use.

b. A human skeleton is made up of 206 bones. There are 54 bones in the hands. Use a benchmark to estimate the percent of a human skeleton's bones found in the hands, and explain how you decided which benchmark to use.

c. The girl's softball team won 25 of the 33 games it played last season. Use a benchmark to estimate the percent of games the team won, and explain how you decided which benchmark to use.

Find each quantity.

39. $\frac{3}{4}$ of 120   **40.** $\frac{7}{9}$ of 54   **41.** $\frac{3}{10}$ of 15

Use the fact that $13 \cdot 217 = 2{,}821$ to find each product *without* using a calculator.

42. $1.3 \cdot 217$ **43.** $13 \cdot 2.17$ **44.** $0.013 \cdot 21.7$

45. $0.13 \cdot 0.217$ **46.** $1{,}300 \cdot 2{,}170$ **47.** $13 \cdot 0.0217$

Find the least common multiple of each pair of numbers.

48. 7 and 37   **49.** 38 and 190   **50.** 21 and 35

51. You can make circle diagrams, called *Venn diagrams,* to group the factors of numbers.

   a. The factors of 24 have been placed in the correct part of the Venn diagram below. Copy the diagram, and write the remaining factors of 36 and 54 where they belong.

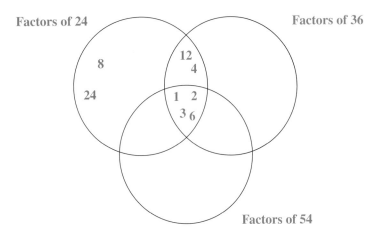

   b. What does the overlap of all three circles represent?

4.2 Finding a Percent of a Quantity

You often read and hear statements that mention the "percent of" a particular quantity.

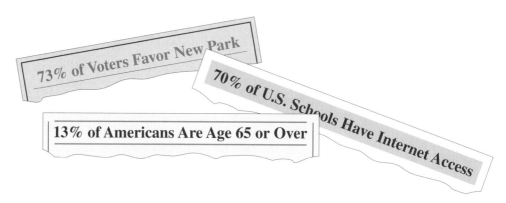

73% of Voters Favor New Park

70% of U.S. Schools Have Internet Access

13% of Americans Are Age 65 or Over

No matter what the quantity, 100% is all of it, and 50% is half of it. The specific amount a given percent represents depends on the quantity. For example, 50% of 10 tree frogs is 5 tree frogs, while 50% of 100 tree frogs is 50 tree frogs.

Think & Discuss

Use what you know about fractions and percents to answer each of these questions.

- What is 100% of 500?
- What is 50% of 500?

- What is 25% of 500?
- What is 1% of 500?

- What is 100% of 40?
- What is 50% of 40?

- What is 25% of 40?
- What is 1% of 40?

Investigation Modeling with a Grid

Imagine that this 100-grid represents a value of 200 and that this value is divided evenly among the 100 small squares.

2	2	2	2	2	2	2	2	2	2
2	2	2	2	2	2	2	2	2	2
2	2	2	2	2	2	2	2	2	2
2	2	2	2	2	2	2	2	2	2
2	2	2	2	2	2	2	2	2	2
2	2	2	2	2	2	2	2	2	2
2	2	2	2	2	2	2	2	2	2
2	2	2	2	2	2	2	2	2	2
2	2	2	2	2	2	2	2	2	2
2	2	2	2	2	2	2	2	2	2

Value: 200

Think & Discuss

What is the value of each small square in the grid? What percent of 200 does each small square represent?

What percent of 200 do 10 small squares represent? What is the value of 10 small squares?

How could you use the grid to find 20% of 200?

MATERIALS
sheet of 100-grids

Problem Set A

1. Refer to the grid above.

   a. What is 15% of 200?   **b.** What is 50% of 200?

2. Imagine that a 100-grid has value 300 and that this value is divided evenly among the small squares. In other words, each small square is worth 3. Use a new grid for each part of this problem, and label each grid "Value: 300."

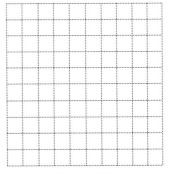

Value: 300

   a. Shade 25% of a grid. What is 25% of 300?

   b. Shade 10% of a grid. What is 10% of 300?

   c. Shade 17% of a grid. What is 17% of 300?

   d. Shade 75% of a grid. What is 75% of 300?

   e. Shade 120% of a grid. What is 120% of 300?

3. Imagine that a 100-grid has value 50 and that this value is divided evenly among the small squares. Use a new grid for each part of this problem, and label each grid "Value: 50."

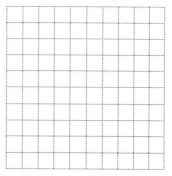

Value: 50

a. Shade 1% of a grid. What is 1% of 50?

b. Shade 10% of a grid. What is 10% of 50?

c. Shade 50% of a grid. What is 50% of 50?

4. Imagine that a 100-grid has value 24 and that this value is divided evenly among the small squares. Use a new grid for each part of this problem, and label each grid "Value: 24."

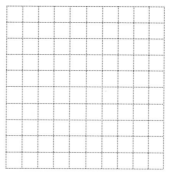

Value: 24

a. Shade 1% of a grid. What is 1% of 24?

b. Shade 10% of a grid. What is 10% of 24?

c. Shade 80% of a grid. What is 80% of 24?

5. Look back at your work in Problems 1–4. Describe a shortcut for finding a percent of a number without using a grid, and give an example to show how it works.

Problem Set B

Find each result without using a grid.

1. 3% of 400

2. 15% of 600

3. 44% of 50

4. 12% of 250

5. In a recent year, about 8,000,000 new cars were sold in the United States. About 16% of these cars were luxury cars.

    a. About how many new luxury cars were sold?

    b. About 10% of the new luxury cars sold were green. How many green luxury cars were sold?

6. You know that 10% of 360 is 36.

    a. Use this fact to find 20% of 360 and 40% of 360. Explain the calculations you did.

    b. Use the fact that 10% of 360 is 36 to find 5% of 360 and 15% of 360. Explain your calculations.

7. The Chang family went out for dinner. The total cost for the meal was $40. Calculate a 15% tip in your head, and explain what you did.

Share & Summarize

Describe two methods for finding 15% of 400.

Investigation 2 — Using a Shortcut

In the last investigation, you used a grid to help find a given percent of a number. You probably discovered a shortcut or two for finding a percent of a number without using a grid. The Example shows how Conor and Rosita thought about finding 23% of 800.

EXAMPLE

In the next problem set, you will practice Rosita's method. When you do a calculation with percents, it is a good idea to estimate the answer by using benchmark fractions and percents. This will help you check whether your answer is reasonable.

For example, 23% of 800 is a little less than 25%, or $\frac{1}{4}$, of 800, so the result should be a little less than 200. Conor and Rosita's answer of 184 is reasonable.

Problem Set C

Estimate each result using benchmarks. Then find the exact value using Rosita's shortcut.

1. 30% of 120   **2.** 45% of 400

3. 9% of 600   **4.** 72% of 1,100

Find each result using any method you like.

5. 3% of 45   **6.** 15% of 64

7. 44% of 125   **8.** 2% of 15.4

9. 125% of 40   **10.** 12.5% of 80

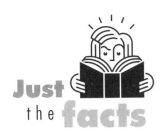

Just the facts

The number of Internet users grew from 3 million in 1994 to over 200 million in 2000— a 6,667% increase!

11. In 2000, there were 105,480,000 households in the United States.

    a. 51.0% of U.S. households had at least one computer. How many U.S. households had a computer?

    b. 41.5% of U.S. households had access to the Internet. How many U.S. households had Internet access?

When stores have sales, they often advertise that items are a certain "percent off." In the rest of this investigation, you will practice calculating the sale price when a percent discount is taken.

Problem Set D

A department store is having their annual storewide sale.

    1. All athletic shoes are on sale for 20% off the original price. Caroline bought a pair of cross-trainers with an original price of $50. What was the sale price for this pair of shoes? Explain how you found your answer.

    2. All fall jackets are on sale for 35% off the original price. Miguel bought a jacket originally priced at $60. How much did he pay? Explain how you found your answer.

    3. A CD player, originally priced at $120, is on sale for 25% off. What is the sale price?

Caroline and Miguel have different ways of calculating the sale price for the items they bought.

As you work on the next problem set, try both of their methods to see which you prefer.

Problem Set E

1. At K.C. Nickel's back-to-school sale, everything is 40% off the price marked on the tag. Find the sale price of each item.

a.

b.

2. At Celebrate! card store, birthday cards are on sale for 25% off. What is the sale price for a card originally marked $1.60?

3. Zears and GameHut usually both charge $75 for Quasar-Z, a hand-held electronic game. This week Quasar-Z is on sale for $60 at GameHut and for 25% off at Zears. At which store is the game less expensive? Explain.

Problem Set F

MATERIALS
- slips of paper numbered 1–20
- paper bag

The If the Shoe Fits shoe store is having a "Draw a Discount" sale. For the sale, the numbers 1 through 20 are put in a box. Each customer draws two numbers and adds them. The result is the percent the customer will save on his or her purchases.

1. To test how the sale works, pretend to be five customers. The price of each customer's purchase is given in the table. Place slips of paper numbered 1 through 20 in a bag. For each customer, draw two slips of paper, record the numbers, and return the two slips to the bag.

Customer	Original Price	First Number	Second Number	Percent Off	Sale Price
1	$37.00				
2	20.50				
3	12.98				
4	45.79				
5	79.99				

2. Complete the table by finding the percent off and the sale price for each customer's purchase.

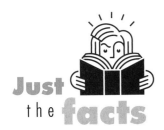
3. What is the total amount the five customers paid for their purchases?

4. Now figure out the price each customer would have paid if, instead of the "Draw a Discount" sale, the store had offered 20% off all purchases.

Customer	Original Price	Percent Off	Sale Price
1	$37.00	20%	
2	20.50	20%	
3	12.98	20%	
4	45.79	20%	
5	79.99	20%	

5. What is the total amount the five customers would have paid during a 20% off sale? How does this compare to the total for the "Draw a Discount" sale?

6. If you were the store manager, which type of sale would you hold? Explain.

Share & Summarize

1. Describe a method for calculating a given percent of a number. Demonstrate your method by finding 67% of 320.

2. Write a problem about a "percent off" sale. Explain how to solve your problem.

On Your Own Exercises

Practice
Apply

1. Imagine that a 100-grid has value 150 and that this value is divided evenly among the small squares.

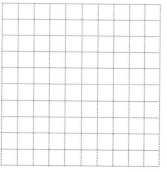

Value: 150

   a. What is the value of 25 small squares?

   b. What is the value of 1% of the grid?

   c. What is the value of $\frac{1}{10}$ of the grid?

   d. What is 40% of 150?

   e. What is 17% of 150?

   f. What is 150% of 150?

Find each result without using a grid.

Just
t h e **facts**

There are more than 300,000 fast-food restaurants in the United States.

2. 22% of 700	**3.** 90% of 120
4. 30% of 15	**5.** 65% of 210

6. In a recent year, Americans spent about $313 billion dollars on food prepared away from home. Of this total, almost 48% was spent on fast food.

   a. About how much money did Americans spend on fast food? Round your answer to the nearest billion dollars.

   b. Of the total dollars spent on fast food, about 64% was spent on takeout food. About how many fast-food dollars were spent on takeout food?

7. **Economics** Mr. Diaz took his mother out for dinner. The total for the items they ordered was $20.

   a. Mentally calculate the 5% sales tax on the order.

   b. Mr. Diaz wants to leave a 20% tip on the food cost plus the sales tax. Mentally calculate how much the tip should be.

impactmath.com/self_check_quiz

Estimate each result using benchmarks. Then find the exact value.

8. 75% of 80   **9.** 60% of 90

10. 65% of 60   **11.** 57% of 80

Find each result using any method you like.

12. 19% of 43   **13.** 45% of 234   **14.** 67% of 250

15. 112% of 70   **16.** 0.55% of 100   **17.** 72% of 3.7

18. **Nutrition** If a 64-ounce carton of fruit juice contains 10% real fruit juice, how many ounces of fruit juice does the carton contain?

19. **Sports** A hockey arena has a seating capacity of 30,275. Of these seats, about 31% are taken by season ticket holders. About how many seats are taken by season ticket holders?

20. Of the 2,000 students at Franklin High School, 28% are freshmen. How many Franklin students are freshmen?

21. **Economics** At Sparks electronic store, all CD players are reduced to 66% of the original price.

     a. What is the sale price for a CD player that originally cost $90?

     b. How much money would you save on a $90 CD player? What "percent off" is this?

22. The Fountain of Youth health-products store is going out of business. To help clear out their remaining merchandise, they are having a "Save Your Age" sale. Each customer saves the percent equal to his or her age on each purchase.

     a. Marcus bought a case of all-natural soda originally priced at $18. If Marcus is 12 years old, how much did he pay for the case of soda?

     b. Marcus' father bought some soap and shampoo originally priced at $24. If he is 36 years old, how much did he pay for the items?

     c. Marcus' grandmother is 63 years old. She bought a juicer originally priced at $57. How much did she pay?

23. You can use 100-grids to model real-world problems involving percents. In Parts a and b, show how you could use a 100-grid to model and solve the problem.

a. The sixth grade has raised $880 of the $2,000 they need for a class trip. What percent of the $2,000 do they still need to raise?

b. Challenge This year, tickets to the dance cost $15. This is 125% of last year's cost. How much did tickets cost last year?

24. This problem will give you practice thinking about percents greater than 100%.

a. Could there be a 125% chance of rain tomorrow? Explain.

b. Could a fund-raiser raise 130% of its goal? Explain.

c. Could a drink be 110% fruit juice?

d. Could a candidate get 115% of the votes in an election?

e. Could prices in a store increase by 120%?

25. Economics You have learned two ways to compute the sale price when a percent discount is taken. You can use similar methods to solve problems involving a percent increase.

a. Last year, Jahmal bought an antique radio for $28. Since then, the radio's value has increased by 25%. By how many dollars did the value increase? What is the new value of the radio?

b. In Part a, you computed the value of the radio in two steps: you calculated the number of dollars the value increased, and then you added the increase to the original value.

How could you calculate the value in one step? Explain why your method works, and show that it gives the same answer you found in Part a.

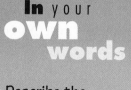

In your **own words**

Describe the difference between finding a "percent of" a given price and the "percent off" a given price. Give examples to show how to do each calculation.

26. Which is greater: 300% of 8, or 250% of 10?

27. Tinley's department store is having a 25% off sale. Jing has a $15 Tinley's gift certificate she wants to use toward a chess set with an original price of $32. She is unsure which of the following methods the sales clerk will use to calculate the amount she must pay:

- Method 1: Subtract $15 from the price and take 25% off the resulting price.

- Method 2: Take 25% off the original price and then subtract $15.

   a. Do you think both methods will give the same result? If not, predict which method will give a lower price.

   b. For each method, calculate the amount Jing would have to pay. Show your work.

   c. Which method do you think stores actually use? Why?

Mixed Review

Evaluate each expression.

28. $\frac{5}{6} + \frac{11}{12}$ **29.** $\frac{6}{11} - \frac{1}{2}$ **30.** $\frac{14}{19} \cdot \frac{38}{77}$

31. $\frac{3}{5} \div \frac{12}{13}$ **32.** $2\frac{3}{4} \div \frac{3}{8}$ **33.** $2\frac{1}{3} \cdot 1\frac{5}{6}$

Describe the pattern in each sequence, and use the pattern to find the next three terms.

34. 2, 7, 12, 17, 22, . . .   **35.** 12, 6, 3, $\frac{3}{2}$, $\frac{3}{4}$, $\frac{3}{8}$, . . .

36. a, c, e, g, i, . . .   **37.** 1, 2, 4, 7, 11, 16, . . .

Geometry Draw a polygon matching each description, if possible. If it is not possible, say so.

38. a concave pentagon

39. a triangle with exactly two lines of symmetry

40. a quadrilateral that is not regular and that has two lines of symmetry

4.3 Percents and Wholes

Think about these mathematical sentences:

$$44\% \text{ of } 125 = 55 \qquad 15\% \text{ of } 200 = 30$$

Both sentences are in this form:

> a percent **of** the whole **=** the part

In the last lesson, you were given a percent and the whole, and you found the part. This is like filling in the blank in sentences like these:

$$44\% \text{ of } 125 = \underline{\hspace{1cm}} \qquad 15\% \text{ of } 200 = \underline{\hspace{1cm}}$$

In this lesson, you will fill in the blanks in sentences like these:

$$\underline{\hspace{1cm}}\% \text{ of } 125 = 55 \qquad 15\% \text{ of } \underline{\hspace{1cm}} = 30$$

As you work, you will find it helpful to use benchmark fractions, decimals, and percents. The *Percent Bingo* game below will help refresh your memory about equivalent fractions, decimals, and percents.

Explore

Choose nine of the numbers listed below, and write one in each square of a grid like this one.

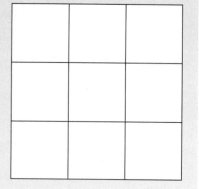

$\frac{1}{8}$	$\frac{1}{4}$	$\frac{1}{2}$	$\frac{1}{3}$	$\frac{1}{5}$	$\frac{1}{6}$
$\frac{2}{3}$	$\frac{2}{5}$	$\frac{3}{4}$	1	$\frac{1}{10}$	$\frac{5}{8}$

0.125	0.25	0.5	$0.\overline{3}$	0.2	$0.1\overline{6}$
$0.\overline{6}$	0.4	0.75	0.1	0.625	

When your teacher calls out a percent, look for an equivalent decimal or fraction on your grid. If you find one, circle it. If your grid contains both a fraction and a decimal equal to the given percent, circle both.

If you circle three numbers in a row—horizontally, vertically, or diagonally—call out "Bingo!" The first student who gets bingo wins.

Investigation 1 Finding the Percent

The question "What percent of 75 is 20?" gives you the part, 20, and the whole, 75, and asks you to find the percent. Answering this question is like filling in the blank in this sentence:

$$\_\_\_\% \text{ of } 75 = 20$$

You already solved several problems like this in Lesson 4.1.

> ## EXAMPLE
>
> In a sports survey, 14 out of 160 students said they like watching ice hockey best. What percent of the students is this? In other words, what percent of 160 is 14?
>
> In this case, the part is 14 and the whole is 160. To find the percent, write "14 out of 160" as a fraction and change the fraction to a percent.
>
> $$14 \text{ out of } 160 = \frac{14}{160} = 0.0875 = 8.75\%$$
>
> Estimating with benchmarks can help you make sure your answer is reasonable. Since $\frac{14}{160}$ is a little less than $\frac{16}{160}$, or $\frac{1}{10}$, the percent should be a little less than 10%. Therefore, 8.75% is reasonable.

Problem Set A

1. Your teacher will give you a page from a telephone book. Quickly scan the last four digits of the phone numbers on your page. Which digit do you think appears most often?

2. Starting with a phone number near the top of the page, analyze 30 phone numbers in a row. Use a table like this one to keep a tally of the last four digits. For example, if one of the numbers you choose ends 2329, make two tally marks next to the 2, one next to the 3, and one next to the 9.

Digit	Tally	Number of Tallies	Fraction of Tallies	Estimated Percent	Exact Percent
0					
1					
2					
3					
4					
5					
6					
7					
8					
9					

3. Count the number of tally marks for each digit and record the results in your table. You should have a total of 120 tally marks.

4. Find the fraction of the 120 tally marks each digit received, and record the results.

5. Use benchmarks to estimate the percent of the 120 tally marks each digit received, and record your results.

6. Calculate the exact percent of the 120 tally marks each digit received, and record your results.

7. Choose one of the digits, and explain how you found the estimated percent and the exact percent for that digit.

8. Which digit occurred most often? What percent of the 120 digits does this digit account for?

9. Which digit occurred least often? What percent of the 120 digits does this digit account for?

The next problem set will give you more practice finding percents when you know the part and the whole.

Problem Set B

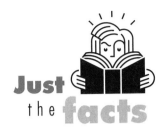
For these problems, first write a fraction and then calculate the percent. Round your answers to the nearest whole percent.

1. What percent of numerals from 1 through 40 are formed at least partially with curved lines? Assume the digits are written like these:

   0 1 2 3 4 5 6 7 8 9

2. Consider the whole numbers from 1 through 80.

   a. What percent of these numbers have two digits?

   b. What percent are multiples of 9?

   c. What percent are even and prime?

   d. What percent are greater than 9?

   e. What percent are factors of 36?

3. Now think about the whole numbers from 1 through 26.

   a. What percent contain only even digits?

   b. What percent contain only odd digits?

   c. What percent contain one even digit and one odd digit?

4. Consider the whole numbers from 1 through 52.

   a. What percent are common multiples of 2 and 3?

   b. What percent are common factors of 24 and 42?

Problem Set C

Fill in the blanks. Round each answer to the nearest whole percent.

1. ___% of 65 = 45

2. ___% of 23 = 17

3. ___% of 9 = 4.5

4. ___% of 93 = 75

5. ___% of 45 = 60

6. ___% of 250 = 500

7. What percent of the area of the large rectangle is the area of the small rectangle? Explain how you found your answer.

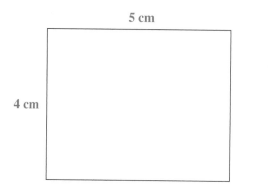

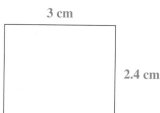

Remember

The area of a rectangle is its length times its width.

Share & Summarize

Describe a method for finding a percent when you are given the part and the whole. Demonstrate your method by finding what percent 25 is of 60.

Investigation 2 Finding the Whole

The question "50% of what number is 8?" gives you the percent, 50%, and the part, 8, and asks you to find the whole. Answering this question is like filling in the blank in this sentence:

$$50\% \text{ of } \underline{\hspace{2em}} = 8$$

To find the whole in this problem, try using what you know about fraction equivalents for percents.

Think & Discuss

Find each missing whole, and explain how you found your answer.

$$50\% \text{ of } \underline{\hspace{1cm}} = 8 \qquad 20\% \text{ of } \underline{\hspace{1cm}} = 10$$

Problem Set D

Remember

$33\frac{1}{3}\% = \frac{1}{3}$

In Problems 1–6, find the missing whole.

1. 50% of ___ = 23

2. 100% of ___ = 65

3. 25% of ___ = 3

4. 10% of ___ = 7

5. 25% of ___ = 20

6. 150% of ___ = 9

7. $33\frac{1}{3}\%$ of what number is 30?

8. 75% of what number is 30?

Problem Set E

It's a Party! caterers rents equipment for parties. The cost to rent an item for one week is 25% of the price the caterers paid to purchase the item. The rental costs are listed at right.

Find the purchase price of each item, and explain how you found your answers.

Item	Rental Cost
Punchbowl	$10.00
Cup and saucer	0.75
Table	12.00
Folding chair	4.00
Sound system	250.00

Problem Sets D and E involved familiar percents whose fraction equivalents are easy to work with. For more complicated problems, you can use what you know about the relationship between multiplication and division.

EXAMPLE

Find the missing whole in this sentence:

$$55\% \text{ of } \underline{\quad} = 20$$

First make an estimate. Since 55% is close to 50%, or $\frac{1}{2}$, the missing number must be close to 40.

To find the exact answer, rewrite the sentence as a multiplication problem:

$$0.55 \cdot \underline{\quad} = 20$$

You know that this multiplication problem is equivalent to the following division problem:

$$20 \div 0.55 = \underline{\quad}$$

Now you can just divide to find the answer:

$$20 \div 0.55 = 36.\overline{36}$$

So, the missing whole is $36.\overline{36}$, or about 36.

Problem Set F

In Problems 1–3, first estimate the missing whole. Then calculate the exact value.

1. 40% of ___ = 70

2. 12% of ___ = 3

3. 124% of ___ = 93

4. 90% of what number is 99.9?

5. 4% of what number is 30?

Problem Set G

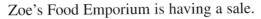

Zoe's Food Emporium is having a sale.

1. Spices are on sale for 35% of their original price.

    a. A jar of cinnamon costs $3 on sale. What was the original price?

    b. Salt is on sale for $0.20 per pound. What was the original price?

    c. Pepper costs $4 per package. What was the original price?

2. Fruit is on sale for 45% of the original price.

    a. Apples are on sale for $4.00 per bag. What was the original price?

    b. Dried apricots cost $2.50 per bag. What was the original price?

3. Snacks are to be marked down to 75% of their original price. Zoe might have marked some incorrectly. For each item, tell whether the sale price is correct. If not, give the correct price.

a.

b.

c.

d.

Share & Summarize

Describe a method for finding the whole when you are given the part and the percent. Demonstrate your method by answering this question: "54 is 6% of what number?"

Investigation ▶ Playing *Percent Ball*

In this lab investigation, you will play a game in which scores are recorded as percents. You will explore how your score for each turn affects your cumulative score for the game.

MATERIALS

- trash can or bucket
- 6 sheets of paper
- score sheets

The Game

Play this game in pairs. Crumple each sheet of paper into a ball. On your turn:

- Try to toss six paper balls into a trash can from about 5 feet away.

- Record your score for the turn as a number out of 6 and as a percent. Round to the nearest whole percent. For example, if 2 of your shots go in, record $\frac{2}{6}$ and 33%.

- Record your *cumulative score* for the game so far. For example, if you made 4 shots on the first turn and 2 on the second, you have made 6 shots out of 12. So, record $\frac{6}{12}$ and 50%.

Player 1's Score Sheet

	Turn Score		Cumulative Score	
Turn 1	$\frac{2}{6}$	33%	$\frac{2}{6}$	33%
Turn 2	$\frac{4}{6}$	67%	$\frac{6}{12}$	50%
Turn 3				
Turn 4				

The game ends when each player has taken 10 turns. The player with the highest cumulative score is the winner.

Try It Out

Play one game with your partner. Then, with your partner, look closely at your scoresheets for the game.

1. Look at the turn-score percents for you and your partner. What pattern do you see? (Hint: Are there certain scores that occur over and over?) Explain why the pattern makes sense.

2. Now look at the cumulative-score percents for you and your partner. Do you see the same pattern you found in Question 1? Explain why or why not.

3. Look at the turns for which your cumulative score went up. Can you see a pattern that could help you predict whether your score for a particular turn will make your cumulative score rise? If so, describe it.

Try It Again

Play the game again. After you record each turn score, predict whether your cumulative score will go up, go down, or stay the same. Keep playing the game until you have a prediction method that works every time.

4. How can you predict, based on your turn score, whether your cumulative score will go up, go down, or stay the same?

Take It Further

Play a new version of the game in which the object is to make your cumulative score alternate between going up and going down. As you play the game, record a D next to a cumulative score if it went down from the previous turn and a U if it went up. Here's how to determine the winner:

- A player earns 1 point each time his or her score changes from D to U or from U to D.

- Subtract 1 point for every turn score of 0 out of 6.

- The player with the greatest point total wins.

For this score-sheet, the cumulative score changes between U and D six times, but the player got 0 out of 6 twice. So, the final score is $6 - 2 = 4$.

	Turn Score		Cumulative Score	
Turn 1	$\frac{2}{6}$	33%	$\frac{2}{6}$	33%
Turn 2	$\frac{4}{6}$	67%	$\frac{6}{12}$	50% U
Turn 3	$\frac{0}{6}$	0%	$\frac{6}{18}$	33% D
Turn 4	$\frac{5}{6}$	83%	$\frac{11}{24}$	46% U
Turn 5	$\frac{2}{6}$	33%	$\frac{13}{30}$	43% D
Turn 6	$\frac{1}{6}$	17%	$\frac{14}{36}$	39% D
Turn 7	$\frac{2}{6}$	33%	$\frac{16}{42}$	38% D
Turn 8	$\frac{3}{6}$	50%	$\frac{19}{48}$	40% U
Turn 9	$\frac{0}{6}$	0%	$\frac{19}{54}$	35% D
Turn 10	$\frac{3}{6}$	50%	$\frac{22}{60}$	37% U

What Did You Learn?

5. Suppose you are playing the game with the original rules. In your first six turns, you've made 27 shots. What is the fewest number of shots you must make on your seventh turn to make your cumulative score rise?

6. In which situation below does making 1 out of 6 cause a greater change in your cumulative score? Why?

- After two turns, you have a cumulative score of 50%. On your third turn, you make 1 out of 6 shots.

- After nine turns, you have a cumulative score of 50%. On your tenth turn, you make 1 out of 6 shots.

On Your Own Exercises

Practice & Apply

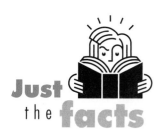

Just the facts

1. Consider the whole numbers from 1 to 64. In Parts a–c, round your answer to the nearest tenth of a percent.

   a. What percent of the numbers are greater than 56?

   b. What percent have an even tens digit?

   c. What percent are multiples of 6?

2. Theo has three dogs, two cats, three parakeets, and eight fish. In Parts a–c, round your answer to the nearest tenth of a percent.

   a. What percent of his pets have four legs? What percent have no legs?

   b. What percent of Theo's pets have beaks? What percent have fins?

   c. What percent of the total number of pet legs belong to birds? What percent belong to cats?

Fill in the blanks. Round each answer to the nearest whole percent.

3. ___% of 15 = 3   4. ___% of 120 = 17   5. ___% of 41 = 20

6. ___% of 132 = 80   7. ___% of 45 = 60   8. ___% of 16 = 2.4

9. Of the 27 girls on the varsity soccer team, 18 are seniors. What percent of the players are seniors?

10. Last year, 235 seniors out of 346 in the graduating class went on to college. What percent went to college?

Find each missing whole.

11. 50% of ___ = 342   12. 100% of ___ = 9   13. 25% of ___ = 5

Estimate each missing whole. Then find the value to the nearest hundredth.

14. 12% of ___ = 7   15. 28% of ___ = 20   16. 98% of ___ = 85

17. **Economics** Rosita and her friends went out to lunch. They left a $5 tip, which was 20% of the bill. How much was the bill?

18. **Life Science** Scientists have named about 920,000 insect species. This is about 85% of all known animal species. How many known animal species are there?

19. About 2,600 bird species live in the rain forest. This is about $33\frac{1}{3}\%$ of the world's bird species. About how many bird species are there?

Connect & Extend

20. An aquarium with an original price of $95 is on sale for $80.

   a. What *percent of* the original price is the sale price? Explain how you found your answer.

   b. What *percent off* the original price is the sale price? Explain how you found your answer.

 impactmath.com/self_check_quiz

21. Preview The students in Ms. Uhura's class were asked how many siblings they had. The results are shown in this plot. An X over a number indicates one student with that number of siblings. For example, the three X's over the 4 indicate that three students have four siblings.

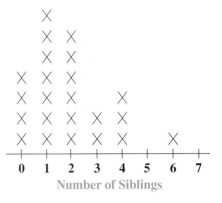

Number of Siblings

a. How many students are in the class? Explain how you know.

b. What percent of the students are only children?

c. What percent of the students have more than five siblings?

d. What percent of the students have fewer than three siblings?

e. What percent of the students are from a three-child family?

In Exercises 22–24, solve the problem if possible. If it's not possible, tell what additional information you would need to solve it.

22. The Tour de France bicycle race has been won by a French cyclist 36 times. What percent of the races have been won by a French cyclist?

23. Of the new trucks and vans sold in the United States in a recent year, 22.5% were white and 11.5% were black. How many of the trucks and vans were black or white?

24. Nutrition A serving of asparagus contains 2 grams of protein. What percent of the asparagus's weight is protein?

Just the facts

In 1999, 2000, 2001, and 2002, American Lance Armstrong won the 2,287-mile Tour de France. Just three years before his first victory, Armstrong was diagnosed with cancer and given a 50% chance of survival.

In y o u r
own
words

Write a problem that requires finding the whole when you know the part and the percent of the whole that part is. Then explain how to solve your problem.

25. Andre is a lifeguard at the local pool. The pool is crowded, and Andre thinks there may be more than 40 swimmers, the maximum number allowed. He starts to count, but Althea says, "Just count the swimmers with red bathing suits. I've already figured out that 20% of the people in the pool have red bathing suits." Andre counts 8 people wearing red bathing suits. How many people are in the pool?

26. Economics The label on a bottle of shampoo says, "20% More Than Our Regular Size!" The bottle contains 18 ounces of shampoo. How many ounces are in a regular-sized bottle? Explain how you found your answer.

27. Geometry Conor's family has a plot in the community garden that measures 9 feet by 12 feet.

   a. A section measuring 6 feet by 2 feet is devoted to tomatoes. What percent of the garden's area is planted in tomatoes?

   b. The green-bean section has 75% of the area of the tomato section. What is the area of the green-bean section?

   c. The green-bean section is 90% of the area of the squash section. What is the area of the squash section?

Mixed Review

Number Sense Order each set of numbers from least to greatest.

28. $^-4, 3, ^-3.99, 2, \frac{3}{2}, \frac{7}{3}, ^-\frac{2}{3}$

29. $\frac{1}{10}, 20\%, \frac{3}{4}, \frac{9}{11}, 0.77, 5\%, 0.\overline{6}$

Measurement Convert each measurement to meters.

30. 43 cm   **31.** 43 mm   **32.** 0.3 cm   **33.** 47,343 mm

Write a rule that fits all the input/output pairs in each table.

34.

Input	1	3	11	7	2	19
Output	$^-1$	1	9	5	0	17

35.

Input	1	2	4	6	10	11
Output	4	7	13	19	31	34

36.

Input	2	4	6	3	7	5
Output	3	11	19	7	23	15

37. Geometry Copy the coordinate grid at left.

   a. Plot the points (2, 5) and (5, 5) on the grid.

   b. Add two more points to the grid so that the four points are vertices of a square. Give the coordinates of the two points.

VOCABULARY
percent

Chapter Summary

In this chapter, you learned that—like a fraction or a decimal—a percent can be used to represent a part of a whole. You used the fact that percent means "out of 100" to convert fractions and decimals to percents and to convert percents to fractions and decimals.

You saw that percents are useful for comparing parts of different groups, even when the groups are of very different sizes. Then you learned how to find a given percent of a quantity and to compute a sale price when a percent discount is taken. Finally, you solved problems that involved finding the percent when you know the part and the whole, and finding the whole when you know the part and the percent that part represents.

Strategies and Applications

The questions in this section will help you review and apply the important ideas and strategies developed in this chapter.

Converting among fractions, decimals, and percents

1. Explain how to convert a decimal to a percent and how to convert a fraction to a percent. Give examples to illustrate your methods.

2. Explain how to convert a percent to a fraction and to a decimal. Give examples to illustrate your methods.

Using a percent to represent part of a whole

3. Estimate the percent of the square that is shaded. Explain how you made your estimate.

4. Of the 20 students in Dulce's ballet class, 17 took part in the spring recital. What percent of the students participated in the recital? Explain how you found your answer.

5. The school band held a carnival to raise $750 for new uniforms. When the carnival was over, they had raised $825. What percent of their goal did they reach? Explain how you found your answer.

Using percents to compare groups of different sizes

6. This summer, the 96 fifth graders and 72 sixth graders at Camp Poison Oak were asked this question:

Which is your favorite camp activity? Choose one.

❏ *swimming* ❏ *hiking* ❏ *arts and crafts*

❏ *volleyball* ❏ *canoeing* ❏ *other*

The results are given in the table.

Activity	Fifth Grade Votes	Sixth Grade Votes
Swimming	34	24
Hiking	5	10
Arts and crafts	18	6
Volleyball	14	12
Canoeing	21	16
Other	4	4

a. Dante said arts and crafts is three times as popular among fifth graders as among sixth graders. Is he correct? Explain.

b. Maya said volleyball is more popular among sixth graders than among fifth graders. Is she correct? Explain.

c. Sameka said that, among sixth graders, swimming is four times as popular as arts and crafts. Is she correct? Explain.

d. Omar said the choice "other" was equally popular among the two grades. Is he correct? Explain.

Calculating a percent of a whole

7. Describe a method for computing a given percent of a quantity. Demonstrate your method by finding 72% of 450 and 125% of 18.

8. A CD originally priced at $15.99 is on sale for 20% off. Calculate the sale price, and explain the method you used.

Finding the whole from the part and the percent

9. Last year, Fran's hourly wage was 75% of her hourly wage this year. She made $12 per hour last year. How much does she make this year? Explain how you found your answer.

10. Write a problem that requires finding the whole when you know the percent and the part. Explain how to solve your problem.

Demonstrating Skills

Convert each fraction or decimal to a percent.

11. 0.56   **12.** $\frac{7}{8}$ **13.** 0.3   **14.** $\frac{90}{125}$

15. 7.25   **16.** $\frac{67}{20}$ **17.** $\frac{2}{3}$ **18.** 0.008

Convert each percent to a decimal and a fraction or mixed number in lowest terms.

19. $33\frac{1}{3}\%$ **20.** 99%   **21.** 25%   **22.** 7.6%

23. 0.4%   **24.** 325%   **25.** $66\frac{2}{3}\%$ **26.** 1,000%

Fill in the blanks.

27. ___% of 25 = 10   **28.** 34% of 650 = ___

29. 10% of ___ = 5.3   **30.** ___% of 54 = 81

31. 77% of 77 = ____   **32.** 45% of ___ = 230

33. What is 83% of 320?

34. What percent of 65 is 26?

35. 80% of what number is 40?

36. What is 175% of 160?

37. A computer originally priced at $950 is on sale for 30% off. What is the sale price?

38. A tricycle originally priced at $76 is on sale for 20% off. What is the sale price?

CHAPTER 5

Exploring Graphs

Real-Life Math

Let It Snow, Let It Snow, Let It Snow Do you remember what you were doing on February 17, 2003? If you lived in New York City, it probably involved snow. The graph below shows the snowfall amounts for five snowstorms that hit New York City in the 2002–2003 season.

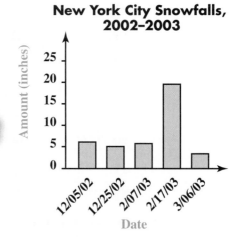

New York City Snowfalls, 2002–2003

Think About It A graph is a useful tool for showing how quantities are related. You can tell at a glance that much more snow fell on February 17, 2003 than on any other date. About how many times more snow fell on February 17, 2003 than on December 25, 2002?

Family Letter

Dear Student and Family Members,

You see graphs everywhere—on the sports page of the newspaper, in advertisements, in your science or social studies books! Most of these graphs are line graphs, bar graphs, or circle graphs. In this chapter, you'll learn about graphs that use points and lines to show patterns and relationships in data.

Here's an example. This graph has a horizontal axis and a vertical axis. It shows the prices of different bags of sugar. From this graph, you can determine facts such as these:

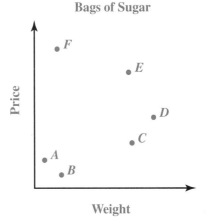

- *D* is the heaviest.

- *B* and *F* are the same weight.

- *C*, *E*, and *D* are heavier than *B*.

- *E* and *F* cost more than *D*.

- *C* would give you a better value for the money than *B*, because you get more sugar for only a little more cost.

In this chapter, you'll draw graphs for many types of situations. Some graphs, like the one above, will not have numerical values. Others will have scales, or sequences of numbers, along each axis.

Vocabulary As you learn about these new kinds of graphs, you'll also learn some new vocabulary terms:

axes	**line graph**	**origin**
coordinates	**ordered pair**	**variable**

What can you do at home?

During the next few weeks, your student might show interest in graphs. You might help him or her find examples of how graphs are used in everyday life by looking in newspapers and magazines. Encourage your student to determine what the graphs show as well as the values for specific points in the graphs.

5.1 Interpreting Graphs

You can find graphs in lots of places—in your school books, on television, and in magazines and newspapers. Graphs are useful for displaying information so it can be understood at a glance. They are also a wonderful tool for making comparisons.

Explore

Choose a topic you are interested in, such as cars, dogs, pizza, or music. Think about some aspects of your topic that would be interesting to display in a graph. Describe what the graph might look like.

For example, if you choose pizza for your topic, you might graph the number of pizzas of different sizes sold at a local pizza parlor. You could list the sizes (small, medium, large) along the bottom and then draw bars to show how many pizzas of each size are sold in a day.

You have seen graphs that use bars, sections of circles, and pictures to display data. In this lesson, you will focus on graphs that use points and curves to show information.

Investigation ▶1 Using Points to Display Information

Here are the front views of four buildings:

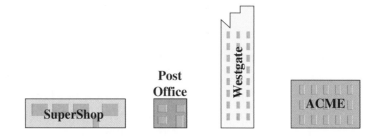

VOCABULARY
variable

Some of the buildings are taller than others, and some are wider than others. In other words, the heights and widths of the buildings *vary*. Quantities that vary, or change, are called **variables.**

Graphs are a convenient way to show information about two variables at the same time. This graph displays information about the heights and widths of the buildings. Each point represents one of the buildings.

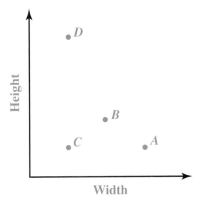

The horizontal line (representing the variable *width*) and the vertical line (representing the variable *height*) are called **axes.** The point where the axes meet is called the **origin.** The origin of a graph is usually the 0 point for each axis.

The arrow on each axis shows the direction in which the values of the variable are increasing. For example, points on the right side of the graph represent wider buildings than points on the left. Points near the top represent taller buildings than points near the bottom.

This graph does not have numbers along the axes, so it doesn't tell you the actual heights or widths of the buildings. However, it does show you how the heights and widths compare.

Think & Discuss

Which point represents a wider building, *A* or *B*? Explain how you know.

Which building does Point *A* represent? How do you know?

Boston, Massachusetts

MATERIALS
- drawing of the buildings
- graph of the buildings

Just the facts

One of the world's largest hotels is the MGM Grand in Las Vegas, Nevada. The hotel has 5,034 guest rooms and covers 112 acres.

Problem Set A

1. Identify the buildings represented by Points *B*, *C*, and *D* on the graph of the buildings.

2. Which letters represent buildings that are less wide than the building represented by Point *B*?

3. Which letters represent buildings that are shorter than the building represented by Point *B*?

4. On a copy of the graph, add two more points—one to represent a skyscraper and the other to represent a doghouse. Explain how you decided where to put the points.

5. Now imagine a building with a different height and width from the four in the drawing.

   a. On a copy of the drawing, make a sketch of the building you are imagining.

   b. Add a point to the graph to represent your building.

   c. Show your graph to a partner, and ask him or her to sketch or describe your building.

In the graph you used in Problem Set A, the variable *width* is represented by the horizontal axis and the variable *height* is represented by the vertical axis. You could instead label the axes the other way.

Problem Set B

1. In this graph, the horizontal axis shows height, and the vertical axis shows width.

   a. Plot a point for each of the four buildings on a copy of this set of axes.

   b. How does your graph compare to the graph on page 279?

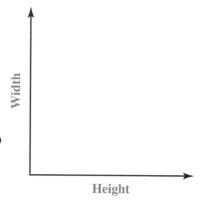

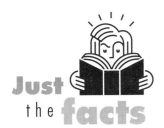
2. The points on the graph below represent the height and weight of the donkey, dog, crocodile, and ostrich shown in the drawing.

a. What are the two variables represented in the graph?

b. Tell which point represents each animal, and explain how you decided.

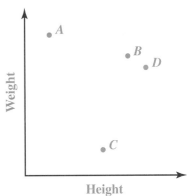

3. The two graphs below compare Car A and Car B. The left graph shows the relationship between age and value. The right graph shows the relationship between size and maximum speed.

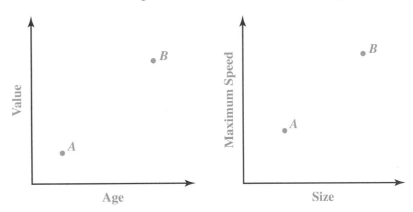

Use the graphs to determine whether each statement is true or false, and explain how you know.

a. The older car is less valuable.

b. The faster car is larger.

c. The larger car is older.

d. The faster car is older.

e. The more valuable car is slower.

This graph shows the heights and widths of four chimneys. Sketch four chimneys that the points could represent. On your sketch, label each chimney with the correct letter.

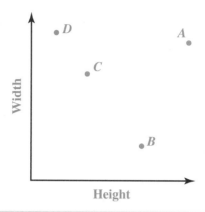

Investigation 2 Interpreting Points

It takes practice to become skillful at reading graphs. When you look at a graph, you need to think carefully about what the variables are and where each point is located.

EXAMPLE

What information does this graph show?

- The two variables are speed and age.

- *Y* is faster than *Z*. *Z* is older than *Y*.

What might *Y* and *Z* represent?

- *Y* and *Z* could be computers, since a newer computer usually processes information faster than an older computer.

- *Y* could represent a boy and *Z* could represent his grandfather. The grandfather is older than the boy, and the boy runs faster than the grandfather.

Problem Set C

Complete Parts a and b for each graph.

a. Tell what two variables the graph shows.

b. Describe what the graph tells you about the things represented by the points. Then try to come up with an idea about what the points could represent.

1.

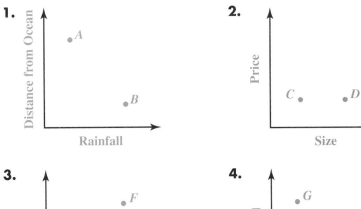

2.

3.

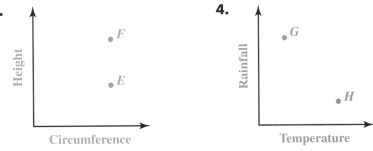

4.

5.

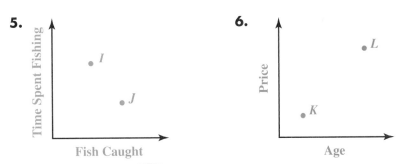

6.

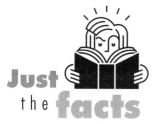

Common white sugar is made from both sugar cane and sugar beets. Although these two sources of sugar look very different, the sugar they produce is identical.

Problem Set D

1. Each point on this graph represents one bag of sugar.

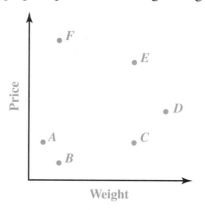

a. Which bag is heaviest? Which bag is lightest?

b. Which bags are the same weight? Which bags are heavier than *B*?

c. Which bags are the same price? Which bags cost more than *D*?

d. Which bag is heavier than *B* and costs more than *D*?

e. Assuming all the bags contain sugar of the same quality, is *C* or *F* a better value? How can you tell?

2. Gina is careless with her felt pens, often losing some and then finding them again. This graph shows how many pens were in her pencil case at noon each day last week. Use the information in the graph to write a story about Gina and her pens over the course of the week.

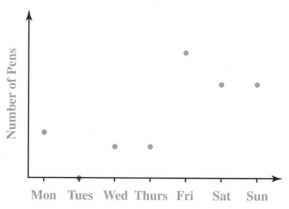

The graphs you have been working with have had only a few points. Graphs of real data often contain many points. Although looking at individual points gives you information about the data, it is also important to consider the overall *pattern* of points.

Problem Set E

Ms. Dimas surveyed two of her classes to find out how much time each student spent watching television and reading last weekend. She made this graph of her results:

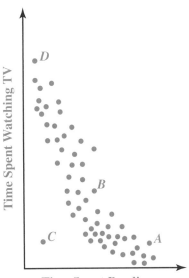

1. Choose one of the four students represented by Points *A*, *B*, *C*, and *D*. Write a sentence or two describing the time that student spent reading and watching television. Do not mention the student's letter in your description.

2. Exchange descriptions with your partner. Try to figure out which student your partner wrote about.

3. Now think about all the points in the graph, not just the four labeled points. Which of these statements best fits the graph?

   a. There is not much connection between how much time the students spent reading and how much time they spent watching TV.

   b. Most students spent about the same amount of time reading as they did watching TV.

   c. The more time students spent reading, the less time they spent watching TV.

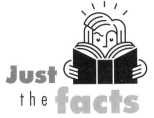

Just the facts

By the time the average American child graduates from high school, he or she will have spent 13,000 hours in school and 18,000 hours watching television!

You have seen several uses for graphs.

- Graphs can help *tell a story*—for example, the graph that shows how the number of pens Gina had changed throughout the week.

- Graphs can be used to *make comparisons*—for example, the graph that shows weights and prices of bags of sugar.

- Graphs can *show an overall relationship*—for example, the graph that shows how time spent watching TV is related to time spent reading.

Create your own graph, and explain what it shows.

Investigation ▶3 Interpreting Lines and Curves

The graphs you have looked at so far have been made up of separate points. When a graph shows a line or a curve, each point on the line or curve is part of the graph. The skills you developed for interpreting graphs with individual points can help you understand information given by a line or a curve.

This graph shows the noise level in Ms. Washington's classroom one Tuesday morning between 9 A.M. and 10 A.M.:

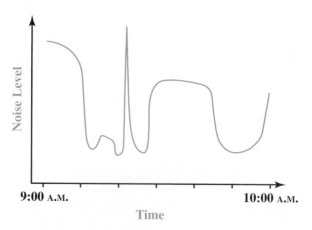

The variable on the horizontal axis is *time*. If you read this graph from left to right, it tells a story about how the noise level in the classroom changed over time.

Think & Discuss

• At about what time did the room first get suddenly quiet? How is this shown on the graph?

• At one point during the hour, the class was interrupted for a very short announcement on the public address system. At about what time did this happen? How do you know?

• When were the students the noisiest? (Ignore the PA announcement.)

• During part of the hour, students worked on a problem in small groups. When do you think this happened? Explain why you think so.

• Ms. Washington stopped the group activity to talk about the next day's homework. When did this happen? Explain how you know.

Problem Set F

1. This graph shows the audience noise in a school auditorium on the evening of a school play. The graph shows only the noise made by the audience, not the noise created by the actors on stage.

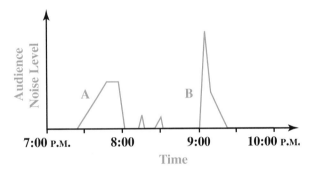

   a. At 7:00, the auditorium was empty. At about what time do you think people started entering the auditorium? Explain why you think so.

   b. What time do you think the play started? How is this shown on the graph?

   c. At the end of the performance, the audience burst into applause. At what time do you think this happened? Why?

   d. There are some small "bumps" in the graph between 8:00 and 9:00. What might have caused these bumps?

   e. A and B mark sections that show an increase in noise level. In which section does the noise level increase more quickly? How can you tell?

2. This graph shows how Jing's hunger level changed over one Saturday. Write a story about Jing's day that fits the information in the graph. Your story should account for all the increases and decreases in her hunger level.

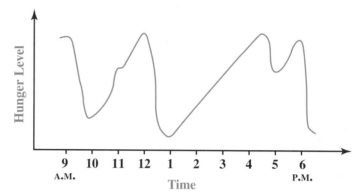

The graphs in this investigation have shown how the values of a variable change over a period of time. In some cases, you will have information about a variable only at certain times during a time period. You can often use what information you *do* have to estimate what happens between those times.

MATERIALS

Susan's height graph

Problem Set G

As Susan was growing up, her father measured her height on each of her birthdays. She recorded the results in a graph in the family scrapbook.

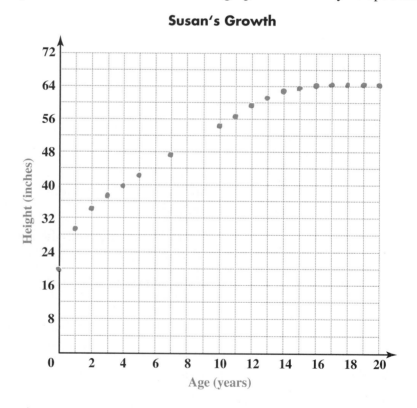

On Susan's sixth birthday, her family was on vacation and her father forgot to measure her. Susan's family moved right before her eighth birthday, and the scrapbook was misplaced. A few weeks before Susan turned 10, her father found the scrapbook and started filling it in again.

Susan wonders what her height was in the missing years.

1. From the graph, what do you know about Susan's height when she was 6 years old?

Susan thought connecting the points might help her estimate her height in the missing years.

2. On a copy of Susan's graph, draw line segments to connect the points in order. Use the segments to estimate Susan's heights at ages 6, 8, and 9.

3. Do you think the values you found in Problem 2 give Susan's exact heights at ages 6, 8, and 9? Explain.

4. Estimate how tall Susan was at age $1\frac{1}{2}$.

5. At what age did Susan's height begin to level off?

VOCABULARY
line graph

Connecting the points in Susan's height graph allowed you to make predictions. It also helped you to see *trends,* or patterns, in the data. Graphs in which points are connected with line segments are called **line graphs.**

In Susan's height graph, it makes sense to connect the points because Susan continues to grow between birthdays. In other words, there are values between those plotted on the graph. For some graphs, there are no values between the plotted values. For example, Susan's father keeps a graph of how many fish he catches each year on his annual fishing trip.

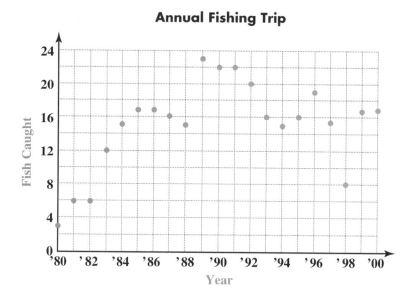

Annual Fishing Trip

Since the number of fish caught does not change between fishing trips, there are no values between the values shown in the graph. In this case, connecting the points does not make sense.

Problem Set H

Tell whether it would make sense to connect the points in each graph described below, and explain why you think so.

1. a graph showing the number of tickets sold for each football game during the season

2. a graph showing the speed of a race car every 10 minutes during a race

3. a graph of the sun's perceived height at each hour during the day

4. a graph of the amount of Carmen's weekly paycheck for 10 weeks

Share & Summarize

These graphs show how something changes over time. For each graph, write a sentence or two describing the change. Then try to think of something that might change in this way.

For example, this graph shows something increasing slowly at first and then more quickly. It could represent the speed of someone running the cross-country race. The runner starts out slowly and then moves faster and faster as she sprints toward the finish line.

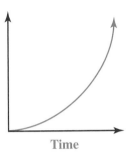

1.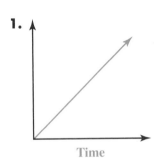

2.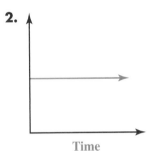

3.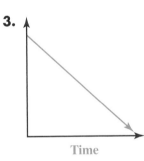

4.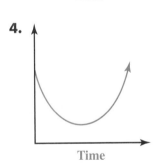

5.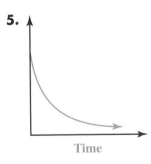

6.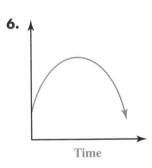

7. Describe a graph involving change over time for which it would make sense to connect the points.

8. Describe a graph involving change over time for which it would not make sense to connect the points.

On Your Own Exercises

Practice & Apply

Just the facts

One of the world's tallest apartment buildings is the John Hancock Center in Chicago, Illinois. The building is 1,127 feet tall and has 100 stories.

1. This drawing shows the front view of several buildings:

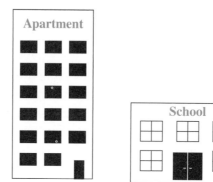
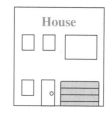

a. Copy this set of axes, and plot a point to represent the height and width of each building.

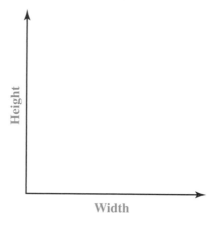

b. Add a new point to your graph, and write a brief description of the building it represents. Describe how the height and width of your building compare to the height and width of at least two of the other buildings.

2. This graph shows the relationship between effort and test results for five students.

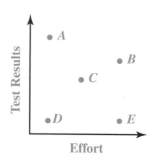

---

impactmath.com/self_check_quiz

The teacher wrote these comments on the report cards for these students:

- Alex's poor attendance this term has resulted in an extremely poor test performance.

- Nicola is a very able pupil, as her test mark clearly shows, but her concentration and behavior in class are very poor. With more effort she could do even better.

- Hoang has worked very well and deserves his marvelous test results. Well done!

- Adrienne has worked reasonably well this term, and she has achieved a satisfactory test mark.

a. Match each student's performance to a point on the graph.

b. Write a comment about the student represented by the point you didn't mention in Part a.

3. Complete Parts a and b for each graph.

a. Tell what two variables the graph shows.

b. Describe what the graph tells you about the things represented by the points. Then try to come up with an idea about what the points could represent.

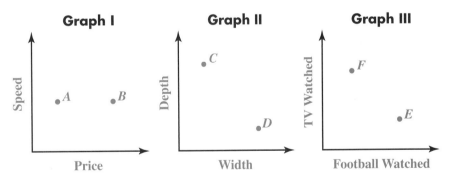

4. The age and height of each person in the cartoon are represented by a point on the graph. Going from left to right, match each person to a point.

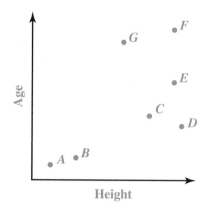

5. This graph shows the height of a hay crop over a summer:

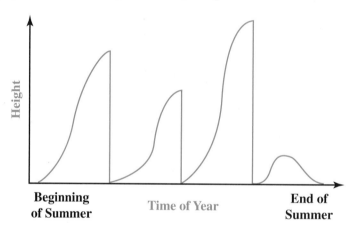

a. How many times was the crop harvested during the summer? How can you know this by looking at the graph?

b. Before which harvest was the hay tallest?

c. Describe the change in the hay's height after the third harvest. Why do you think the height changed this way?

6. This graph shows Jahmal's mood during one Saturday:

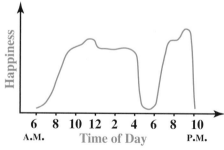

a. What two variables does this graph show?

b. Write a short story about what might have happened during Jahmal's day. Your story should account for all his mood changes.

7. The De Marte family went on a picnic last Sunday. This graph shows how far they were from home at various times of the day:

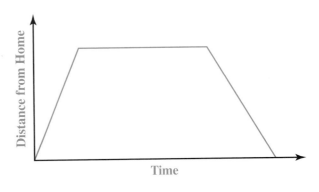

Hannah and Rosita each told a story about the graph. Which story best fits the graph? What is wrong with the other story?

Hannah: "The family drove up a tall mountain. They stayed on a level area for several hours. Then they came down the mountain on a road that was less steep than the first."

Rosita: "The family drove fairly quickly to the picnic spot. They stayed at the picnic area most of the day, and then drove home more slowly."

8. Science As part of his science project, Conor was supposed to record the temperature every hour one Saturday from 6 A.M. to midnight. At noon, he was eating lunch and completely forgot. At 8:00 P.M., his favorite show came on and he forgot again. He recorded the data he had collected in a graph and connected the points.

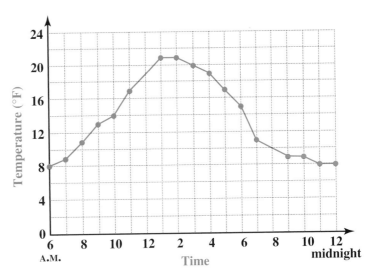

a. Why does it make sense to connect the points in this situation?

b. Describe the overall trend, or pattern, in the way the temperature changes over the time period shown on the graph.

c. Estimate the temperature at noon and 8 P.M.

Connect & Extend

9. Social Studies This graph shows the percent of U.S. students in grades 10–12 who dropped out of high school on even-numbered years from 1982 to 2000. The point labeled *A* shows that, in 1986, about 4.7% of students dropped out.

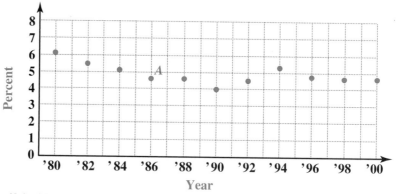

U.S. High School Dropout Rates, 1980–2000

Source: National Center for Education Statistics

a. In which year was the dropout rate highest? In which year was it lowest?

b. When did the percent of students who dropped out of high school first fall below 5%?

c. About what percent of students dropped out of high school in 2000? About what percent of students stayed in high school in 2000?

d. If the U.S. high school population was about 13,000,000 in 2000, approximately how many students dropped out that year?

10. In Problem 3 of Problem Set B, you used these graphs to compare two cars:

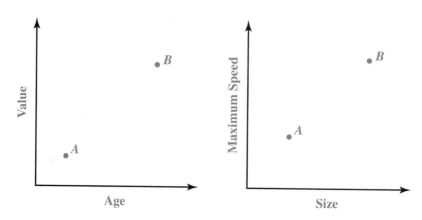

This graph gives some additional information about the cars:

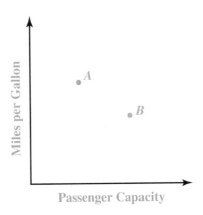

a. *True or false?* The car that holds more passengers gets fewer miles per gallon.

b. Copy each set of axes below. Mark and label points to represent Car A and Car B.

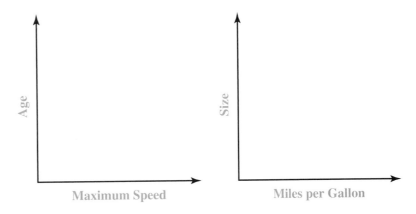

11. Complete Parts a and b for each graph below.

    a. Copy the graph, and make up a variable for each axis.

    b. Describe what the graph tells you about the things represented by the points. Then try to come up with an idea about what the points could represent.

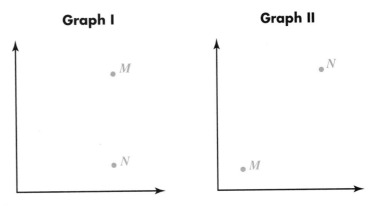

12. Elena thinks her math teacher, Mr. Malone, gives too much homework! To prove it, she asked a group of students how much time they spent doing math homework each week and how much time they spent doing homework for other subjects. She plotted points for Mr. Malone's students in green and points for students in other math classes in blue.

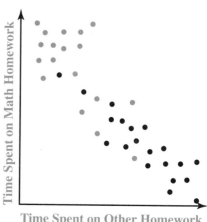

a. Look at the overall pattern of points in the graph. In general, what is the relationship between the time spent on math homework and the time spent on homework for other subjects?

b. In general, how are the points for Mr. Malone's students different from the points for the other students?

c. Write an argument Elena could use to try to convince Mr. Malone to give less homework.

13. In an experiment, the heights of 192 mothers and their adult daughters were measured.

Based on this graph, does there appear to be a connection between the heights of the mothers and the heights of their daughters? Explain your answer.

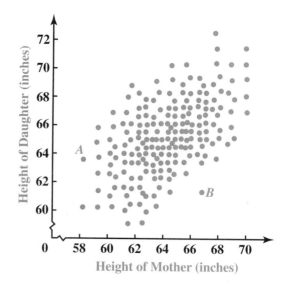

14. Science This graph was created to show how the amount of air in a balloon changed as it was blown up and then deflated. What is wrong with this graph?

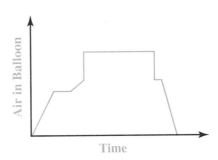

15. A firefighter walked from the fire truck up the stairs to the second floor of the firehouse. Several minutes later, the alarm rang. The firefighter slid down the pole and ran back to the fire truck. Sketch a graph to show the firefighter's height above the ground from the time she left the fire truck to the time she returned to it.

Mixed
Review

Change each fraction or decimal to a percent.

16. $\frac{2}{5}$　　　　　　**17.** 0.78　　　　　　**18.** $\frac{1}{3}$

19. 0.7　　　　　　**20.** $\frac{112}{70}$　　　　　**21.** 3.06

Change each fraction to a decimal.

22. $\frac{2}{3}$　　　　　　**23.** $\frac{7}{9}$　　　　　　**24.** $\frac{4}{11}$

25. $\frac{5}{111}$　　　　　**26.** $\frac{29}{15}$　　　　　**27.** $\frac{11}{12}$

28. Whenever Hannah has a party, she makes a batch of her famous salsa. To figure out how much to make, she uses this rule:

Make $\frac{2}{3}$ cup per person, plus 1 extra cup just in case.

a. Last year, there were 6 people at Hannah's birthday slumber party. How much salsa did she make?

b. This year, Hannah is having a pool party. She is expecting 20 people. How much salsa should she make?

c. Challenge At one holiday party, Hannah made $6\frac{1}{3}$ cups of salsa. How many people were at the party?

You have already had lots of experience describing the information shown by graphs. In this investigation, you will practice making your own graphs.

Just the facts

When an Argentinean child has a birthday, it is customary to tug his or her earlobe once for each year of age. In Israel, the birthday child sits in a chair while adults raise and lower it once for each year, plus once for good luck.

Explore

Althea and Miguel wanted to make a graph to show how the noise level at a typical birthday party is related to time. Here is the first graph they drew:

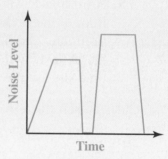

Althea said the graph wasn't quite right, since it isn't usually completely quiet at the start of a party. Miguel said that most people arrive on time for a party, and the noise level would go up more quickly. They also realized that the party would probably never be completely silent. They drew a new graph:

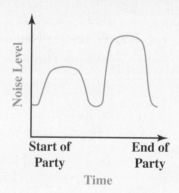

Think about birthday parties you have been to. What would you change to improve Althea and Miguel's graph?

Another pair of variables that might be useful for describing what happens at a birthday party is the amount of food and the time. Draw a graph showing how these variables might be related. Explain what your graph shows.

Investigation 1 Drawing Your Own Graphs

In this investigation, you will sketch graphs to fit various situations.

Problem Set A

For each problem, create your own graph and then compare and discuss graphs with your partner. Then work together to create a final version of your graph. Be sure to label the axes.

1. At a track meet, the home team won a relay race, and the crowd roared with excitement. Make a graph to show how the noise level might have changed from just before the win to a few minutes after the win.

2. A child climbs to the top of a slide, sits down, and then starts to slide, gaining speed as he goes. At the bottom, he gets up quickly, runs around to the ladder, and climbs up again. Make a graph to show how the height of the child's feet above the ground is related to time.

3. At Computer Cafe, customers are charged a fixed price plus a certain amount per minute for using a computer. Create a graph that shows how the amount a customer is charged is related to the time he or she spends using a computer.

4. Smallville is a town surrounded by farms. The number of people in Smallville changes a lot during a typical school day. During the day, many children come to town for school, and adults drive in for business and shopping. In the evenings, people come to town to eat dinner or attend social events. Draw a graph to show how the number of people in town might be related to time on a typical school day.

5. Make a graph showing how the number of hours of daylight is related to the time of year. Assume the time axis starts in January and goes through December.

6. Make a graph showing approximately how the temperature outside has changed over the past three days.

Just the facts

In the summer in Barrow, Alaska, the sun does not set for more than 80 consecutive days! In the winter, the town experiences about 64 days of darkness.

Share & Summarize

Write a description of something that changes over time. Then sketch three graphs, one that matches your description and two that don't. See if your partner can guess which graph is correct.

Investigation ▶ 2 ▶ Plotting Points

When you made your graphs in the last investigation, you had to think about the overall shape of the graph, not about exact values. To draw a graph that shows exact values, you need to plot points.

This graph shows a map of an island just off the coast of a continent. The point labeled *S* represents a major city on the coast. The distance between grid lines represents 1 mile.

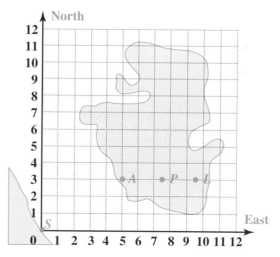

VOCABULARY
coordinates
ordered pair

Point *A* represents a resort that is located 5 miles east and 3 miles north of Point *S*. The values 5 and 3 are the **coordinates** of Point *A*. The coordinates can be given as the **ordered pair** (5, 3), where 5 is the horizontal coordinate and 3 is the vertical coordinate.

As you might guess, the *order* of the numbers in an ordered pair is important—the first number is always the horizontal coordinate, and the second number is always the vertical coordinate.

Problem Set B

1. On a copy of the map, mark the point that is 3 miles east and 5 miles north of Point *S* and label it *B*. Is Point *B* in the water or on the island? Is Point *B* in the same place as Point *A*?

2. Mark the point that is 7 miles east and 5 miles north of Point *S* and label it *C*. Then mark the point that is 5 miles east and 7 miles north of Point *S* and label it *D*. Are Points *C* and *D* in the same place? Give the coordinates of Points *C* and *D*.

3. Which point is in the water, (2, 7) or (7, 2)? Mark the point on your map and label it *E*.

4. Developers want to build another resort on the island. Which would be the better location, (6, 11) or (11, 6)? Why?

5. Give the coordinates of two points on the island that are exactly 2 miles from Point *A*.

6. Coordinates are not always whole numbers. For example, Point *L*, the island lighthouse, has coordinates (9.5, 3). Point *P* represents the swimming pool. What are the coordinates of Point *P*? How far is the lighthouse from the pool?

7. Give the coordinates of the point that is halfway between Points *L* and *P*.

8. Give the coordinates of the point that is halfway between Points *A* and *P*.

9. List three points on the island with a first coordinate greater than 8.

10. List three points on the island with a second coordinate equal to 8.

11. List three points on the island with a second coordinate less than 4.

In the next problem set, you will use what you know about plotting points to make a graph.

Just the facts

The system of naming points with ordered pairs is called the Cartesian coordinate system. The system was invented by René Descartes in the early 1600s. It is said that Descartes came up with the idea while staring at a fly on his ceiling. He discovered that he could describe the fly's position by giving its distance from two adjacent walls (the axes).

Problem Set C

Rosita wants to use some of the money she earns from her paper route to sponsor a child in a developing country. Sponsors donate money each month to help pay for food, clothing, and education for a needy child.

Rosita learned that sponsoring a child costs $48 a month. This is more than she can afford, so she wants to ask some of her friends to share the cost.

1. If two people divide the monthly sponsorship cost, how much will each person pay? If three people divide the cost, how much will each pay?

2. Copy and complete the table to show how much each person would pay if the given number of people split the cost.

Number of People	1	2	3	4	6	8	12	16	24	48
Cost per Person (dollars)	48							3		

3. On a copy of the grid below, plot and label points for the values from your table. For example, for the first entry, plot the point with coordinates (1, 48). Two of the points have been plotted for you.

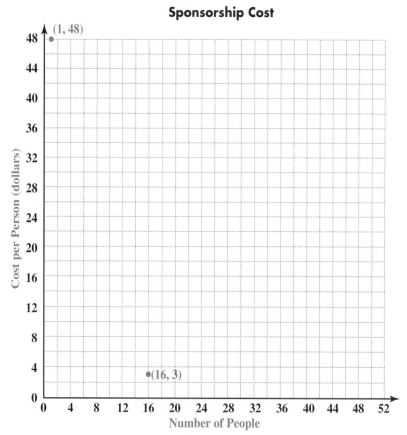

Sponsorship Cost

4. Describe the overall pattern of points in your graph.

5. Luke wanted to connect the points on his graph with line segments. Marcus said, "That wouldn't make sense. How can $1\frac{1}{2}$ people share the cost?" What do you think?

6. As the number of people increases, the amount each person pays decreases.

   a. If you continued to plot points for more and more people, what would the graph look like?

   b. Could there ever be so many people that each person would pay nothing? Explain.

MATERIALS
grid paper

Share & Summarize

1. Give the coordinates of the point halfway between (2, 7) and (2, 3). Then make a graph showing all three points. Label the points with their coordinates.

2. Give the coordinates of two more points on the same vertical line. Plot and label the points.

3. Give the coordinates of two points that, along with (2, 7) and (2, 3), form the vertices of a rectangle.

4. Is your answer for Question 3 the only one possible? Explain.

Investigation 3 Choosing Scales

When you created a graph showing the relationship between the number of sponsors and the amount each sponsor would have to pay, you were given a set of axes labeled with *scale values*. Often when you make a graph, you have to choose the scales yourself.

Think & Discuss

Conor and Jing are baking ginger snaps for the school bake sale. They need half a cup of molasses for each batch of cookies.

Sketch a rough graph showing the relationship between the number of batches and the number of cups of molasses. Your graph does not need to show precise points.

Conor and Jing made a table to show how much molasses they would need for different numbers of batches.

Batches	Molasses (cups)
1	$\frac{1}{2}$
12	6
20	10
3	$1\frac{1}{2}$
8	4
40	20

They each decided to graph the data. Here are their graphs:

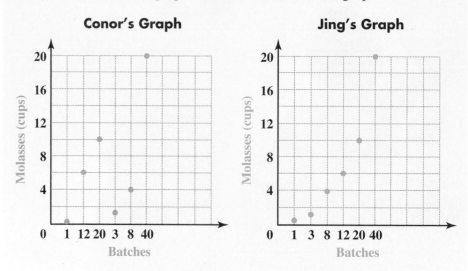

Discuss each graph. Do you think the graph is correct? If not, explain what is wrong and tell how you would fix it.

When you draw a graph, you need to think about the greatest value you want to show on each axis. For the cookie data, you need to show values up to 40 on the "Batches" axis and up to 20 on the "Molasses (cups)" axis.

You also need to consider the scale to use on each axis. The *scale* is the number of units each equal interval on the grid represents. You want to choose a scale that will make your graph easy to read but will not make it so large that it won't fit on your paper. Here are two possibilities for the cookie data:

- Let each interval represent 4 batches on the horizontal axis and 2 cups on the vertical axis. Then the graph would fit on a 10-by-10 grid.

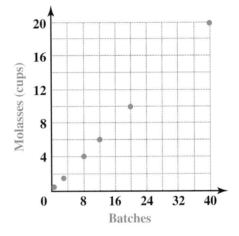

- Let each interval represent 2 batches on the horizontal axis and 2 cups on the vertical axis. Then the graph would fit on a 20-by-10 grid.

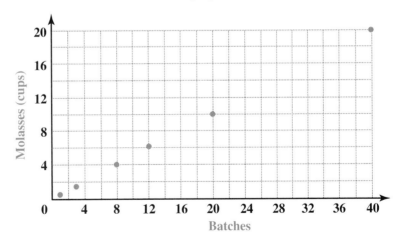

Problem Set D

Miguel wants to buy a birthday present for his sister. He has $10 to spend on the present and the gift wrap.

1. Copy and complete this table to show some of the ways the $10 Miguel has to spend can be distributed between the cost of the present and the cost of the gift wrap.

Cost of Present (dollars)	10	9	8	7	6			3
Cost of Gift Wrap (dollars)	0	1	2			5	6	

2. Suppose you want to graph these data with the cost of the present on the horizontal axis and the cost of the gift wrap on the vertical axis. What is the greatest value you need to show on each axis?

3. What scale would you use on each axis? In other words, how many dollars would you let each interval on the grid represent?

4. Make a graph of the data. Be sure to do the following:
   • Label the axes with the names of the variables.
   • Add scale labels to the axes.
   • Plot a point for each pair of values in the table.

5. Describe the pattern of points on your graph.

6. In this situation, either variable could have a value that is not a whole number. For example, Miguel could spend $6.50 on the present and $3.50 on the wrapping. Plot the point (6.5, 3.5) on your graph. Does this point follow the same pattern as the others?

7. Find two more pairs of non-whole-number values that fit this situation. Plot points for these values. Check that the points follow the same pattern as the others.

8. You can use your graph to make predictions.
   a. Connect the points with line segments.
   b. Choose a point on the graph that is not one of the points you plotted. Use the coordinates of the point to predict a pair of values for the present cost and the wrapping cost. Then check your prediction by verifying that the values add to $10.

Just the facts

The birthday cake with lit candles that we know today was a tradition started by Germans in the Middle Ages who adapted it from an ancient Greek custom.

When you make the graphs in the next problem set, you will need to choose a scale that gives a good view of the overall pattern in the data.

Problem Set E

In this problem set, make each graph on a 10-by-10 grid.

1. Conor is planning a party at a local pizza parlor. The party will cost $6 per person.

    a. Complete the table to show the cost for various numbers of people.

People	1	2	3	4	5	6	7	8
Cost (dollars)	6	12						

    b. Conor wants to graph these data on a 10-by-10 grid, with the number of people on the horizontal axis. What scale do you think he should use on each axis?

    c. Graph the values in the table using an appropriate scale.

    d. Do the points on your graph form a pattern? If so, describe it.

    e. Use your graph to predict the cost of the pizza for nine people. Check that your prediction is correct.

    f. Would it make sense to connect the points on this graph? Explain.

Just the facts

Americans eat 100 acres of pizza every day!

2. The social committee needs streamers for decorations. Material for the streamers costs 20¢ per yard.

   a. Complete the table to show the cost of various lengths of material.

Length (yards)	1	2	3	4	5	6	7	8	9
Cost (cents)	20	40							

   b. Suppose you graphed these data with the length on the horizontal axis and the cost on the vertical axis. What would happen if you let each interval on the vertical axis represent 1¢?

   c. What scale would be appropriate for each axis?

   d. Make a graph of the data with length on the horizontal axis.

   e. Do the points on your graph form a pattern? If so, describe it.

   f. Would it make sense to connect the points on this graph? Explain.

   g. Use your graph to predict the cost for $4\frac{1}{2}$ yards. Check your prediction by multiplying the number of yards by the cost per yard.

   h. Draw another graph using the same data, but this time put the cost on the horizontal axis.

   i. Describe how your two graphs are alike and different.

Share & Summarize

1. When you make a graph, how do you decide on the scale for each axis?

2. After you have plotted points from a table, how do you decide whether to connect the points?

On Your Own Exercises

Practice & Apply

1. **Sports** A weight lifter grips a barbell and struggles with it for several seconds. She suddenly lifts it part of the way, and then steadily raises it until it is fully above her head. She holds it for a few seconds and then drops it. Sketch a graph to show how the height of the barbell changes from the time the weight lifter first grips the barbell until just after she drops it.

2. Goin' Nuts sells mixed nuts by the pound. Draw a graph that shows how the cost for cashews is related to the number of pounds purchased.

3. For this problem, use a grid with horizontal and vertical axes from 0 to 14. Plot the points below in order, reading across the rows, and connect the points as you go with straight line segments. The line segments should form a picture.

(7, 3)	(13, 3)	(10, 1)	(2, 1)	(1, 2)	(1, 3)	(7, 3)
(7, 12)	(13, 4)	(7, 4)	(8, 5)	(8, 10)	(7, 12)	(6, 10)
(6, 5)	(7, 4)	(0, 4)	(7, 12)	(5, 13)	(7, 13)	(7, 12)

4. **Geometry** The area of a square is the product of the lengths of two sides.

   a. What is the area of a square with side length 1? What is the area of a square with side length 2?

   b. Complete the table to show areas of squares with the given side lengths.

Side Length	1	2	3	4	5	6
Area	1				25	

   c. On a copy of the axes at right, plot and label points for the values from your table. For example, for a square with side length 5, plot the point (5, 25).

   d. Describe the overall pattern of points in your graph.

   e. Would it make sense to connect the points on this graph? Explain.

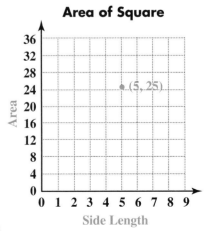

Area of Square

on your graph? Explain.

5. Jahmal has only 30 minutes to finish his homework, which includes practicing the violin and studying for a history quiz.

a. Make a table showing at least six ways Jahmal can split up the time between the two activities.

b. Suppose you want to graph your data with the time spent practicing the violin on the horizontal axis. What is the greatest value you need to show on each axis?

c. What scale would you use on each axis?

d. Make a graph of your data. Be sure to add labels and scale values to the axes.

e. Describe the general pattern of points on your graph.

f. Connect the points on your graph with line segments. Then choose a point on the graph that is not one of the points you plotted. Use the coordinates of the point to predict a pair of values for the violin time and the study time.

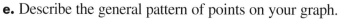

Time (hours)	0	1	2	3	4	5	6	7	9	10
Cost (dollars)	5	6.20								

6. Economics Sparks Internet Service charges $5 per month plus $0.02 for each minute a customer is on line.

a. Complete the table to show the cost for various amounts of time on line.

b. Graph the data from the table on a 10-by-10 grid. Show time on the horizontal axis, and use appropriate scales.

c. Do the points on your graph make a pattern? If so, describe it.

d. Does it make sense to connect the points on this graph? Explain.

e. Use your graph to predict the cost for 8 hours. Check that your prediction is correct.

7. A family is driving to visit some relatives a few hundred miles away. They begin driving at a moderate pace along back roads, and then travel for several hours on a major highway. When they get to the city where their relatives live, they are slowed by heavy traffic and lights at intersections.

a. Sketch a graph showing the time since the trip began on the horizontal axis and the family's speed on the vertical axis.

b. Sketch a graph showing the time on the horizontal axis and the family's distance from the starting point on the vertical axis.

c. Sketch a graph showing the time on the horizontal axis and the distance from the destination on the vertical axis.

8. Althea said, "When my schoolwork is much too easy, I don't learn very much. But I also don't learn very much when it is much too hard. I learn the most when the difficulty level is somewhere between 'too easy' and 'too hard.'" Draw a graph to illustrate Althea's ideas.

9. You could play tic-tac-toe on a grid like this. Instead of writing X's and O's in the squares, you would write them at points where the grid lines meet. For example, you could mark (0, 0) or (2, 1).

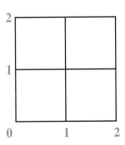

a. How can you tell from just the coordinates whether three points are on the same horizontal line?

b. How can you tell from just the coordinates whether three points are on the same vertical line?

c. How can you tell from just the coordinates whether three points are on the same diagonal line?

10. In Chapter 1, you looked at this toothpick pattern:

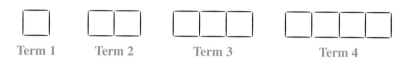

Term 1   Term 2   Term 3   Term 4

a. Imagine that this pattern continues. Complete the table to show the number of toothpicks in the first six terms.

Term	1	2	3	4	5	6
Toothpicks	4			13		

b. Make a graph with the term number on the horizontal axis and the number of toothpicks on the vertical axis. Make the horizontal axis from 0 to 10 and the vertical axis from 0 to 30.

c. Describe the pattern of points in your graph.

d. Use the pattern in your graph to predict the number of toothpicks in Terms 7 and 8. Check your answers by drawing these stages.

e. Would it make sense to connect the points on this graph? Explain.

11. Consider this input/output table:

Input	1	2	4	5	7
Output	2	5	11	14	20

a. Describe a rule relating the input and output values.

b. Graph the values from the table. Make the "Input" axis from 0 to 8 and the "Output" axis from 0 to 24.

c. Use your graph to predict the outputs for inputs of 3, 6, and 8. Use your rule to check your predictions.

12. Here are the first seven rows of Pascal's triangle:

```
                1                    Row 0
              1   1                  Row 1
            1   2   1                Row 2
          1   3   3   1              Row 3
        1   4   6   4   1            Row 4
      1   5  10  10   5   1          Row 5
    1   6  15  20  15   6   1        Row 6
```

a. Complete a table to show the sum of the numbers in each row.

Row	0	1	2	3	4	5	6
Sum							

b. Make a graph of these values. Put the row number on the horizontal axis and the sum on the vertical axis.

c. Describe the pattern of points in your graph.

Mixed Review

Replace each ● with $<$, $>$, or $=$ to make a true statement.

13. $\frac{3}{7}$ ● $\frac{2}{5}$

14. $\frac{17}{23}$ ● $\frac{5}{9}$

15. $3\frac{8}{9}$ ● $\frac{15}{4}$

16. $\frac{3}{11}$ ● $\frac{3}{10}$

17. $\frac{33}{165}$ ● $\frac{41}{200}$

18. $\frac{21}{15}$ ● $\frac{22}{16}$

19. $\frac{-2}{3}$ ● $\frac{-1}{2}$

20. $\frac{-5}{2}$ ● -2

21. $\frac{-7}{8}$ ● $\frac{-4}{5}$

22. Economics Marcus bought a used guitar for $70 plus 6% sales tax. What total amount did he spend on the guitar?

23. Economics Six months ago, Ms. Donahoe bought a computer system for $1,300. Since then, the price of the system has dropped by 27%. How much does the system cost now?

24. Geometry Rosita's math test included a drawing of a triangle with sides labeled 8 cm, 3 cm, and 4 cm. She knew right away that her teacher had made a mistake. How do you think Rosita knew?

Using Graphs to Find Relationships

In the graphs you made in the last lesson, the points followed a definite pattern—either a line or a curve—indicating a relationship between the variables. When data are collected, it is often not obvious how the variables are related, or even if they are related at all. One way to start searching for a relationship is by plotting the data on a graph and looking for trends.

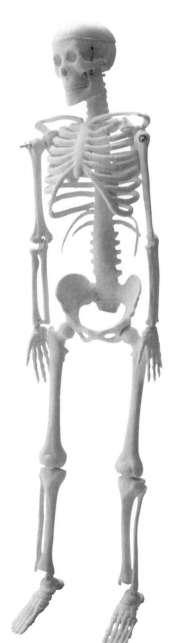

These data were collected by a team of archeologists. They show the length of the femur, or thighbone, and the height for several ancient human skeletons.

If the archeologists can discover a relationship between femur length and height, they will be able to predict the heights of other skeletons even if they know only the femur length.

Femur Length (centimeters)	Height (meters)
10.0	0.50
20.0	0.55
21.8	0.61
25.0	0.67
28.0	0.80
29.2	0.67
29.8	0.80
33.0	0.97
36.0	1.20
37.0	1.00
38.5	1.48
39.5	1.10
42.5	1.10
42.5	1.51
44.0	1.30

Explore

Graph the archeologists' data on a grid like this one:

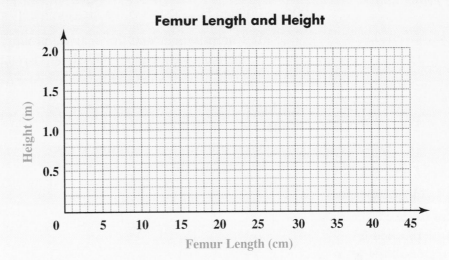

Does there appear to be an overall relationship between femur length and height? If so, describe the relationship.

Hannah said that, in general, longer femurs belong to taller skeletons. Miguel said that doesn't make sense because there are two skeletons with femur length 42.5 centimeters but very different heights. Who do you think is correct? Explain.

Suppose an archeologist uncovers portions of two skeletons. One has femur length 15 centimeters and the other has femur length 35 centimeters. Use your graph to estimate the heights of the skeletons. Explain how you made your estimates.

Althea said, "The skeleton with femur length 38.5 centimeters has height about 1.5 meters. The skeleton with femur length 42.5 centimeters has height 1.1 meters. It seems that as femur length increases, height decreases." Explain the flaw in Althea's reasoning.

The human femur

Investigation 1 ▶ Looking for a Connection

In this investigation, you will graph data you and your classmates collect to see whether you can discover a relationship.

- measuring tape, or string and ruler
- graph paper

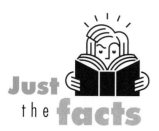

Just the facts

In the book *Gulliver's Travels*, the Lilliputians make clothes for Gulliver based on a single measurement: the distance around his thumb. From this measurement, they are able to predict the distances around Gulliver's wrist and waist.

GREAT ILLUSTRATED CLASSICS
GULLIVER'S TRAVELS
Jonathan Swift

Problem Set A

In this problem set, you will measure the distances around your partner's wrist and neck. Before you begin, decide as a class what units of measure to use and how accurate the measurements should be.

1. Carefully measure the distance around your partner's wrist. Measure the narrowest part of the wrist. Also measure the distance around your partner's neck. Record the data in a class table.

Initials	Distance around Wrist	Distance around Neck

2. Now you will draw a graph of the class data. Put the distance around the wrist on the horizontal axis.

    a. Look at the data and decide on a good scale for each axis. Draw and label the axes, including the scale values.

    b. Carefully plot the data.

3. Describe the relationship between the distance around the wrist and the distance around the neck for students in your class.

4. Measure the distance around your teacher's wrist. Use your graph to predict the distance around your teacher's neck. Then measure your teacher's neck to see how close your prediction is.

Share & Summarize

If you have data for two variables, you can graph the data and look for a relationship between the variables.

1. How would you describe the relationship between the variables if the points form a pattern that slants upward from left to right?

2. How would you describe the relationship if the points fall almost in a horizontal line?

Investigation How Some Things Grow

When you first think about growing up, you probably think about getting taller—but growing up isn't only about height. Your brain, heart, and other organs also grow. Even the number of teeth in your mouth increases from infancy to adulthood. Sometimes a graph can be a useful tool for showing how things grow over time.

MATERIALS
graph paper

Just the facts

The brain makes up only 2% of an adult's body weight but uses 20% of the blood's oxygen supply.

Problem Set B

Each table below contains data about the way something grows. Complete Parts a–c for each table.

   a. Graph the data with the time variable on the horizontal axis. Be sure to choose a suitable scale for each axis.

   b. Connect the points with line segments.

   c. Write a sentence or two describing the growth shown in your graph.

1. This table shows average heights of girls at different ages:

Age (years)	2	4	6	8	10	12	14	16
Avg. Girl's Height (cm)	85	100	114	127	138	151	161	162

2. This table shows average heights of boys at different ages:

Age (years)	2	4	6	8	10	12	14	16
Avg. Boy's Height (cm)	87	102	115	128	139	148	163	173

3. This table shows how the weight of an average human brain changes from a person's birth to age 18. Each column shows a person's age and the percent of the brain's adult weight that the brain has reached at that age.

Age (years)	0	1	2	3	4	5	10	15	18
Percent of Adult Weight	27	65	75	82	88	94	98	100	100

4. This table shows how a typical chicken grew. Each column shows the chicken's age and the percent of its adult weight the chicken had reached at that age.

Age (days)	21	50	75	100	125	150	175	200
Percent of Adult Weight	4	7	20	38	55	65	78	82

5. This table shows the mass of a plant at various times after planting:

Weeks after Planting	2	4	6	8	10	12	14	16
Mass (grams)	20	100	210	350	560	700	800	900

6. A biologist grew a sample of bacteria. This table shows how many millions of bacteria there were per milliliter at various times:

Hours	0	12	24	36	48	60	72
Millions of Bacteria	4	5	8	70	99	103	104

Share & Summarize

1. What variables are used in Problem Set B to describe growth? What are some other variables that could describe growth?

2. You have seen growth data given in tables and in graphs. What types of information are easier to get from a graph? What types of information are easier to see in a table?

Investigation ▶3 Making Predictions from Graphs

You have seen that when the points in a graph form a pattern, you can use the graph to make predictions. Connecting the points on a graph with line segments can help you find in-between values. Extending the pattern can help you make predictions about values beyond those shown in the graph.

MATERIALS
copy of the graph

Problem Set C

This graph shows the cost of various numbers of decks of playing cards.

1. How much do 5 decks of cards cost?

2. How many decks of cards can you buy for $9?

3. Can you use the graph to find the cost for 4 decks of cards? For 10 decks of cards? Why or why not?

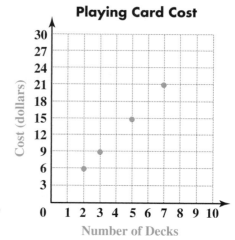

Playing Card Cost

Here is how Hannah thought about Problem 3 in Problem Set C:

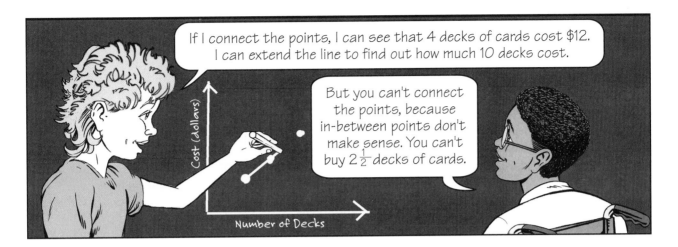

Think & Discuss

What do you think about Hannah's method? Does it make sense?
Is Jahmal correct?

Sometimes connecting the points in a graph can help you find information
or to see a pattern, even when all the in-between points don't make sense.
In cases like this, people often use dashed segments to connect the points.

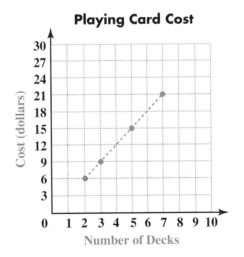

The dashed segments in the graph above make it easy to find information
and to see that all the points fall on a straight line. The dashes also indicate that not every point on the line makes sense.

Problem Set D

1. This table shows the cost for different numbers of packs of bubble gum:

Packs	3	5	8	15
Cost	$1.50	$2.50	$4.00	$7.50

a. Make a graph of the data. Be sure to choose appropriate scales and to label the axes with the variable names.

b. Use your graph to find the cost of 6 packs of gum and the cost of 9 packs of gum. Explain how you found your answers.

c. Use your graph to find the cost of 19 packs of gum. Explain how you found your answer.

d. Use your graph to find how many packs of bubble gum you can buy for $6.00. Explain how you found your answer.

2. This graph shows the masses of different lengths of a certain type of copper pipe:

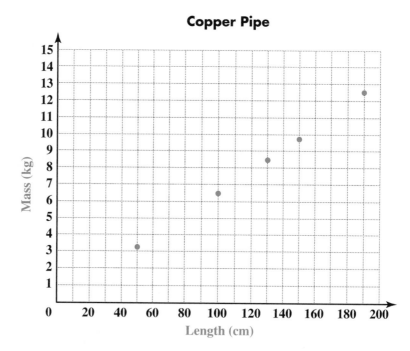

Copper Pipe

a. What is the mass of 100 centimeters of copper pipe?

b. Would you connect the points on this graph? If so, would you use dashed or solid segments? Explain your answers.

Just the facts

The Statue of Liberty contains 200,000 pounds of copper.

c. Estimate the mass of a copper pipe with length 180 centimeters. Explain how you found your answer.

d. Estimate the length of a copper pipe with a mass of 5 kilograms. Explain how you found your answer.

Share & Summarize

1. If you were to make a graph of these data, would you connect the points? If so, would you use dashed segments or solid segments? Explain your answers.

Tickets	Price
1	$3
5	15
8	24
15	45
21	63

2. If you were to make a graph of these data, would you connect the points? If so, would you use dashed segments or solid segments? Explain your answers.

Time	Temp. (°F)
6 A.M.	17
8 A.M.	20
10 A.M.	25
noon	32
2 P.M.	30
4 P.M.	28
6 P.M.	15

Lab Investigation ▶ Graphing with Spreadsheets

In this lab investigation, you will use a spreadsheet to create a graph. Spreadsheet software allows you to create accurate graphs quickly and easily. And, once you have created a graph, a spreadsheet makes it easy to make adjustments—such as changing scales and labels or adding and deleting data points.

MATERIALS

computer with spreadsheet software (1 per group)

The Situation

Because of safer living conditions, improved nutrition and health care, and an emphasis on wellness, people are living longer than ever. This table shows the expected life span of people born in the United States in various years.

1. The table shows that for the birth year 1990, the expected life span is 75.4 years. Explain what this means.

2. By how many years has the expected life span increased from 1920 to 1996?

Birth Year	Expected Life Span (years)
1920	54.1
1930	59.7
1935	61.7
1940	62.9
1945	65.9
1950	68.2
1955	69.6
1960	69.7
1965	70.2
1970	70.8
1972	71.2
1974	72.0
1976	72.9
1978	73.5
1980	73.7
1982	74.5
1984	74.7
1986	74.7
1988	74.9
1990	75.4
1992	75.8
1994	75.7
1996	76.1

Source: National Center for Health Statistics.

Just the facts

The greatest age a human has ever reached is 122 years 164 days. (This is the greatest recorded age. It is possible that someone has lived longer.)

Enter the Data

Before you use a spreadsheet to make a graph, you need to enter the data.

Begin by typing the column headings into Cells A1 and B1.

	A	B	C	D
1	Birth Year	Expected Life Span (years)		
2				
3				
4				
5				
6				
7				

Each box is called a *cell*. A cell is named for the column letter and row number. This is Cell A2.

Now enter the data. You can enter data down each column or across each row. When you are finished, check that you have entered the correct values.

	A	B	C	D
1	Birth Year	Expected Life Span (years)		
2	1920	54.1		
3	1930	59.7		
4	1935	61.7		
5	1940	62.9		
6	1945	65.9		
7	1950	68.2		

Create a Graph

Just as when you graph by hand, you need to think about the type of graph you want to make and how the axes and scales should be labeled. Since a spreadsheet makes it easy to change your graph, you don't have to get it exactly right the first time.

3. What variable do you want to show on each axis? What scales do you think would work well for this graph?

Now follow these steps to create your graph:

- Highlight the data you want to include in your graph. (For your first graph, you will probably want to include all the data. You can modify your graph later.)

- Enter information about what the graph should look like. This includes your answers to Question 3, along with other things like the shape and color of the points. For some spreadsheets, you type this information into a dialog box; other spreadsheets ask you a series of questions.

- Look at your graph carefully. Are the axes set up how you want them? Are the scales appropriate? Can you read the information easily? Make changes if necessary until the graph looks the way you want it to.

Analyze the Graph

4. Use the graph to help you describe the relationship between the year a person was born and his or her expected life span.

5. Write two or three sentences comparing the change in life span from 1920 to 1930 with the change from 1986 to 1996.

6. Use your graph to predict the expected life span of a person born in 1925.

7. Use your graph to predict the expected life span of a person born in 2000.

8. Add your prediction for 2000 to the spreadsheet, and make a new graph. Does this new data point seem to fit the pattern of the rest of the graph? Explain why or why not. If the point doesn't fit the pattern, make a new prediction and redraw the graph.

What Did You Learn?

9. Use what you have learned in this lab to graph another set of data. You can choose a data set from this chapter or create your own. Print your table and graph, and write a short paragraph describing what your graph shows.

On Your Own Exercises

Practice & Apply

1. **Sports** This table shows the number of field goals attempted and the number made by 12 members of a professional women's basketball team. In this problem, you will make a graph of these data.

   a. Decide on a good scale for each axis. Draw and label the axes, including the scale values.

   b. Carefully plot the data.

   c. Describe the overall relationship between the number of field goals attempted and the number made.

   d. The table shows that one player attempted 10 field goals and made 4. Another attempted 13 field goals and made 3. Does this information fit the relationship you described in Part c? If not, do you need to change your answer to Part c? Explain.

   e. Suppose another player on the team attempted 250 field goals. About how many would you expect her to have made?

   f. Suppose another player on the team made 175 field goals. About how many field goals would you expect her to have attempted?

Field Goals Attempted	Field Goals Made
458	212
489	226
339	142
177	76
164	71
10	4
117	52
58	24
35	13
40	14
13	3
4	4

2. Social Studies This table shows the percent of U.S. workers who were farmers from 1820 to 1994.

Year	Percent of U.S. Workers in Farming Occupations
1820	71.8
1840	68.6
1860	58.9
1880	57.1
1900	37.5
1920	27.0
1940	17.4
1960	6.1
1980	2.7
1994	2.5

Source: World Almanac and Book of Facts 2000. Copyright © 1999 Primedia Reference Inc.

a. Make a graph of the data with years on the horizontal axis. Be sure to choose appropriate scales and to label the axes with the variable names.

b. Describe how the percent of U.S. workers in farming occupations has changed over the years shown in your graph. Be sure to mention any trends or patterns you observe.

c. Use your graph to estimate the percent of U.S. workers in farming occupations in 1830, 1850, and 1965.

d. What percent of U.S. workers do you estimate were in farming occupations in 2000?

Just the facts

The Great Dane was developed 400 years ago to hunt wild boar. The breed originated in Germany, not Denmark as the name suggests.

3. Life Science Conor kept a record of his Great Dane Roscoe's height.

Age (weeks)	Height (inches)
7	12
9	14
11	18
12	20
14	21
16	21.5
18	25
20	25.5
21	26
23	27
30	30

a. Graph the data with age on the horizontal axis. Choose a suitable scale for each axis.

b. Connect the points with line segments.

c. Write a sentence or two describing the dog's growth.

4. Life Science In addition to measuring the length and weight of a baby, doctors often measure the distance around the baby's head. The chart shows how the distance around a particular baby's head changed for the first 36 months after birth.

Age (months)	0	4	8	12	16	20	24	28	32	36
Distance Around Head (cm)	34	40.2	43	45.5	46.6	47.8	48.5	49.4	49.8	50

a. Graph the data with age on the horizontal axis. Choose a suitable scale for each axis.

b. Connect the points with line segments.

c. Write a sentence or two describing the growth in the distance around the baby's head.

5. Geometry
Circumference is the distance around a circle. The table lists the approximate circumferences of circles with given diameters.

Diameter (cm)	Circumference (cm)
0.5	1.6
1.5	4.7
2	6.3
3	9.4
3.5	11.0
5	15.7

a. Make a graph of the data. Be sure to choose appropriate scales and to label the axes with the variable names.

b. Use your graph to estimate the circumference of a circle with diameter 4 cm. Explain how you found your answer.

c. Use your graph to estimate the diameter of a circle with circumference 8 cm. Explain how you found your answer.

6. Earth Science This table shows the normal monthly temperatures for Washington, DC.

a. Graph the data with the month on the horizontal axis and the temperature on the vertical axis.

b. In which month is the temperature highest? In which month is the temperature lowest?

c. Would you connect the points on your graph? If so, would you use dashed or solid segments? Explain your answers.

**Washington, DC
Normal Monthly Temperatures**

Month	Temperature (°F)
January	31
February	34
March	43
April	
May	62
June	71
July	76
August	74
September	67
October	55
November	
December	35

Source: *World Almanac and Book of Facts 2000.* Copyright © 1999 Primedia Reference Inc.

d. Use your graph to estimate the temperatures for April and November. Explain how you made your estimates.

Connect & Extend

7. Life Science When real data fall in a pattern that is almost a straight line, you can *fit* a line to the data and use it to make predictions. Here is a graph showing the height and weight of 21 Great Danes. The points form a pattern that is close to a line, so a line has been fit to the data.

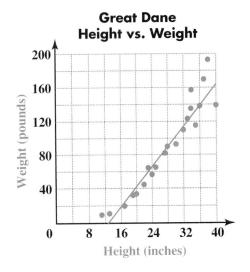

Great Dane Height vs. Weight

The line does not pass through all the points, but it comes close to most of them and gives a good approximation of the relationship.

a. Describe the overall relationship between height and weight for the dogs in the table.

b. Roscoe, Conor's Great Dane, is 31 inches tall. Use the line to predict Roscoe's weight.

c. Duke, Ms. Pinsky's Great Dane, weighs 50 pounds. Use the line to predict Duke's height.

8. This graph shows the shoe sizes and midterm grades for 20 of the students in Ms. Washington's fourth-period math class.

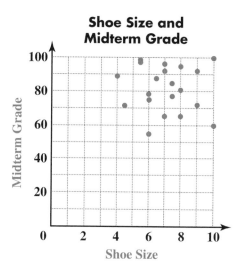

Shoe Size and Midterm Grade

a. Is there an overall relationship between shoe size and midterm grade? Explain.

b. Donny, another student in the class, wears size 7.5 shoes. Can you use the graph to predict his midterm grade? Explain.

9. Physical Science Caroline has two cylindrical jars. The heights of the jars are the same, but the base of one jar has twice the area of the base of the other.

Caroline poured exactly 1 cup of water into each jar and measured the height of the water. She then poured another cup into each jar and measured the water height. She repeated this process until she had poured 10 cups of water into each jar. She recorded her data in a table.

a. Use the pattern in the data to complete the table.

Cups	Water Height in Small Jar (cm)	Water Height in Large Jar (cm)
0	0	0
1	2	1
2	4	2
3	6	3
4	8	4
5		
6		
7		
8		
9		
10		

b. Draw a graph showing how the water height changed in the small jar. Put the number of cups on the horizontal axis and the water height on the vertical axis. Choose appropriate scales, and label the axes.

c. Describe the pattern of points in your graph.

d. *On the same axes,* draw a graph showing how the water height
changed in the large jar. Use a different color or point shape so
you can distinguish the two sets of points.

e. How is the pattern of points for the small jar similar to that for the
large jar? How are the patterns different?

f. Caroline also has a jar with a base area three times
that of the small jar. If she repeats the filling process
with this jar, how do you think the new graph would
compare to the other two graphs?

10. Mr. Takai drove his car on a mountain path. He reached the highest
point of the mountain, continued to drive on the top of the peak for a
while, and then drove down the other side. This table summarizes the
car's height and distance from his starting point at various times:

Time (hours)	Distance from Start (km)	Height from Start (km)
0	0	0
$\frac{1}{4}$	7.5	0.75
$\frac{1}{2}$	15	1
$\frac{3}{4}$	22.5	1.5
1	30	1.5
$1\frac{1}{4}$	37.5	1.5
$1\frac{1}{2}$	45	0.75
$1\frac{3}{4}$	52.5	0

a. Plot the data for distance and time. Use the horizontal axis for
time and the vertical axis for distance. Connect the points with
line segments.

b. Describe how the distance changed over time.

c. Plot the data for height and time. Use the same scale for the time
axis that you used for the first graph. Connect the points with line
segments.

d. Describe how the height changed over time.

e. Estimate the car's height and distance after $\frac{7}{8}$ hour and after
$1\frac{5}{8}$ hours.

f. Explain how it is possible for the height to be decreasing after
$1\frac{1}{4}$ hours while the distance is increasing.

Remember

- A right angle measures 90°.
- A right triangle is a triangle with a right angle.
- The sum of the angle measures of any triangle is 180°.

11. Geometry Marcus drew 10 right triangles. For each triangle, he measured the two non-right angles and recorded the measures in a table.

Measure of Angle 1 (degrees)	Measure of Angle 2 (degrees)
10	80
15	75
20	70
30	60
36	54
40	50
47	43
53	37
75	15

a. Make a graph of his data. Connect the points with dashed or solid segments if it makes sense to do so.

b. If you connected the points, explain why and tell how you decided whether to use dashed or solid segments. If you did not connect the points, explain why you didn't.

c. Write a sentence or two describing the relationship between the measures of Angle 1 and Angle 2.

d. Use your graph to estimate the measure of Angle 2 when Angle 1 measures 5° and when Angle 1 measures 55°. Check your answers by verifying that the sum of the three angles in each case is 180°.

Mixed Review

Find each sum or difference.

12. $\frac{7}{10} - \frac{3}{15}$

13. $3\frac{4}{7} + 2\frac{18}{21}$

14. $\frac{7}{8} - \frac{5}{12}$

15. $\frac{37}{13} - 2\frac{1}{2}$

16. $\frac{3}{8} + 10\frac{4}{5}$

17. $\frac{1}{15} - \frac{1}{40}$

Find the greatest common factor of each pair of numbers.

18. 84 and 72

19. 47 and 63

20. 54 and 108

Find the prime factorization for each number.

21. 62

22. 248

23. 702

24. Here are the first three terms of a sequence:

Term 1   Term 2   Term 3

a. Describe the pattern in this sequence.

b. Draw the next two terms in the sequence.

c. Draw Term 15, and explain how you know you are correct.

25. Here are the first four terms of a sequence:

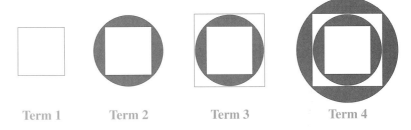

Term 1   Term 2   Term 3   Term 4

a. Draw Term 5.

b. Write a rule for creating each term from the previous term. Then describe what Term 55 would look like.

26. Think of your own sequence of shapes or numbers.

a. Write the first five terms of your sequence.

b. Write a rule for finding each term from the previous term.

27. The number of students in the science club is 87.5% of the number on the math team. If there are 21 students in the science club, how many are on the math team?

28. Of the 21 students in the science club, 8 are also on the math team. What percent of the science club is this?

VOCABULARY

VOCABULARY
axes
coordinates
line graph
ordered pair
origin
variable

Chapter Summary

In this chapter, you interpreted and created graphs. You started by looking at graphs with only a few points and figuring out what information was revealed by the positions of the points. Then you looked at graphs made up of lines and curves. You saw that when the variable on the horizontal axis is time, the line or curve tells a story about how the other variable changes.

You then made your own graphs. For some graphs, you drew a line or curve to fit a story or description, without worrying about specific values. For other graphs, you made a table of values, chose scales for the axes, and plotted points. You saw that sometimes it makes sense to connect the points on a graph and that connecting points can help you see patterns and make predictions.

You also learned that plotting collected data and looking for trends, or patterns, can help you determine whether the variables are related. When the plotted points do show a pattern, you can make predictions about values not on the graph.

MATERIALS
graph for Question 7

Strategies and Applications

The questions in this section will help you review and apply the important ideas and strategies developed in this chapter.

Interpreting graphs

1. These graphs give information about Hannah's dogs. Use them to determine whether each statement that follows is true or false, and explain how you decided.

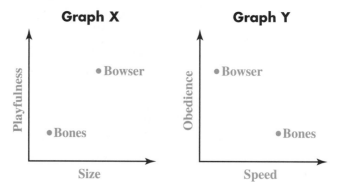

Graph X — Playfulness (vertical axis) vs Size (horizontal axis): • Bowser, • Bones

Graph Y — Obedience (vertical axis) vs Speed (horizontal axis): • Bowser, • Bones

   a. The smaller dog is less playful.

   b. The larger dog is more obedient.

   c. The faster dog is more playful.

   d. The slower dog is smaller.

impactmath.com/chapter_test

2. This graph shows how Althea's distance from home changed one Saturday morning. Write a story about Althea's morning. Your story should account for all the changes in the graph.

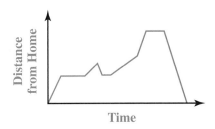

3. The graph below shows the percent of the U.S. labor force that was unemployed in even-numbered years from 1980 to 2000.

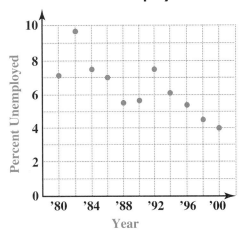

U.S. Unemployment

Source: *World Almanac and Book of Facts 2003*. Copyright © 2003 Primedia Reference Inc.

a. In which year was the unemployment rate highest? In which year was it lowest?

b. Over which two-year period did the unemployment rate decrease most? By about how much did it decrease?

c. Over which two-year period did the unemployment rate increase most? By about how much did it increase?

d. Over which two-year period did the unemployment rate change least?

e. Predict what the unemployment rate was in 1993.

Creating graphs

4. Fran went diving to explore a sunken ship. She descended quickly at first and then more slowly. As she was descending, she noticed an interesting fish. She ascended a little and stayed at a constant depth for a while as she watched the fish. Then she slowly descended until she reached the sunken ship. She explored the ship for several minutes and then returned to the surface of the water at a slow, steady pace. Draw a graph that shows how Fran's depth changed during her dive.

5. A plumber charges $40 for a house call, plus $15 for each half hour he works.

    a. Copy and complete the table to show how much the plumber charges if he works the given numbers of hours.

Time (hours)	0	0.5	1	1.5	2	2.5	3	3.5	4	4.5	5
Charge (dollars)	40	55									

    b. Choose an appropriate scale, and graph the data.

    c. Does it make sense to connect the points on your graph? Explain.

    d. Suppose the plumber charges by fractions of a half hour. Use your graph to estimate how much the plumber would charge if he worked 3 hours 15 minutes. Explain how you found your answer.

Using graphs to find relationships and make predictions

6. The *Greenville Tribune* asked 17 of its newspaper carriers about how long they spent delivering papers each weekday. For each carrier, the graph shows the number of addresses on the carrier's route and the number of minutes it takes him or her to complete the route.

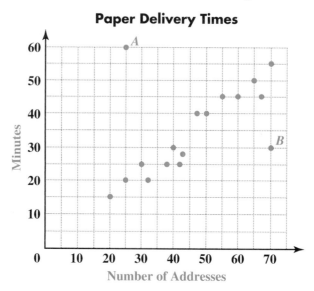

Paper Delivery Times

a. Does there appear to be an overall relationship between the number of addresses and the time it takes to complete the route? If so, describe the relationship.

b. Points *A* and *B* are far from most of the other points. Give some reasons why the data for the carriers represented by these points might be different from the others.

c. A new *Greenville Tribune* carrier has been assigned a route with 37 addresses. Estimate how long it will take her to finish her route each day.

d. Chandra spends about 35 minutes delivering papers each day. Estimate the number of addresses on her route.

7. Rey's Market sells pistachios by the pound. The graph shows the weight and price of pistachios purchased by five customers.

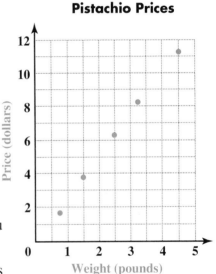

Pistachio Prices

a. Would it make sense to connect the points on this graph? If so, would you use dashed or solid segments? Explain your answers.

b. Estimate the cost of 1 pound of pistachios and of 5 pounds of pistachios. Explain how you found your answers.

c. Estimate the number of pounds of pistachios you could buy for $7.50. Explain how you found your answer.

Demonstrating Skills

8. Plot each point on one set of axes. Label each point with its coordinates.

   a. (0, 5)   **b.** (1.5, 5.5)   **c.** (9.5, 8)

   d. (7, 7)   **e.** (7, 0)   **f.** (3.3, 6.5)

9. Give the coordinates of the point halfway between (3, 5) and (7, 5).

10. Give the coordinates of the point halfway between (6, 2) and (6, 7).

11. Plot the points (1, 4) and (5, 4). Plot two more points so that the four points can be connected to form a square. Give their coordinates.

CHAPTER 6

Analyzing Data

Real-Life Math

The Average American You have probably read or heard reports that give information about the "average American." In these reports, a single number is often used to describe what is typical about the entire U.S. population. Usually, such a number is found by calculating an "average" value from a large set of data. As you will discover in this chapter, there are several types of averages—and for some sets of data, they may be very different from one another.

Think About It How might you find the height of the "average" student in your school?

Family Letter

Dear Student and Family Members,

Did you know that the "average American" eats 24 quarts of ice cream and 68 quarts of popcorn every year? Did you ever wonder how people come up with numbers like these? Our next chapter is about analyzing data—numbers, facts, or other measurable information.

You'll explore three ways of measuring what is typical—the median, mode, and mean. Here's an example. Suppose you ask 15 of your friends to tell you how many movies they've seen in the past month. You organize the data into a list.

0, 0, 0, 0, 0, 2, 2, 2, 2, 3, 3, 3, 3, 4, 6

- The median is the middle number in the list, 2.
- The mode is the number that appears the most often, 0.
- The mean is the number you get if the total, 30 movies, is divided evenly among all your friends. The mean is $30 \div 15$ or 2.

So, you might conclude that your friends saw an average of 2 movies in the last month.

Along with analyzing data, it's important to display the data. You're already familiar with bar graphs and line graphs, but in this chapter you'll learn about histograms, stem-and-leaf plots, double bar graphs, and line plots too. Here's a line plot of the data above.

Graphs can make it easier to see patterns in data and to draw conclusions. What is obvious in the line plot that wasn't when the data was shown in the list?

Vocabulary Throughout the chapter you'll see these new vocabulary terms:

distribution	mean	outlier
histogram	median	range
line plot	mode	stem-and-leaf plot

What can you do at home?

During the next few weeks, your student may show an interest in data outside of the classroom. You might help him or her to think of different kinds of data people collect and why—purchases in the grocery story, telephone opinion surveys, political exit polls, or hits at a Web site.

6.1

Using Graphs to Understand Data

Just the facts

The word *data* is plural and means "bits of countable or measurable information." The singular form of data is *datum.* When you talk about data, you should use the plural forms of verbs—for example *are, were,* or *show.*

People in many professions use data to help make decisions. When data are first collected, they are often just lists of numbers and other information. Before they can be understood, data must be organized and analyzed. In this chapter, you will investigate several tools for understanding data.

In many activities in this chapter, your class will play the role of a company called Data Inc.—a consulting group that specializes in organizing and analyzing data. Various people and organizations will come to Data Inc. for advice and suggestions.

In Chapter 5, you saw how graphs can help you discover patterns and trends. In this lesson, you will look at several types of graphs and compare the kinds of information each tells you about a set of data.

Explore

These graphs have no labels or scale values. Tell whether one of these graphs could describe each situation below. If it could, tell what the axes would represent and what the graph would reveal about the situation. If neither graph could describe the situation, describe or sketch a graph that could.

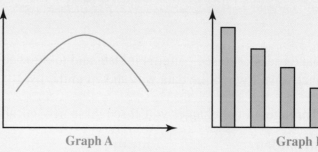

Graph A   Graph B

• The number of visitors to a zoo over the past year

• The weight of a young hippo from birth to age one

• The distance from a ball to the ground after the ball is dropped

• The number of minutes of daylight each day during a year

• The number of school days remaining on the first of each month, from February through June

• The number of children born each month in one year in Canada

Investigation ▶1 Using Line Graphs to Solve a Mystery

The Smallville police are investigating the disappearance of a man named Gerald Orkney. Here is what they know so far:

- Mr. Orkney lives alone with his pet iguana, Agnes.

- Mr. Orkney didn't show up for work on December 15. When his friends came to check on him, he and Agnes were gone.

- An atlas and the graphs below, which have no scales or labels, were found in Mr. Orkney's apartment.

The police have asked Data Inc. to help them figure out what happened to Gerald Orkney.

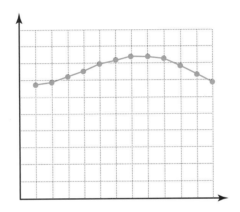
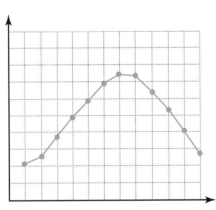

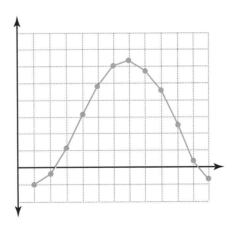

Think & Discuss

Look carefully at the three graphs the police found. Think about the shape of each graph and about the axes. How are the graphs alike? How are they different?

Since the graphs don't have labels, it is impossible to know exactly what they represent or how they are related to Gerald Orkney's disappearance. However, you may be able to make a *hypothesis,* or an educated guess, based on the information you do have.

Try to think of some ideas about what the graphs might show. Consider both what the graphs look like and how they might be related to the other information the police found.

Using the graphs and the other information given, try to come up with at least one hypothesis about what might have happened to Gerald Orkney.

MATERIALS
copies of the "clues"

Problem Set A

1. During a search of Gerald Orkney's office, the police found more detailed versions of the graphs. Your teacher will give you copies of the new graphs. Look at them closely.

   a. What new information do the graphs reveal? Does this information fit any of the ideas your class had in the Think & Discuss? Now what do you think the graphs might show?

   b. Does the new information support any of the hypotheses your class made about what happened to Gerald Orkney? Explain.

   c. Make a new hypothesis, or make changes to an earlier hypothesis, to fit all the information you have so far.

2. The police have just discovered more clues: a list and a note. Your teacher will give you a copy of this new information.

   a. What might the list have to do with the graphs? Do you have new ideas about what each graph might show? If so, add appropriate titles and axis labels to the graphs.

   b. Does this new information support your hypothesis about what happened to Gerald Orkney? Explain.

   c. Now what do you think happened to Gerald Orkney? Make a new hypotheses, or make changes to an earlier hypothesis, to fit this new information.

Share & Summarize

Write a letter to the Smallville police summarizing your group's investigation and presenting your hypothesis about what happened to Gerald Orkney.

Investigation 2 Using Bar Graphs to Analyze Data

The environmental group Citizens for Safe Air has asked Data Inc. to analyze some data about *hydrocarbons*. These compounds are part of the emissions from cars and other vehicles that pollute the air. The group wants to know how the total amount of hydrocarbon emitted by vehicles has changed over the past several decades—and how it might change in the future.

Just the facts

Hydrocarbons react with nitrogen oxides and sunlight to form ozone, a major component of smog. Ozone causes choking, coughing, and stinging eyes, and it damages lung tissue.

The table shows estimates of the typical amount of hydrocarbon emitted per vehicle for each mile driven in the United States for years from 1960 to 2015. The values from 2000 to 2015 are predictions.

Source: "Automobiles and Ozone," Fact Sheet OMS-4 of the Office of Mobile Sources, the U.S. Environmental Protection Agency.

Year	Average Per-Vehicle Emissions (grams of hydrocarbon per mile)
1960	17
1965	15.5
1970	13
1975	10.5
1980	7.5
1985	5.5
1990	3
1995	1.5
2000	1
2005	0.75
2010	0.5
2015	0.5

Problem Set B

1. For which 5-year period is the decrease in per-vehicle emissions greatest? For which 5-year period is it least?

2. On a set of axes like this one, draw a bar graph showing the typical per-vehicle emissions for each year given in the table.

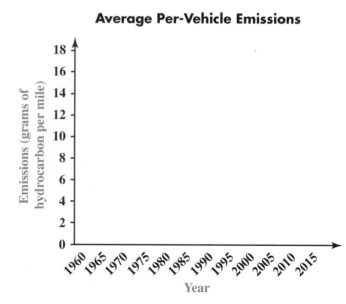

Average Per-Vehicle Emissions

3. Describe what your graph indicates about the change in per-vehicle emissions over the years. Discuss high and low points, periods of greatest and least change, and any other patterns you see.

4. To determine when the greatest decrease in per-vehicle emissions occurred, is it easier to use the table or the graph? Explain.

You have seen that the amount of hydrocarbon *each vehicle* emits *per mile* has decreased over the years, but this is not enough information to conclude that the *total amount* of hydrocarbon emitted by *all vehicles* is decreasing. You also need to consider the total number of miles driven by all vehicles.

This table shows estimates of the number of miles driven, or expected to be driven, by all vehicles in the United States for various years between 1960 and 2015.

Year	Vehicle Miles Traveled (billions)
1960	750
1965	950
1970	1,150
1975	1,250
1980	1,500
1985	1,500
1990	2,000
1995	2,300
2000	2,600
2005	2,850
2010	3,150
2015	3,400

Source: "Automobiles and Ozone," Fact Sheet OMS-4 of the Office of Mobile Sources, the U.S. Environmental Protection Agency, Jan 1993.

Problem Set C

1. Look at the table on page 347. During which 5-year period does the number of miles driven increase most? During which 5-year period does it increase least?

2. On a set of axes like the one below, draw a bar graph showing the billions of vehicle miles traveled for each year given.

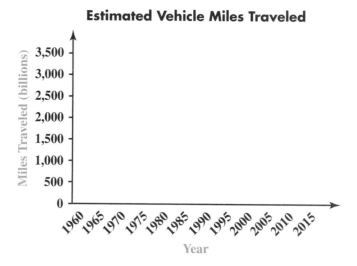

3. Describe what your graph indicates about the change in the total number of miles driven over the years. Discuss high and low points, periods of greatest and least change, and any other patterns you see.

You now know that, over time, the typical amount of hydrocarbon emitted per vehicle has decreased. You also know that more miles are driven each year. In the next problem set, you will combine this information to answer this question: Is the *total amount* of hydrocarbon emitted from vehicles increasing or decreasing?

Think & Discuss

How could you use the data in the two previous tables to calculate estimates of the total amount of hydrocarbon emitted by all vehicles each year?

Problem Set D

1. Copy and complete the table to show the total amount of hydro-carbon emitted by U.S. vehicles each year.

Year	Estimated Total Emissions (billions of grams of hydrocarbon)
1960	
1965	
1970	
1975	
1980	
1985	
1990	
1995	
2000	
2005	
2010	
2015	

2. Make a bar graph of the data in the table.

3. Describe what your graph indicates about the change in the total hydrocarbon emissions over the years. Discuss high and low points, periods of greatest and least change, and any other patterns you see.

Share & Summarize

Write a letter to Citizens for Safe Air. Describe Data Inc.'s investigation of hydrocarbon emissions, summarizing your findings about how total emissions have changed over the past few decades and how they might change in the future.

Investigation 3 Making Histograms

There are many types of graphs. The graph that is best for a given situation depends on the data you have and the information you want to convey.

Problem Set E

This bar graph shows the times of some of the participants in the men's 10-kilometer cross-country skiing event at the 1998 Winter Olympics:

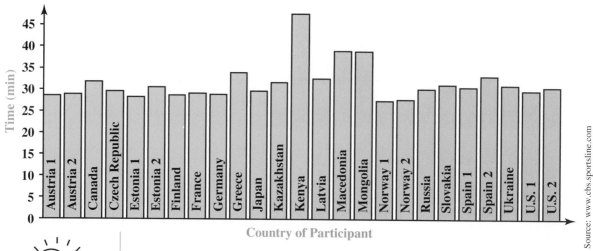

1998 Olympic Cross–Country Ski Times

Source: www.cbs.sportsline.com

Just the facts

V O C A B U L A R Y
histogram

1. How many participants' times are shown in the graph? How many countries are represented?

2. From which country was the gold medalist? What was his time?

3. How many participants completed the race with a time between 31 minutes and 32 minutes 59 seconds?

4. The bars in the graph are arranged alphabetically by country. Think of another way the bars could be ordered. What kinds of questions would be easier to answer if they were ordered that way?

It probably took you some time to figure out the answer to Problem 3 of Problem Set E. Although it is easy to use the bar graph to find the time for each participant, it is not as easy to find the number of skiers who finished within a particular time interval.

You will now use a *histogram* to display the ski times. In a **histogram**, data are divided into equal intervals, with a bar for each interval. The height of each bar shows the number of data values in that interval. There are no gaps between intervals.

Problem Set F

In this problem set, you will make a table of frequencies. *Frequencies* are counts of the number of data values in various intervals. You will use your frequency table to create a histogram.

1. Copy this table. Use the bar graph in Problem Set E to count the number of participants who finished in each time interval. Record this information in the "Frequency" column.

Time (minutes:seconds)	Frequency
27:00–28:59	
29:00–30:59	
31:00–32:59	4
33:00–34:59	
35:00–36:59	
37:00–38:59	
39:00–40:59	
41:00–42:59	
43:00–44:59	
45:00–46:59	
47:00–48:59	

2. Copy the axes below. Create a histogram by drawing bars showing the number of participants who finished in each time interval. The bar for the interval 31:00–32:59 has been drawn for you.

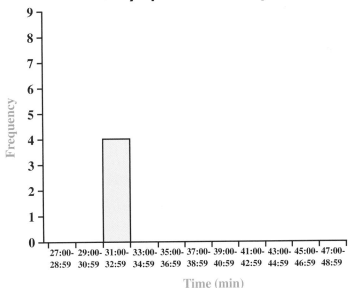

1998 Olympic Cross–Country Ski Times

Just the **facts**

Relative frequency is the ratio of the number of data in an interval to the total number of data in all intervals. For example, the relative frequency of the data in the 31:00–32:59 interval is $\frac{4}{25}$ or 0.16.

3. In which 2-minute interval did the greatest number of skiers finish?

4. The shape of a histogram reveals the **distribution** of the data values. In other words, it shows how the data are spread out, where there are gaps, where there are many values, and where there are only a few values. What can you say about the distribution of times for this event?

V O C A B U L A R Y
distribution

Rather than showing the *number* of values in each interval, some histograms show the *percent* of values in each interval. For example, this histogram shows how the test scores for Mr. Lazo's math exam were distributed. The maximum possible score was 75 points.

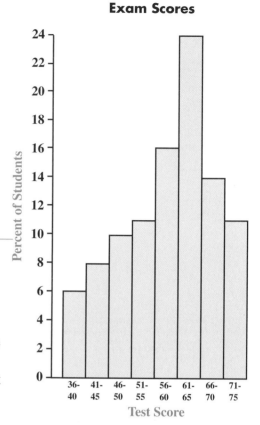

Exam Scores

Problem Set G

1. Describe the shape of the histogram. Tell what the shape indicates about the distribution of test scores.

2. Which interval includes the greatest percent of test scores? About what percent of scores are in this interval?

3. Which interval includes the least percent of test scores? About what percent of scores are in this interval?

4. Suppose 64 students took Mr. Lazo's test. How many of them received a score from 66 to 70?

5. If you were to add the percents for all the bars, what should the total be? Why?

Just the facts

Cumulative frequency is the total number of all data values less than the upper limit of a certain interval. This is found by adding together the frequencies of the interval and all other intervals that come before it. For example, the cumulative frequency of the data less than 50 is 6 + 8 + 10 or 24.

Cumulative relative frequency is the ratio of the cumulative frequency for an interval to the total number of data in all intervals. For example, the cumulative relative frequency of the 46–50 interval is $\frac{24}{100}$ or 0.24.

Share & Summarize

1. What type of information does a histogram display? Give an example of a situation for which it would make sense to display data in a histogram.

2. In this investigation, you looked at a bar graph and a histogram of Olympic ski data. What are some things the bar graph shows better than the histogram? What are some things the histogram shows better than the bar graph?

On Your Own Exercises

Practice & Apply

1. In Parts a–d, tell which graph could represent the situation.

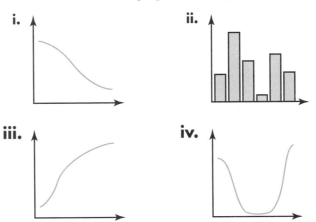

i. ii.

iii. iv.

a. a child's activity level from before a nap until after a nap

b. the populations of six cities

c. the change in water level from high tide to low tide

d. the change in the weight of a cat from birth until age 2

Chicago, Illinois

2. **Earth Science** This line graph shows the monthly normal temperatures for a U.S. city. Compare this graph to the temperature graphs for Miami, Chicago, and Fairbanks from Problem Set A.

   a. Which city's graph is this graph most similar to?

   b. In which of the following regions do you think this city is located: North, South, East Coast, Midwest, or West?

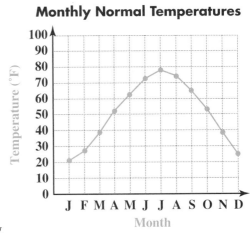

Monthly Normal Temperatures

Temperature (°F)

J F M A M J J A S O N D
Month

3. Social Studies This table shows the number of people who visited the United States from other countries in the years from 1990 to 2001:

Visitors to United States

Year	Visitors (millions)
1990	39.4
1991	42.7
1992	47.3
1993	45.8
1994	44.8
1995	43.3
1996	46.5
1997	47.8
1998	46.4
1999	48.5
2000	50.9
2001	45.5

Source: *World Almanac and Book of Facts 2003.* Copyright © 2003 Primedia Reference Inc.

a. The change in the number of visitors is greatest from 2000 to 2001. Between which two years is the change in the number of visitors least?

b. By what percent did the number of visitors increase from 1990 to 2001?

c. Make a bar graph showing the number of visitors to the United States during the years shown in the table.

d. Describe what your graph indicates about the change in the number of visitors to the United States. Discuss high and low points, periods of greatest and least change, and any other patterns you see.

The Coliseum in Rome, Italy

4. Ecology The bar graph on the left shows the number of farms in the United States from 1940 to 2000. The bar graph on the right shows the size of the average farm, in acres, for the same years.

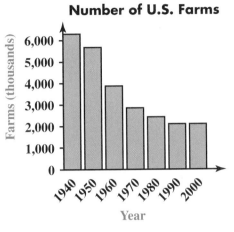

Number of U.S. Farms

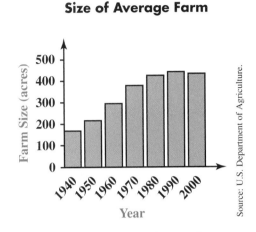

Size of Average Farm

Source: U.S. Department of Agriculture.

a. Describe how the number of U.S. farms has changed over the years.

b. Describe how the size of the average U.S. farm has changed over the years.

c. Jing used the data from the two farm graphs to create the bar graph below. Her graph shows the total amount of U.S. land devoted to farms from 1940 to 2000. How do you think Jing calculated the values for her graph?

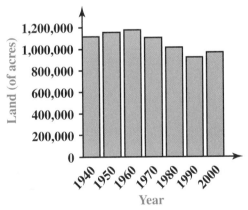

Total Land Devoted to U.S. Farms

d. Describe how the total amount of land devoted to U.S. farms has changed over the years.

5. Think about the multiplication facts from 0×0 to 12×12. You can group the products into intervals of 10. For example, a product can be between 0 and 9, between 10 and 19, between 20 and 29, and so on.

a. Do you think the products are evenly distributed among the intervals of 10, or do you think some intervals contain more products than others?

b. Copy and complete this multiplication table:

×	0	1	2	3	4	5	6	7	8	9	10	11	12
0													
1													
2													
3													
4													
5													
6													
7													
8													
9													
10													
11													
12													

c. Make a table, like that at right, showing the number of products that fall in each interval of 10.

d. Make a histogram that shows the number of products in each interval of 10. Be sure to include axes labels and scale values.

e. What does the shape of your histogram reveal about the distribution of the products?

f. Now make another histogram showing the number of products that fall into intervals of 20—that is, 0–19, 20–39, 40–59, and so on.

g. Describe the similarities and differences in the two histograms.

Product	Frequency
0–9	
10–19	
20–29	
30–39	
40–49	
50–59	
60–69	
70–79	
80–89	
90–99	
100–109	
110–119	
120–129	
130–139	
140–149	

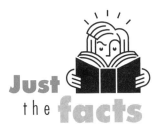

Connect & Extend

Remember

To make a line graph, plot the data points and connect them with line segments.

6. **Sports** This table shows the number of U.S. girls of various ages who played soccer in leagues recognized by the American Youth Soccer Organization in a recent year:

a. Create a histogram showing these data. The first bar, which includes 5- and 6-year-old girls, has been drawn for you.

Soccer Players

Ages	Girls
5 and 6	23,805
7 and 8	45,181
9 and 10	46,758
11 and 12	39,939
13 and 14	26,147
15 and 16	11,518
17 and 18	4,430

Source: American Youth Soccer Organization

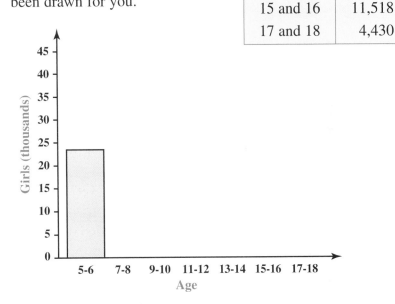

b. Describe the shape of the histogram. Tell what the shape indicates about the distribution of ages.

7. **Economics** Here are data about the number of motor vehicles manufactured in the United States, Europe, and Japan from 1993 to 2001:

a. On the same set of axes, make a line graph of the data for each group. Use a different point shape or line color for each group.

Motor Vehicles Manufactured (millions)

Year	U.S.	Europe	Japan
1993	10.9	15.2	11.2
1994	12.3	16.2	10.6
1995	12.0	17.0	10.2
1996	11.8	17.6	10.3
1997	12.1	17.8	11.0
1998	12.0	16.3	10.1
1999	13.1	17.6	9.9
2000	12.8	17.7	10.1
2001	11.5	17.7	9.8

Source: *World Almanac and Book of Facts 2003.* Copyright © 2003 Primedia Reference Inc.

b. Is there one group that consistently produces more motor vehicles than the others? If so, which group is it?

c. Write two or three sentences comparing the number of vehicles manufactured in the United States to the number manufactured in Japan for the years from 1993 to 2001.

d. Given the trends in these data, which group do you think produced the most motor vehicles in 2002? Which group do you think produced the fewest? Give reasons for your answers.

8. Earth Science The *latitude* of a location indicates how far it is from the equator, which has latitude 0°. The farther from the equator a place is, the greater its latitude. The latitude measure for a location includes the letter N or S to indicate whether it is north or south of the equator.

This table gives the lowest average monthly temperature and the latitude of nine cities:

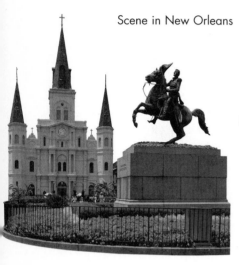
Scene in New Orleans

City	Latitude	Lowest Average Monthly Temp. (°F)
Albuquerque, New Mexico, U.S.A.	35° N	34
Georgetown, Guyana	7° N	79
New Orleans, Louisiana, U.S.A.	30° N	51
Portland, Maine, U.S.A.	44° N	22
Porto Alegre, Brazil	30° S	58
Recife, Brazil	8° S	75
San Juan, Puerto Rico	18° N	72
St. John's, Newfoundland, Canada	48° N	23
Stanley, Falkland Islands	52° S	36

Source: www.worldclimate.com

a. Make a line graph of the latitude and temperature data. When you graph the latitude values, ignore the N and S, and just graph the numbers. This way you will be graphing each city's distance from the equator.

b. Does there appear to be an overall relationship between the latitude of a city and its lowest average monthly temperature? If so, describe the relationship.

c. The island of Nassau in the Bahamas has a latitude of about 25° N. Predict Nassau's lowest average monthly temperature. Explain how you made your prediction.

9. **Preview** A survey asked middle school students how much time they spend with their parents or guardians on a typical weekend. Here are the results:

Time Spent with Parents	Boys (percent)	Girls (percent)
Almost all	39.6	49.6
One full day	18.6	21.8
Half a day	17.5	17.1
A few hours	24.3	11.5

If you wanted to compare the boys' responses with the girls' responses, you could display these data in two circle graphs.

Time Spent with Parents

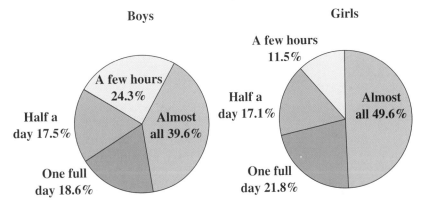

a. You could also show the data in a *double bar graph*. For each time category, the graph will have two bars—one showing the percent of boys in that category and the other showing the percent of girls. Copy and complete the graph below.

Time Spent with Parents

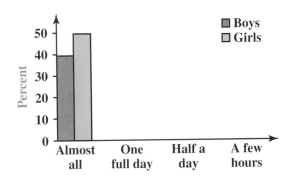

b. Which display do you think makes it easier to compare the two categories of data? Give reasons for your choice.

10. Drake's mother told him he could not play video games after school until his performance in math class improved significantly. Drake's math teacher gives a 20-point quiz each week. Drake made this graph to show his mother how much his scores had improved over the past five weeks:

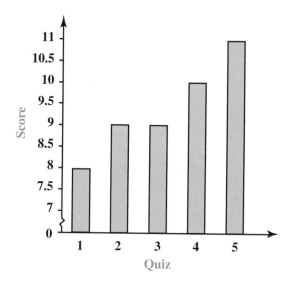

a. The bar for Quiz 5 is four times the height of the bar for Quiz 1. Is Drake's score on Quiz 5 four times his score on Quiz 1?

b. Drake's mother says his graph is misleading because it makes his improvement look more dramatic than it really is. What features of the graph make it misleading?

c. Make a new bar graph that you feel gives a more accurate view of Drake's performance on the weekly quizzes.

Find each product or quotient.

11. $\frac{3}{4} \cdot \frac{4}{3}$ **12.** $\frac{3}{4} \div \frac{4}{3}$ **13.** $\frac{12}{21} \cdot \frac{7}{16}$

14. $\frac{27}{32} \cdot \frac{24}{45}$ **15.** $2\frac{2}{5} \cdot \frac{1}{3}$ **16.** $3\frac{5}{8} \div \frac{1}{4}$

17. $1\frac{3}{8} \cdot 4\frac{1}{2}$ **18.** $4\frac{4}{7} \div 1\frac{1}{2}$ **19.** $5 \div \frac{1}{9}$

Geometry Find each missing angle measure.

20.

21.

22.

23. The 180 sixth-grade girls at Wright Middle School were asked to name their favorite activity in gym class. The results are shown in this circle graph:

Favorite Gym Activities

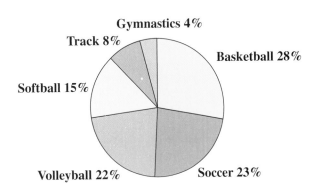

  a. Which activity is most popular? About how many girls chose that activity?

  b. Which activity is least popular? About how many girls chose that activity?

  c. What is the difference in the *percent* of girls who chose volleyball and the percent who chose track? What is the difference in the *number* of girls who chose these sports?

6.2 What Is Typical?

To help people understand a set of data, it is useful to give them an idea of what is *typical,* or average, about the data. In this lesson, you will learn three ways to describe the typical value in a data set. You will also learn about two simple types of graphs that are useful for showing the distribution of values in a set of data.

Think & Discuss

Have each student in your class estimate how many minutes he or she spent doing homework yesterday. Your teacher should record the data on the board. How would you describe to someone who is not in your class what is typical about your class's data?

Investigation 1 ▶ Mode and Median

The Jump Shot shoe store sells basketball shoes to college players. The table shows the brand and size of each pair the store sold one Saturday.

Brand	Size
Swish	13
Dunkers	14.5
Hang Time	8.5
Swish	13
Hang Time	13
Airborne!	15
Swish	12.5
Swish	8
Dunkers	13
Big J	13.5
Hang Time	14
Big J	14.5
Dunkers	14

Brand	Size
Swish	14
Hang Time	10
Airborne!	11
Dunkers	12
Airborne!	14
Big J	12.5
Swish	14
Swish	14.5
Hang Time	10.5
Swish	10.5
Swish	14
Hang Time	13.5

In the next two problem sets, you will learn some ways to summarize the shoe store's data.

Problem Set A

VOCABULARY
line plot

1. You can create a line plot to show the sizes of the shoes sold on Saturday. A **line plot** is a number line with X's indicating the number of times each data value occurs.

   a. To make the plot, copy the number line below. Mark an X above a shoe size each time it appears in the data set. For example, 14.5 appears three times, so put three X's above 14.5.

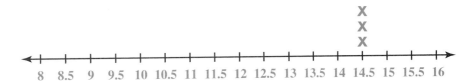

   b. Describe the shape of your line plot. Tell what the shape indicates about the distribution of shoe sizes.

2. When you describe a data set, it is helpful to give the *minimum* (least) and *maximum* (greatest) values.

   a. Give the minimum and maximum shoe sizes in the data set.

   b. How can you find the minimum and maximum values by looking at a line plot?

VOCABULARY
median
mode
range

3. The **range** of a data set is the difference between the minimum and maximum values. Give the range of the shoe-size data.

4. The **mode** of a data set is the value that occurs most often.

   a. Give the mode of the shoe-size data.

   b. How can you find the mode of a data set by looking at a line plot?

5. The **median** is the middle value when all the values in a data set are ordered from least to greatest.

   a. List the shoe-size data in order from least to greatest, and then find the median size.

   b. How can you find the median of a data set by looking at a line plot?

6. Suppose the store discovered five Saturday sales that weren't recorded: 14.5, 14.5, 15, 14.5, and 16. Add these to your line plot.

   a. What is the range of the data now?

   b. What is the mode now?

   c. When a data set has an even number of values, there is no single middle value. In such cases, the median is the number halfway between the two middle values. Find the median of the new data set.

In Problem Set A, you looked at ways to summarize *numerical data*—that is, data that are numbers. You will now look at the brand-name data, which are not numbers. Non-numerical data are sometimes called *categorical data* because they can be thought of as names of *categories,* or groups.

Problem Set B

1. Is it possible to make a line plot to show the distribution of the brand-name data? Explain.

2. Can you find the range of the brand-name data? If so, find it. If not, explain why it is not possible to find the range.

3. Do the brand-name data have a mode? If so, find it. If not, explain why it is not possible to find the mode.

4. Do the brand-name data have a median? If so, find it. If not, explain why it is not possible to find the median.

5. What are some other ways you might summarize the brand-name data?

The mode and the median are two measures of the typical, or average, value of a data set. In some cases, one of these measures describes the data better than the other.

Problem Set C

1. Ms. Washington gave her class a 10-point quiz. Here are her students' scores:

       7 9 10 5 5 8 6 10 6 7 10 2

         7 5 8 8 4 9 10 4 10 7 6

   a. Find the range, mode, and median of the quiz scores.

   b. Do you think the mode or the median is a better measure of what is typical in this data set? Explain.

2. Hannah asked nine of her classmates how many pets they have. Here are the results:

0   4   1   0   0   4   4   0   4

a. Find the range, mode, and median of these data.

b. Do you think the mode or the median is a better measure of what is typical in this data set? Explain.

3. During one afternoon practice, an athlete threw a javelin 13 times. Here are the distances for each throw, rounded to the nearest foot:

257   210   210   255   210   220

275   253   210   255   250   252   200

a. Find the range, mode, and median of the data.

b. If you had to use only one type of average—the mode or the median—to summarize this athlete's performance, which would you choose? Give reasons for your choice.

4. When you are given summary information about a set of data, you can sometimes get an overall picture of how the values are distributed. For example, suppose you know these facts about a data set:

- It has 15 values.
- The minimum value is 50, and the maximum value is 100.
- The mode is 55.
- The median is 57.

a. What do you know about how the data values are distributed?

b. Make up a data set that fits this description.

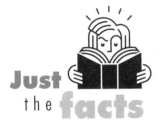
Share & Summarize

1. How is a line plot similar to a histogram? How is it different?

2. Describe what the range, mode, and median tell you about a set of numerical data.

3. Which measure—the range, the mode, or the median—can be used to describe a set of categorical data? Explain.

Investigation 2 Stem-and-Leaf Plots

You have seen that line plots are useful for showing the distribution of numerical data and for locating the mode and median. However, for some data sets, creating a line plot may not be practical.

Think & Discuss

The students in Ms. Washington's class kept track of how many minutes they spent doing homework one evening. Here are their results:

Student's Initials	Time Spent Doing Homework (minutes)	Student's Initials	Time Spent Doing Homework (minutes)
AF	42	RL	90
JB	5	DD	39
RC	60	AG	30
HE	30	RT	49
JL	45	CB	58
MM	47	MC	55
DL	0	FB	75
SK	25	JM	45
FR	67	TK	44
CO	51	MG	37
DW	56	LK	62
PG	20	EL	65

Think about what a line plot of these data would look like. Do you think a line plot is a good way to display these data? Why or why not?

VOCABULARY
stem-and-leaf plot

When a data set contains many different values, or when the values are spread out, a **stem-and-leaf plot** (also called a *stem plot*) may be more helpful than a line plot.

To make a stem-and-leaf plot of the homework-time data, think of each data value as being made up of two parts: a tens digit and a ones digit.

Write the tens digits, from least to greatest, in a column. Draw a vertical line to the right of the digits. The values in this column are the *stems*.

Stem	Leaf
0	
1	
2	
3	
4	
5	
6	
7	
8	
9	

To add the *leaves*, write the ones digit for each data value to the right of the appropriate tens digit. For example, to plot the first time of 42 minutes, write a 2 next to the stem value 4. The first five data values—42, 5, 60, 30, and 45— have been plotted here.

Stem	Leaf
0	5
1	
2	
3	0
4	2 5
5	
6	0
7	
8	
9	

After you have plotted all the data values, redraw the plot, listing the ones digits for each stem in order from least to greatest.

Stem	Leaf
0	0 5
1	
2	0 5
3	0 0 7 9
4	2 4 5 5 7 9
5	1 5 6 8
6	0 2 5 7
7	5
8	
9	0

Key: 4|2=42

Stem plots and histograms both group data into intervals. However, unlike a histogram, a stem plot allows you to read individual values.

For example, if you made a histogram that grouped the data above into the intervals 0–9, 10–19, and so on, it would show that there are four values between 50 and 59, but it would not show that these values are 51, 55, 56, and 58.

M A T E R I A L S
homework-time data
for your class

Problem Set D

1. Look at the completed stem-and-leaf plot in the Example on page 367.

   a. Explain how you can use this graph to find the range of the homework times.

   b. Use the stem plot to find the mode homework time. How did you find it?

   c. Use the stem plot to find the median homework time, and explain how you found it.

   d. Describe the shape of the graph. Explain what the shape tells you about how the data are distributed.

2. In the Think & Discuss at the beginning of this lesson, you collected homework times for your class. Make a stem plot of these data. (If your data include values of 100 minutes or more, your plot will need to include two-digit stem values. For example, a value of 112 would have a stem of 11 and a leaf of 2.)

3. Find the range, median, and mode of your class data.

4. Describe the distribution of your class data.

5. Write a few sentences comparing your class data to the data from Ms. Washington's class.

In the plots in Problem Set D, the tens digits are used as stem values and the ones digits are used as leaves. The stem and leaf values you choose for a given situation depend on the minimum and maximum values in the data set.

Problem Set E

1. This table shows the batting averages for the 2002 Cleveland Indians baseball team:

Just the facts

A player's *batting average* is the number of hits divided by the number of times the player is officially at bat.

2002 Cleveland Indians

Batter	Average	Batter	Average
Bradley	.249	Lawton	.236
Branyan	.205	Magruder	.217
Burkes	.301	McDonald	.250
Diaz	.206	Selby	.214
Fryman	.217	Stevens	.222
Garcia	.297	Thome	.304
Gutierrez	.275	Vizquel	.275

Source: *World Almanac and Book of Facts 2003*. Copyright © 2003 Primedia Reference Inc.

a. The batting averages range from .205 to .304. In this case, you can use the tenths and hundredths digits as stem values and the thousandths digits as leaves.

Make a stem-and-leaf plot using the stem values shown at right.

b. Describe the distribution of data values.

c. Find the mode and median batting average.

2. This table shows the American League RBI leaders for each season from 1973 to 2002:

2002 Cleveland Indians Batting Averages

Stem	Leaf
.20	
.21	
.22	
.23	
.24	
.25	
.26	
.27	
.28	
.29	
.30	

Key: .27|5=.275

American League RBI Leaders

Year	Player	RBI	Year	Player	RBI
1973	Reggie Jackson	117	1988	Jose Canseco	124
1974	Jeff Burroughs	118	1989	Ruben Sierra	119
1975	George Scott	109	1990	Cecil Fielder	132
1976	Lee May	109	1991	Cecil Fielder	133
1977	Larry Hisle	119	1992	Cecil Fielder	124
1978	Jim Rice	139	1993	Albert Belle	129
1979	Don Baylor	139	1994	Kirby Puckett	112
1980	Cecil Cooper	122	1995	A. Belle / M. Vaughn	126
1981	Eddie Murray	78	1996	Albert Belle	148
1982	Hal McRae	133	1997	Ken Griffey, Jr.	147
1983	C. Cooper / J. Rice	126	1998	Juan Gonzalez	157
1984	Tony Armas	123	1999	Manny Ramirez	165
1985	Don Mattingly	145	2000	Edgar Martinez	145
1986	Joe Carter	121	2001	Brett Boone	141
1987	George Bell	134	2002	Alex Rodriguez	142

Source: World Almanac and Book of Facts 2003. Copyright © 2003 Primedia Reference Inc.

a. Make a stem-and-leaf plot of the RBI data.

b. Describe the distribution of data values.

c. Find the minimum, maximum, mode, and median of the RBI data.

Share & Summarize

1. In what types of situations would you use a stem-and-leaf plot, rather than a line plot, to display a set of data?

2. What information can you get from a stem plot that you can't get from a histogram?

Investigation ▶3 The Meaning of *Mean*

The median and the mode are two ways to describe what is typical, or average, about a set of data. These values are sometimes referred to as *measures of central tendency,* or simply *measures of center,* because they give an idea of where the data values are centered. In this investigation, you will explore a third measure of center: the *mean.*

Problem Set F

1. Althea's scouting troop went strawberry picking. They decided to divide the strawberries they picked equally, so each girl would take home the same amount. The drawing shows how many quarts each girl picked.

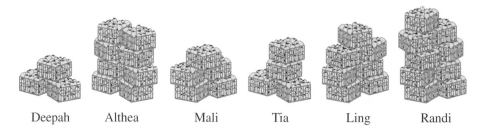

Deepah   Althea   Mali   Tia   Ling   Randi

How many quarts did each girl take home? How did you find your answer?

2. Another group of friends went blueberry picking. If they divided their berries equally, how many quarts did each friend get?

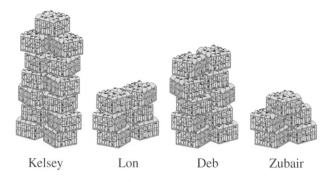

Kelsey   Lon   Deb   Zubair

3. A group of 10 friends picked the following numbers of quarts of blackberries:

5   10   4   5   7   9   9   6   8   7

If they divided the blackberries equally, how many quarts did each friend get?

Just the facts

Strawberries were originally known as "strewberries" because they appeared to be strewn among the leaves of the strawberry plant.

In each situation in Problem Set F, you redistributed the quarts to give each person the same number. The result was the *mean* of the number of quarts picked. The **mean** of a set of values is the number you get by distributing the total evenly among the members of the data set. You can compute the mean by adding the values and dividing the total by the number of values.

The mean is another measure of the typical, or average, value of a data set. In everyday language, the word *average* is often used to mean *mean*. However, it is important to remember that the mean, median, and mode are *all* types of averages.

Problem Set G

1. The astronomy club is selling calendars to raise money to purchase a telescope. The 10 club members sold the following numbers of calendars:

    3   5   7   10   5   3   4   6   9   8

    a. Find the mean, median, and mode of the numbers of calendars the club members sold.

    b. Suppose two very motivated students join the club. One sells 20 calendars and the other sells 22. Find the new mean, median, and mode.

    c. In Part b, suppose that instead of 22 calendars, the 12th club member had sold 100 calendars. What would the new mean, median, and mode be?

    d. How does the median in Part c compare to the median in Part b? How do the two means compare? Explain why your answers make sense.

2. Luke asked 12 students in his class how many books they had read (other than school books) in the past six months. Here are the responses given by 11 of the students:

    3   5   7   10   5   3   4   6   9   8   20

    a. Suppose you know that the mean number of books read by the 12 students is 10. Is it possible to find the number of books the 12th student read? If so, explain how. If not, explain why not.

    b. Suppose you know that the median number of books read by the 12 students is 5.5. Is it possible to find the number of books the 12th student read? If so, explain how. If not, explain why not.

Problem Set H

1. At her party, Ines had a contest to see who could pick up the most jelly beans in a handful. The nine people at her party reached into a bag and pulled out these numbers of jelly beans:

22   23   24   28   32   32   35   37   37

a. Find the mean and median of the data set.

b. Add two values to the data set so the median remains the same but the mean decreases. Give the new mean.

c. Start with the original data set. Add two values to the set so the median remains the same but the mean increases. Give the new mean.

d. Start with the original data set. Add two values to the set so the mean remains the same but the median changes. Give the new median.

Create a data set with 10 values that fits each description below.

2. The minimum is 45, the maximum is 55, and the median and mean are both 50.

3. The minimum is 10, the maximum is 90, and the median and mean are both 50.

4. The range is 85, the mean is 50, and the median is 40.

5. The range is 55, the mean is 40, and the median is 50.

Share & Summarize

1. Jing said that the students in her class have an average of three pets each.

a. If Jing is referring to the mode, explain what her statement means.

b. If Jing is referring to the median, explain what her statement means.

c. If Jing is referring to the mean, explain what her statement means.

2. Suppose you have a data set for which the mean and median are the same. If you add a value to the set that is much greater than the other values in the set, would you expect the median or the mean to change more? Explain.

Investigation ▶4▶ Mean or Median?

This investigation will help you better understand what the mean and median reveal about a set of data.

Problem Set ▮

1. Lee and Arturo collected the heights of the students in their math classes. They found that the median height of students in Arturo's class is greater than the median height of students in Lee's class.

   Tell whether each statement below is *definitely true,* is *definitely false,* or *could be true or false* depending on the data. In each case, explain why your answer is correct. (Hint: For some statements, it may help to create data sets for two small classes, with three or four students each.)

   a. The tallest person is in Arturo's class.

   b. Lee's class must have the shortest person.

   c. If you line up the students in each class from shortest to tallest, each person in Arturo's class will be taller than the corresponding person in Lee's class. (Assume the classes have the same number of students.)

   d. If you line up the students in each class from shortest to tallest, the middle person in Arturo's class would be taller than the middle person in Lee's class. (Assume the classes have the same odd number of students.)

2. Marta and Grace collected height data for their math classes. They found that Marta's class has a greater mean height than Grace's class.

   Tell whether each statement below is *definitely true,* is *definitely false,* or *could be true or false* depending on the data. In each case, explain how you know your answer is correct.

   a. The tallest person is in Marta's class.

   b. Grace's class must have the shortest person.

   c. If you line up the students in each class from shortest to tallest, each person in Marta's class will be taller than the corresponding person in Grace's class. (Assume the classes have the same number of students.)

   d. If you line up the students in each class from shortest to tallest, the middle person in Marta's class would be taller than the middle person in Grace's class. (Assume the classes have the same odd number of students.)

Books, news reports, and advertisements often mention average values.

You have learned about three types of average: the mode, the median, and the mean. The average reported in a particular situation depends on many factors. Sometimes, one measure is "more typical" than the others. Other times, a measure is selected to give a particular impression or to support a particular opinion.

Problem Set J

Career Connections is a small company that helps college graduates find jobs. They are creating a brochure to attract new clients and would like to include the average starting salary of their recent clients. They have asked Data Inc. to help them determine which type of average to use.

Listed below are the starting salaries of the clients they have helped in the past three months.

$30,000	$25,000	$60,000	$40,000	$25,000
$50,000	$70,000	$50,000	$25,000	$60,000
$25,000	$1,000,000	$60,000	$25,000	$40,000
$50,000	$25,000	$50,000	$25,000	$25,000

1. You know how to compute three types of averages: the mode, the median, and the mean.

   a. Find the mode, median, and mean for these data.

   b. Which average do you think best describes a typical value in this data set? Explain.

2. One of the salaries, $1,000,000, is much greater than the rest. A value that is much greater than or much less than most of the other values in a data set is called an **outlier.** One Data Inc. analyst suggested that this outlier should not be included when determining the average salary.

   a. Remove $1,000,000 from the data set, and recompute the mode, median, and mean.

   b. How does removing the outlier affect the three measures of center?

3. Write a brief letter to Career Connections telling them what value you recommend they report as the average starting salary of their clients. Give reasons for your choice. Consider all the averages you have computed for the salary data, both including and not including the $1,000,000 salary.

In the next problem set, you will use a single set of data to support two very different points of view.

Problem Set K

The Hillsdale School District will hold its annual girls' basketball banquet next Friday night. The head of athletics will present an award to the best scorer in the two high schools. Below are the points per game earned for each school's best offensive player.

Points per game for Westside Wolves' best player: 30, 61, 10, 0, 28, 48, 55, 12, 23, 55, 6, 25, 39, 18, 55, 31, 30

Points per game for Eastside Eagles' best player: 22, 35, 12, 37, 19, 36, 39, 13, 13, 36, 11, 37, 13, 38, 21, 37, 35

Each coach wants to be able to argue that his player deserves the award. Both coaches have come to Data Inc. for help.

1. Use your knowledge of statistics to argue that the Westside Wolves' player deserves the award.

2. Use your knowledge of statistics to argue that the Eastside Eagles' player deserves the award.

Just the **facts**

The first women's inter-collegiate basketball game was played in 1896 in San Francisco. The game pitted Stanford University against the University of California at Berkeley. Stanford won the game by a score of 2 to 1. Male spectators were not allowed at the game.

The next problem set will help you better understand what the mean and median tell you about the distribution of a data set.

Problem Set L

1. Create a data set with eight values from 1 to 20 that fits each description. Give the median and mean of each data set you create.

   a. The median is greater than the mean.

   b. The mean is greater than the median.

   c. The mean and median are equal.

   d. The mean is 3 more than the median.

   e. The median is 3 more than the mean.

2. The data set 1, 2, 3, 4, 5 has a mean of 3. Change two values so the new data set has a mean of 4.

Share & Summarize

1. Reports of the typical income of a city, state, or country often use the median rather than the mean. Why do you think this is so?

2. What might cause the mean of a data set to be much greater than the median?

3. What might cause the median of a data set to be much greater than the mean?

4. What might cause the median of a data set to be equal to the mean?

Investigation 5 ▶ Analyzing Data

Data Inc. has just received a letter:

Dear Data Inc.,

We are a sports company specializing in protective gear for kids. We have an idea for a new product called the ForArm—an armband that would protect the area from the wrist to the elbow. Skaters and skateboarders would find this protective gear very useful. We would like to make a few samples to test with middle school students, but we need your help to determine how long middle school students' forearms are.

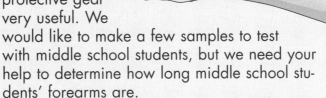

We would like you to start by analyzing the forearm lengths for students in your class. We are also gathering data from another middle school class and will send it to you as soon as it is available.

Thank you for your help.

Sincerely,
SportSafe

MATERIALS
measuring tape or yardstick

Problem Set M

1. Measure the forearm of each member of your group to the nearest quarter inch. Measure from just below the knuckles to just above the elbow. Check your results by having two people do each measurement. Record the data for your group in a class table.

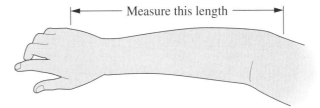

Measure this length

2. Work with your group to organize, analyze, and summarize your data. Prepare a short report of your findings. Include the range, mode, median, and mean, and any other information or graphs you think SportSafe would find useful.

Save your class data. You will need it for the On Your Own Exercises.

Another letter just arrived from SportSafe.

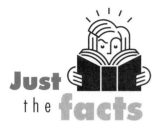

Just the facts

A *vambrace* is a protective armband worn as part of a suit of armor in the 14th century.

Dear Data Inc.,

We just received the forearm-length data from the other middle school class.

Please send us a report that includes the following:

- An analysis of the data for this class

- A comparison of these data with the data for your class

- An analysis of the combined data for the two classes

Thank you for all your hard work.

Sincerely,
SportSafe

Name	Forearm Length (inches)
Albert	$15\frac{3}{4}$
Ali	$13\frac{1}{2}$
Anna	13
Bianca	$13\frac{1}{2}$
Jackie	$12\frac{3}{4}$
Mariel	$12\frac{3}{4}$
Michael	$13\frac{3}{4}$
Nick	14
Olivia	13
Pablo	$13\frac{1}{2}$
Powell	$13\frac{1}{2}$
Sumi	$13\frac{1}{2}$
Susan	$13\frac{3}{4}$
Tomas	$13\frac{1}{2}$

Problem Set N

1. Analyze the data SportSafe has sent. Find the mean, median, mode, and range, and create a plot of the data.

2. Write a few sentences comparing data from your class with the data SportSafe collected.

3. Combine your class data with the data sent by SportSafe. Analyze the combined data set, and describe how it is similar to and different from your class data.

4. You have now analyzed three sets of forearm data: your class data, SportSafe's data, and the combined set of data. Which results do you think SportSafe should use to determine the sizes of their sample products? Why?

Share & Summarize

Create a set of forearm lengths containing at least 10 measurements, a range from 12 to 15, and a median between the medians you found for your class and for the SportSafe data. (If the medians of the two sets were the same, make the median of your new data set 1 inch greater.)

On Your Own Exercises

Practice & Apply

1. The table shows the style and size of all the hats sold at the Put a Lid on It! hat shop last Thursday.

   a. Make a line plot of the hat-size data.

   b. Describe the shape of your line plot. Tell what the shape indicates about the distribution of the hat sizes sold.

   c. Find the range, mode, and median of the hat-size data.

   d. Find the mode of the hat-style data.

   e. Is it possible to find the median and range of the hat-style data? If so, find them. If not, explain why it is not possible.

Style	Size
Cap	$6\frac{5}{8}$
Beret	$7\frac{3}{8}$
Fedora	$7\frac{1}{4}$
Sombrero	7
Cap	$7\frac{1}{4}$
Cap	$7\frac{3}{8}$
Fedora	$7\frac{3}{8}$
Cap	$7\frac{1}{4}$
Beret	$6\frac{7}{8}$
Panama hat	$7\frac{5}{8}$
Fedora	$6\frac{7}{8}$
Cap	$7\frac{1}{4}$
Sombrero	$7\frac{1}{2}$
Fedora	$7\frac{1}{2}$
Fedora	$7\frac{1}{4}$
Chef's hat	$7\frac{1}{8}$
Beret	7
Derby	$7\frac{1}{8}$
Beret	$7\frac{1}{8}$
Top hat	$7\frac{3}{4}$
Panama hat	$7\frac{3}{8}$

2. **Sports** This list shows the number of hits Jing got in each softball game this season:

   0 3 2 7 4 2 3 0 4 0 6 5 5 2 4 0

   a. Find the mode and median of these data.

   b. Do you think the mode or the median is a better measure of what is typical in this data set? Explain.

3. Create a data set with 13 values that satisfies these conditions: the minimum value is 3, the maximum value is 13, the mode is 4, and the median is 8.

 impactmath.com/self_check_quiz

4. Sports Caroline had a bowling party for her birthday. Each person at the party bowled three games. The stem-and-leaf plot shows their scores.

a. What is the lowest score? What is the highest score?

b. Find the mode and median scores.

c. Describe the distribution of scores.

Bowling Party Scores

Stem	Leaf
5	2
6	
7	3 7
8	1 6
9	2 9
10	0 3
11	0 1 4 4 4 9
12	3 3
13	2 6 6
14	
15	
16	
17	
18	5

Key: 9|2=92

5. Measurement Two classes of elementary school students measured their heights in centimeters. Here are the results:

Ms. Cho's class: 117, 117, 119, 122, 127, 127, 114, 137, 99, 107, 114, 127, 122, 114, 120, 125, 119

Mr. Diaz's class: 130, 147, 137, 142, 140, 135, 135, 142, 142, 137, 135, 132, 135, 120, 119, 125, 142

a. For each class, make a stem-and-leaf plot of the height data.

b. Find the range, mode, and median for each class.

c. The two classes are at two different grade levels. Which class do you think is the higher grade?

d. What percent of the students in Mr. Diaz's class are as tall or taller than the median height of the students in Ms. Cho's class?

Just the facts

In the United States, bowling involves 10 pins. In Canada, 5-pin bowling is popular. In the Canadian game, the pins are smaller and the ball weighs only 3.5 pounds.

6. A scientist in a science fiction story finds the mass of alien creatures from four areas of the planet Xenon.

Area of Xenon	Mass (kilograms)
Alpha	6, 21, 12, 36, 15, 12, 27, 12
Beta	18, 36, 36, 27, 21, 48, 36, 33, 21
Gamma	12, 18, 12, 21, 18, 12, 21, 12
Delta	30, 36, 30, 39, 36, 39, 36

a. Find the range, mean, median, and mode of the masses for each area of Xenon. Round to the nearest tenth.

b. The scientist realizes he made a mistake. One of the 39-kilogram creatures in the Delta area actually has a mass of 93 kilograms. Compute the new mean and median for the Delta area.

c. Compare the original Delta mean and median to the mean and median you computed in Part b. Which average changed more? Explain why this makes sense.

d. The scientist realized that one value for the Alpha area is missing from the table. He doesn't remember what the value is, but he remembers that the complete data set has a mean of 19. What value is missing?

7. Listed here are the number of unusual birds spotted by each member of a bird-watching club on a weekend excursion:

$$4 \quad 4 \quad 6 \quad 10 \quad 11 \quad 11 \quad 11 \quad 14 \quad 19$$

a. Find the mean and median of the data.

b. Add two values to the data set so the median remains the same but the mean increases. Give the new mean.

c. Add two values to the original data set so the median decreases but the mean remains the same. Give the new median.

d. Add two values to the original data set so both the mean and median stay the same.

8. Lonnie earns money by tutoring students in algebra. Here are the scores his students received on their most recent algebra tests:

$$0 \quad 60 \quad 78 \quad 79 \quad 90 \quad 95 \quad 95$$

a. Lonnie claims that the students he tutored received an average score of 95 on their tests. Which measure of center is he referring to? Do you think 95 is a good measure of what is typical about these tests scores? Explain.

b. Find the mean and median of the test scores.

c. Lonnie said the score of 0 should not be counted when finding the average because the student didn't even show up for the test. Delete the 0, and find the new mean and median.

d. Which of the averages you computed in Parts b and c do you think best represents the typical test scores for the students Lonnie tutored?

9. Elsa received the following scores on her first four math tests this semester: 81, 79, 90, 70. There is one more test left. Elsa's teacher has told her she may choose to use her mean or her median test score as her final grade, but she must decide *before* she takes the final test.

a. Calculate Elsa's current mean and median test scores.

b. If Elsa is not confident she will do well on the final test, should she choose the mean or the median? Explain.

c. If Elsa is confident she will do well on the final test, should she choose the mean or the median? Explain.

10. Sports Alano and Kate are swimming instructors at the local recreation center. One day, both instructors asked their students to swim as many laps as they could. The results are shown in the table.

Student	Instructor	Laps Swum
Lucinda	Kate	7
Jay	Alano	15
Guto	Kate	9
Deb	Alano	11
Ebony	Kate	6
Darius	Kate	7
Carlos	Alano	9
Carmen	Alano	4.5
Avi	Kate	8
Toku	Kate	7
Gil	Alano	4
Lana	Alano	4

a. Find the range, mean, median, and mode for all 12 swimmers.

b. Find the range, mean, median, and mode for each instructor's students.

c. Alano said his students were stronger swimmers. Kate argued that her students were stronger. Use your knowledge of statistics to write two arguments—one to support Alano's position and one to support Kate's position.

Just the facts

The first public swimming pool in the United States was built in Brookline, Massachusetts, in 1887. There are now more than 200,000 public swimming pools in the United States.

11. Divide your class's forearm data into two sets—one with the girls' measurements and one with the boys' measurements.

    a. Organize, analyze, and summarize each set of data. Include a line plot of each set.

    b. Write a few sentences comparing the data for the girls with the data for the boys.

    c. If SportSafe wants to make different armbands for boys and for girls, what size recommendations would you make based on your class data?

12. The table shows the number of left-handed and right-handed students in each homeroom class at Martin Middle School.

Room Number	Left-handed Students	Right-handed Students
101	3	27
102	4	26
103	2	28
104	5	25
105	2	29
106	6	23

The principal is buying new desks for a 30-student classroom. How many left-handed desks do you think she should buy? Use statistics to defend your answer.

Connect & Extend

13. Zeke's class made a line plot showing the number of people in each student's family.

```
                    X
                    X
                    X
                    X
                    X
                    X   X
                X   X   X
                X   X   X
                X   X   X
            X   X   X   X
            X   X   X   X           X
        ←---+---+---+---+---+---+---+---+---→
            1   2   3   4   5   6   7   8
```

    a. What is the total number of people in all the students' families?

    b. Zeke said, "The plot can't be right! My family has 8 people. If I have the largest family, why is the stack of X's over the 8 the shortest one on the graph?" Answer Zeke's question.

14. **Sports** The students in Consuela's gym class recorded how many times they could jump rope without missing. The results are shown in the table.

Team	Name	Gender	Jumps
Group 1	Jorge	male	1
	Felise	female	1
	Lana	female	5
	Sean	male	7
	Matt	male	8
	David	male	11
	Aaron	male	16
	Karen	female	26
	Brandon	male	26
	Enrique	male	26
	Emma	female	40
	Nicholas	male	50
	Shondra	female	95
	Selena	female	300
Group 2	Lucas	male	4
	Colin	male	4
	Olivia	female	4
	Trent	male	23
	Lauren	female	35
	Tyrone	male	48
	Francisca	female	68
	Elsa	female	83
	Shari	female	89
	Kiran	male	96
	Meela	female	110
	Consuela	female	138
	Tino	male	151

a. Group 1 claims they did better. Use what you have learned about statistics, along with any other information you think is useful, to write an argument Group 1 could use to support their claim.

b. Group 2 says they did better. Use what you have learned about statistics, along with any other information you think is useful, to write an argument Group 2 could use to support their claim.

15. Sports You can create a *back-to-back stem plot* to display and compare two data sets. This plot shows the batting averages for the 2000 New York Yankees and the 2000 New York Mets.*

Leaf values for the Yankees are given to the left of the stem, and leaf values for the Mets are given to the right of the stem. For example, the leaves 7 and 5 in the eleventh row indicate a Yankee batting average of .297 and a Met batting average of .295.

a. Write a few sentences comparing the two teams' batting averages. Be sure to discuss the distribution of values and the mean and mode.

Yankees and Mets 2000 Batting Averages

Yankees	Stem	Mets
	.19	9
	.20	
	.21	5
	.22	
2	.23	
7 7 3 0	.24	0
	.25	4 4 9
8 0	.26	0 6
	.27	
	.28	0
7	.29	5
0	.30	8
4	.31	
	.32	
3	.33	

Key: .21|5 = .215

b. These data give the total number of home runs hit by American League and National League teams in 2002. Make a back-to-back stem plot comparing the number of home runs for the two leagues.*

National League

Team	Home Runs
Arizona	165
Atlanta	164
Chicago	200
Cincinnati	169
Colorado	152
Florida	146
Houston	167
Los Angeles	155
Milwaukee	139
Montreal	162
New York	160
Philadelphia	165
Pittsburgh	142
San Diego	136
San Francisco	198
St. Louis	175

American League

Team	Home Runs
Anaheim	152
Baltimore	165
Boston	177
Chicago	217
Cleveland	192
Detroit	124
Kansas City	140
Minnesota	167
New York	223
Oakland	205
Seattle	152
Tampa Bay	133
Texas	230
Toronto	187

*Source: *World Almanac and Book of Facts 2003.*
Copyright © 2003 Primedia Reference Inc.

c. Write a few sentences comparing the numbers of home runs for the two leagues. Be sure to discuss the distribution of values and the mean and mode.

In y o u r **own words**

Describe what a line plot and a stem plot are. Tell how you can find the median and mode of a data set from each type of graph.

16. If you have a data set that includes only whole numbers, which measures of center—mode, median, or mean—will *definitely be* whole numbers? Which measures of center *may or may not be* whole numbers? Explain your answers.

17. Suppose a data set has a mean and a median that are equal.

    a. What must be true about the distribution of the data values?

    b. Suppose one value is added to the set, and the new mean is much greater than the median. What must be true about the new value? Explain.

    c. Now suppose you start with a new data set in which the mean and median are equal. You add one value to the set, and the new median is much greater than the mean. What must be true about the new value?

Just the facts

It's difficult to think of *The Wizard of Oz* without picturing Dorothy's ruby slippers. However, in the book on which the movie was based, Dorothy wore silver shoes, not ruby slippers.

18. Emelia asked her friends to rate three movies on a scale from 1 to 5, with 5 being terrific and 1 being terrible.

    a. For each movie, make a line plot of the friends' ratings.

    b. Compute the mean and median rating for each movie.

Movie Ratings

Friend	Star Wars	The Sound of Music	The Wizard of Oz
Adam	2	1	4
Ashley	5	1	5
Corey	2	4	3
Emelia	4	4	2
Eric	4	4	4
Hector	3	5	5
Ilene	2	5	3
Jay	3	2	1
Jose	5	1	5
Kareem	3	3	2
Karen	2	5	3
Lauren	4	5	4
Letonya	3	5	4
Lynn	1	2	2
Mai Lei	4	3	3
Maria	3	2	1
Michael	3	1	1
Peter	1	1	2

    c. Do you think reporting the means and medians is a good way to summarize the ratings for the three movies? Explain.

    d. How would you summarize these data if you wanted to emphasize the differences in the ratings among the three movies? Explain why you would summarize the data this way.

19. Life Science As their science project, Althea and Luke compared how two soil mixtures affected the growth of morning glories and zinnias. They planted some seeds of each type in a mixture of peat moss and sand and some in a mixture of top soil, compost material, and sand. They put all the plants by the same window and watered them at the same time with the same amount of water. After 20 days, they measured the height of each plant.

Peat Moss and Sand		Topsoil, Compost, and Sand	
Morning Glory Height (mm)	Zinnia Height (mm)	Morning Glory Height (mm)	Zinnia Height (mm)
145	47	157	27
156	49	155	42
139	52	167	50
142	42	149	54
154	45	87	51
0	43	127	38
151	47	116	36
147	50	145	0
143	56	4	39
145	12	15	32
168	47	143	35
129	52	105	23
148	5	132	33

a. Each column contains data for one type of plant in one type of soil mixture. Althea and Luke decided to ignore the outliers in each column, reasoning that these seeds probably wouldn't grow well in any type of soil. For each column, tell which values you think are outliers.

b. Compute the mean, median, and mode for each column of data. Ignore the outliers.

c. For each column of data, make a display that shows how the values are distributed. Do not include outliers in your display.

d. Do you think the different types of soil affected the growth of the plants? Use your answers from Parts b and c, along with any other information you think is useful, to support your answer.

Mixed Review

Fill in the blanks.

20. 25% of ____ = 14

21. ____ % of 75 = 25

22. 80% of 200 = ____

23. 125% of ____ = 300

24. ____ % of 280 = 238

25. 1% of 30 = ____

Fill in each ● with $<$, $>$, or $=$ to make a true statement.

26. $\frac{6}{5}$ ● 1.2

27. 0.37 ● $\frac{7}{20}$

28. $\frac{13}{18}$ ● $\frac{7}{10}$

29. 1.5 ● $\frac{25}{19}$

30. 0.0375 ● $\frac{3}{80}$

31. $\frac{23}{24}$ ● $\frac{24}{25}$

Geometry In Exercises 32–34, tell whether the given lengths could be side lengths of a triangle.

32. 3 cm, 3 cm, 3 cm

33. 3 in., 4 in., 7 in.

34. 5 ft, 12 ft, 14 ft

35. Jahmal left his apartment and started walking toward school. After walking a couple of blocks, he realized he had forgotten his homework. He walked back home to get it, and then started walking toward school again. He stopped to meet his friend Miguel and had to wait a few minutes while Miguel finished his breakfast. The two boys thought they might be late, so they ran from Miguel's house to the school.

Sketch a graph showing how Jahmal's distance from home might have changed from the time he first left his apartment.

Find the next three terms in each sequence.

36. 99, 98, 96, 93, 89, 84, . . .

37. 729, 243, 81, 27, 9, . . .

38. 1, 1, 2, 6, 24, 120, 720, . . .

39. ❀, ▲, ❀, ▲, ▲, ❀, ▲, ▲, ▲, ❀, ▲, ▲, . . .

6.3 Collecting and Analyzing Data

The editors of *All about Kids!* magazine are researching an article about the activities middle school students participate in. They would like the article to address these questions:

- What activities do middle school students participate in after school and on weekends?

- What percent of students participate in each activity?

- How many hours a week do students typically spend on each activity?

- What are students' favorite activities?

- Do boys and girls like different activities?

- Do students tend to spend the most time on the activities they like best?

The editors have hired Data Inc. to help with the article. They would like you to answer the above questions for the students in your class. They have suggested using the form below to collect your class data.

Are you male or female?

Please fill out this table, listing the time you spend each week in each activity, after school and on weekends.

What is your favorite activity?

Activity	Time Spent Each Week
Doing homework	
Spending time with friends	
Playing sports	
Reading books	
Reading newspapers	
Using a computer	
Taking care of pets	
Watching movies or videos	
Talking on the phone	
Watching TV	
Listening to music	
Playing video or board games	
Shopping	

Think & Discuss

Look over the list of activities. Decide with your class whether to add or delete any activities.

Consider each question the editors want to answer, and think about whether the survey form will collect the information needed to answer it. Decide as a class whether anything should be added to the form.

Each student in your class should fill out a survey form. All the data from the class will be combined later.

Investigation Planning Your Analysis

In this investigation, you will think about what types of statistics and graphs might be useful for reporting the results of the survey. You won't do your analysis until the next investigation, but carefully planning your strategy now will make your analysis much easier.

Problem Set A

Problems 1–6 list the six questions asked by the magazine editors. For each question, complete Parts a and b.

> **a.** Tell which collected data you will need to answer the question.
>
> **b.** Describe a procedure you could use to answer the question. Be sure to indicate any statistical measures—mean, median, mode, or range—you will need to find or computations you will need to do.

1. What activities do middle school students participate in after school and on weekends?

2. What percent of students participate in each activity?

3. How many hours a week do students typically spend on each activity?

4. What are students' favorite activities?

5. Do boys and girls like different activities?

6. Do students tend to spend the most time on the activities they like best?

Save your answers from this problem set for Investigation 2.

Magazine articles often use graphs to present data. In Problem Set B, you will think about what types of graphs might be useful to include in the magazine article.

You are familiar with several types of graphs: line plots, line graphs, stem-and-leaf plots, bar graphs, histograms, pictographs, and circle graphs. You might also consider one of the special types of bar graphs described in the Example.

EXAMPLE

A *double bar graph* compares data for two groups. For example, this double bar graph compares the favorite primary colors of boys and girls in one middle school class:

Favorite Primary Colors

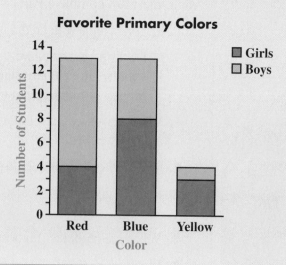

A *stacked bar graph* shows how the data represented by each bar is divided into two or more groups. This graph shows how the number of children who chose each color is divided between boys and girls:

Favorite Primary Colors

Problem Set B

For each question, decide whether it would be useful to include a graph with the answer to the question. If so, describe the graph you would use. Include at least one double bar graph or stacked bar graph.

1. What activities do middle school students participate in after school and on weekends?

2. What percent of students participate in each activity?

3. How many hours a week do students typically spend on each activity?

4. What are students' favorite activities?

5. Do boys and girls like different activities?

6. Do students tend to spend the most time on the activities they like best?

Save your answers from this problem set for Investigation 2.

Share & Summarize

Think of at least one more question you think would be interesting to address in the magazine article. Describe the data you would need to collect to answer the question, and tell what statistics and graphs you would include in your answer.

Investigation ▶2 Carrying Out Your Analysis

You have collected data about the activities the students in your class participate in. Now you will analyze the class data and use your results to answer the questions posed by the magazine editors.

M A T E R I A L S

- results of class survey
- answers for Problem Sets A and B

Problem Set C

The editors' six questions appear in Problems 1–6. Work with your group to analyze the data and answer each question. Include the following information for each problem:

- The results of your computations and the measures you found

- A few sentences answering the question, including statistical measures that support your answers

- A graph, if appropriate, to help illustrate your answer

You can use your answers to Problem Sets A and B as a guide, but you may change your mind about what statistics and graphs to include.

As you work, you may want to create tables to organize your data and calculations. Here is an example you might find useful:

Activity	Number Who Participate			Percent Who Participate			Mean Time Spent			Median Time Spent		
	Boys	Girls	All	Boys	Girls	All	Boys	Girls	All	Boys	Girls	All
Doing homework												
Spending time with friends												
Playing sports												
Reading books												
Reading newspapers												
Using a computer												
Taking care of pets												
Watching movies or videos												
Talking on the phone												
Watching TV												
Listening to music												
Playing video or board games												
Shopping												

1. What activities do middle school students participate in after school and on weekends?

2. What percent of students participate in each activity?

3. How many hours a week do students typically spend on each activity?

4. What are students' favorite activities?

5. Do boys and girls like different activities?

6. Do students tend to spend the most time on the activities they like best?

Share & Summarize

1. Write a few sentences summarizing your group's work on Problem Set C. Discuss how you divided the work among group members, and how you organized the data to make it easier to answer the questions.

2. Do you think a nationwide survey of middle school students would give results similar to your class results? Explain why or why not.

Lab Investigation ▶ Statistics and Spreadsheets

Analyzing data can be time-consuming, especially if you are working with a large data set. In this investigation, you will see how using a spreadsheet program can make analyzing a data set easier.

MATERIALS

computer with spreadsheet software (1 per group)

The Situation

Conor read an article about how difficult it is to hold a book at arm's length with one hand for very long. He tried it himself and was amazed how hard it was.

Conor hypothesized that students who work out in the school weight room would be able hold a book longer than students who don't. To test his hypothesis, he recruited 20 students: 5 girls and 5 boys who work out and 5 girls and 5 boys who don't. He had each student hold a 7-pound book at arm's length, using the weaker arm, for as long as he or she could. He summarized the data in a table.

Name	Work Out?	Time (seconds)	Name	Work Out?	Time (seconds)
Abby	yes	44	Ken	yes	18
Barbara	yes	40	Kiran	no	34
Ben	no	30	Lee	yes	55
Cassie	yes	52	Liz	no	25
Chris	yes	49	Michele	no	30
Eileen	yes	58	Roland	yes	55
Grason	no	20	Sara	no	29
Jackie	no	38	Sydney	yes	38
John	no	36	Tia	no	20
Jon	yes	45	Tomas	no	28

1. What do you think Conor should do to determine whether students who work out did better than those who don't?

Set Up the Spreadsheet

Conor set up a spreadsheet to help him analyze the data. He decided not to include the students' names. Here are the first few rows of his spreadsheet:

	A	B	C	D
1	Work Out?	Time (s)		
2	Y	44		
3	Y	40		
4	N	30		
5	Y	52		

Set up your spreadsheet like Conor's. Enter the column heads shown, and then enter the data.

The results will be easier to analyze if you divide the data values into two groups—data for students who work out and data for students who don't. You can do this using the Sort command.

- Highlight all the cells containing data. (Don't select the column labels.)

- Choose the Sort command.

- Indicate that you want the data sorted according to the values in Column A (the Y or N values) and that you want the values sorted in ascending order. (*Ascending* sorts from least to greatest or, in this case, from A to Z.)

2. Describe what happened to the data.

3. Choose Sort again, but this time choose to sort the data values in descending order. Describe what happened to the data.

Find the Minimum and Maximum

You can use a spreadsheet to find the minimum and maximum values in a data set. For this data set, you can probably find these values fairly quickly by looking at the table. However, when you have a large data set, it is much more efficient to have a spreadsheet do it for you.

You can tell the spreadsheet to display the minimum value in any empty cell. In this example, you will put the result in Cell C2. First, enter the column head "Min. Time" in Cell C1. In Cell C2, type the following:

$$=MIN(B2:B21)$$

The = sign tells the spreadsheet that the entry is a formula to evaluate. MIN means "find the minimum value." B2:B21 tells the spreadsheet to search for the minimum value in Column B from Cell B2 to Cell B21.

	A	B	C	D
1	Work Out?	Time (s)	Min. Time	
2	Y	44	=MIN(B2:B21)	

4. After you enter the formula and press Return, what value appears in Cell C2?

Enter the column head "Max. Time" in Cell D1. In Cell D2, use the MAX command to calculate and display the maximum time.

5. What is the maximum time for these data?

6. Use your spreadsheet to find the minimum and maximum values for each group—for the students who work out and for those who don't.

Find the Median and Mean

For most spreadsheets, the command for computing the median is MEDIAN. So, to compute the median of all the time values, you would enter =MEDIAN(B2:B21).

7. Use your spreadsheet to find the median of all the time values.

8. Now find the median time value for each group.

For most spreadsheets, the command for computing the mean is AVERAGE.

9. Use your spreadsheet to find the mean of all the time values.

10. Now find the mean time value for each group.

Back to the Situation

11. Do the data Conor collected support his hypothesis? Use what you know about statistics to justify your answer.

12. Does Conor's experiment *prove* his hypothesis? Explain.

What Did You Learn?

13. Have each student in your group hold a heavy book at arm's length for as long as he or she can. Record the times for your group in a class table. Enter the times for your class into a spreadsheet, and calculate the minimum value, maximum value, mean, and median of the data.

On Your Own Exercises

Practice & Apply

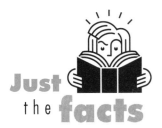

Just the facts

The average college student spends three hours per day on-line.

In Exercises 1 and 2, use this information:

The editors of *All about Kids!* would like to publish an article about teenage Internet users. Here are the questions they would like to answer:

Question 1: How much time do teen Internet users typically spend on the Internet each week?

Question 2: How much time do they typically spend on various Internet activities?

Question 3: What are these teenagers' favorite Internet activities?

The magazine editors have collected data from 15 students who are regular Internet users. The green entries indicate the students' favorite activities.

Weekly Time Spent on the Net (minutes)

Student Initials	Chatting in a Chat Room	Playing Games	Doing Homework	Surfing the Web	E-mail
AB	90	75	80	90	12
BT	75	150	0	150	15
CP	0	75	60	150	5
CT	0	240	0	0	0
GO	120	90	90	60	3
KQ	135	60	40	80	15
LM	75	160	30	45	6
MC	15	0	35	60	15
MH	80	180	30	90	6
NM	90	90	45	90	15
PD	100	150	60	90	0
RL	100	90	45	90	0
SK	90	135	60	240	10
SM	60	135	40	60	22
YS	120	30	60	45	15

1. Complete Parts a and b for Questions 1, 2, and 3 above.

   a. Describe any statistical measures you will need to find or computations you will need to do to answer the question.

   b. Describe a graph that would be appropriate to include with the answer to the question.

2. Refer to the information on page 399. Analyze the given data, and use your results to answer the three questions posed by the magazine editors. Provide the following information for each question:

a. The results of your computations and the measures you found

b. One or more sentences answering the question

c. A graph to help illustrate your answer

Connect & Extend

3. Social Studies This double bar graph shows the number of licensed drivers per 1,000 people and the number of registered vehicles per 1,000 people in seven states:

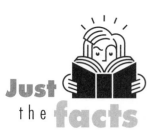

Just the facts

Automobiles were commercially available in the United States beginning in 1896. In 1903, Missouri and Massachusetts adopted the first driver's license laws. In 1908, Rhode Island became the first state to require a driver's test.

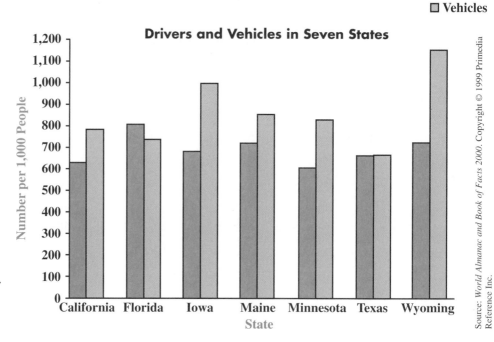

■ Drivers
□ Vehicles

Drivers and Vehicles in Seven States

Number per 1,000 People

States: California, Florida, Iowa, Maine, Minnesota, Texas, Wyoming

State

Source: *World Almanac and Book of Facts 2000.* Copyright © 1999 Primedia Reference Inc.

a. Which state has the greatest difference between the number of registered vehicles and the number of licensed drivers?

b. Which state has about one registered vehicle per licensed driver?

c. Which state has less than one registered vehicle per licensed driver?

d. Which two states have about the same number of registered vehicles?

4. Social Studies This stacked bar graph shows the number of bachelor's degrees awarded to men and women in the United States in various years.

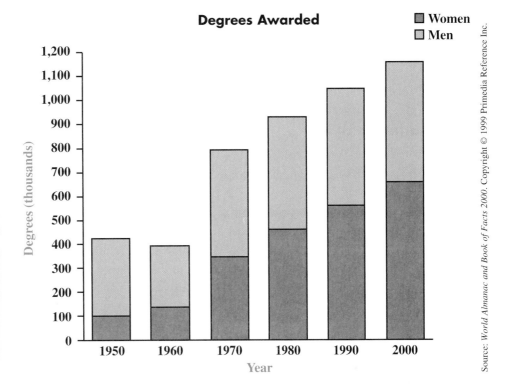

Degrees Awarded

■ Women
□ Men

Degrees (thousands)

Year

Source: *World Almanac and Book of Facts 2000.* Copyright © 1999 Primedia Reference Inc.

a. Describe how the total number of bachelor's degrees awarded changed over the years shown in the graph.

b. Describe how the total number of bachelor's degrees awarded to women has changed over the years shown in the graph.

c. In 1970, about how many bachelor's degrees were received by men? About how many were received by women?

d. In 1950, about what fraction of bachelor's degrees were received by women? In 2000, about what fraction were received by women?

Mixed Review

Number Sense Order each set of numbers from least to greatest.

5. $4, {}^-3, 0, {}^-1.5, 0.5, 2, {}^-4$

6. $0.108, 0.08, 0.1, 0.018, 0.081, 0.801$

7. $\frac{3}{4}, \frac{8}{11}, \frac{21}{33}, \frac{15}{21}, \frac{209}{330}, \frac{41}{52}$

8. $\frac{4}{3}, {}^-\frac{3}{4}, {}^-0.999, {}^-1, 0.009, \frac{2}{3}, {}^-\frac{1}{3}$

List all the factors of each number.

9. 115 **10.** 92 **11.** 71 **12.** 90

Chapter Summary

In this chapter, you learned methods for organizing, analyzing, and displaying data. First you looked at *bar graphs* and *histograms*. You saw that bar graphs are used to show information about individuals or groups, while histograms are used to show the number or percent of data values that fall into various intervals.

You then learned about two more types of displays: *line plots* and *stem-and-leaf plots*. Both are helpful for showing a distribution of data values. Line plots show a stack of X's for each value. Stem plots group values with the same "stem" and are useful when the values are spread out or when there are many different values.

You were also introduced to some statistics used to summarize a data set. The *range* of a data set is the difference between the minimum and maximum values. The *mode* is the value that occurs most often. The *median* is the middle value. The *mean* is the value arrived at by dividing the sum of the data values equally among the data items. The mean, median, and mode are all measures of what is typical, or average, about a set of data.

Finally, you conducted a data investigation in which you had to decide what information, statistics, and graphs to use to answer a series of questions.

A snow-survey crew gathering data in Nevada

impactmath.com/chapter_test

Strategies and Applications

The questions in this section will help you review and apply the important ideas and strategies developed in this chapter.

Interpreting and creating bar graphs and histograms

1. The bar graph shows the number of cassettes and CDs sold in the years from 1992 to 2000.

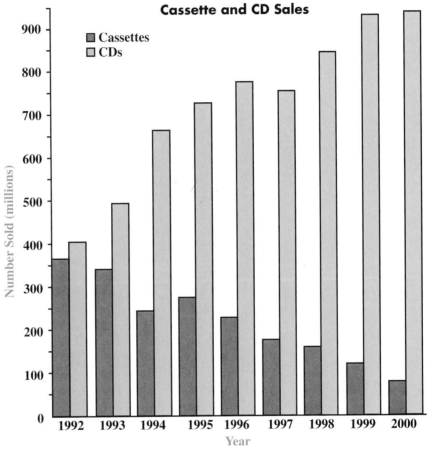

Cassette and CD Sales

Number Sold (millions)

Year

Cassettes / CDs

Source: World Almanac and Book of Facts 2003. Copyright © 2003 Primedia Reference Inc.

  a. Describe what the graph indicates about the change in cassette sales over the years.

  b. Describe what the graph indicates about the change in CD sales over the years.

  c. Describe how the difference between CD sales and cassette sales has changed over the years.

2. The American Film Institute released a list of the best American movies of all time. Below are the years in which the top 50 movies were released.

Movie Ranking	Year Released	Movie Ranking	Year Released	Movie Ranking	Year Released
1	1941	18	1960	35	1934
2	1942	19	1974	36	1969
3	1972	20	1975	37	1946
4	1939	21	1940	38	1944
5	1962	22	1968	39	1965
6	1939	23	1941	40	1959
7	1967	24	1980	41	1961
8	1954	25	1982	42	1954
9	1993	26	1964	43	1933
10	1952	27	1967	44	1915
11	1946	28	1979	45	1951
12	1950	29	1939	46	1971
13	1957	30	1948	47	1976
14	1959	31	1977	48	1975
15	1977	32	1974	49	1937
16	1950	33	1952	50	1969
17	1951	34	1962		

a. Copy and complete the frequency table to show the number of movies released during each decade.

Decade	Frequency
1910–1919	
1920–1929	
1930–1939	
1940–1949	
1950–1959	
1960–1969	
1970–1979	
1980–1989	
1990–1999	

b. Use your frequency table to help you make a histogram showing the number of values in each decade.

c. What does the shape of your histogram reveal about the distribution of years in which the best 50 movies were released?

Interpreting and creating line plots and stem plots

3. The table shows the number of cookies sold by the girls in a particular scouting troop.

Initials	Boxes Sold
JJ	12
PK	80
TT	29
CR	23
HT	90
SI	19
FM	87
VY	72

Initials	Boxes Sold
PN	21
KT	99
FV	84
DY	25
HH	33
SE	85
MI	79
CK	21

a. Do you think a line plot or a stem plot would be better for displaying the "Boxes Sold" values? Explain why.

b. Make the plot you suggested in Part a.

c. Use your plot to find the mode, median, and range of the "Boxes Sold" values.

d. Describe the shape of the plot. Explain what the shape tells you about how the data are distributed.

4. How can you find the mode of a data set by looking at a line plot? How can you find the mode by looking at a stem plot?

Finding and interpreting the mode, median, and mean

5. Marcus asked the students in his class how many first cousins they have. He summarized the data by reporting three averages.

a. Marcus said the mode number of cousins is 6. What does this average tell you about the data?

b. He reported that the median number of cousins is 4. What does this average tell you about the data?

c. Marcus found that the mean number of cousins is 5. What does this average tell you about the data?

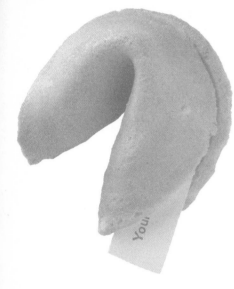

6. Angelia measured the mass of each of the fortune cookies her family brought home from dinner at a Chinese restaurant. Here are her measurements, in grams:

11   13   13   14   15   17   19   20   22

a. Find the mode, median, and mean of these data.

b. Add two values to the data set so the mean decreases but the median and mode remain the same.

c. Start with the original data set. Add two values to the set so the mean remains the same but the median changes.

d. Start with the original data set. Add two values to the set so the median, mean, and mode all increase.

7. You learned that an *outlier* is a value that is much greater or much less than most of the other values in a data set. Which measure of center—the mean or the median—is more influenced by outliers? Explain your answer, and give an example to illustrate.

Choosing the best average for a given situation

8. Hannah's father wants to start giving Hannah a weekly allowance, but he isn't sure how much to give her. Hannah asked 13 of her friends how much weekly allowance they get. Here are the results:

$0 $2.50 $3 $2.50 $5 $15

$7.50 $0 $10 $0 $10 $75 $5

a. Find the mean, median, and mode of the allowance data.

b. Which of the averages you found do you think best represents a typical value in this data set? Give reasons for your answer.

c. The $75 allowance is an outlier, since it is much greater than the other data values. Remove this value and find the new mean, median, and mode.

d. Hannah thinks telling her father the average allowance of her friends will help him decide how much to give her. Which average do you think she should report? Consider the averages for the data set with and without the outlier. Give reasons for your choice.

Demonstrating Skills

Find the range, mean, median, and mode of each set of numbers.

9. 9, 3, 7, 3, 3, 2, 5, 5, 7, 7

10. 144, 120, 196, 95, 180, 5, 136, 175, 114

11. Find the range, median, and mode of the data in this stem plot:

Stem	Leaf
50	0 3 7 7
51	0 5
52	
53	3 5 7
54	
55	3 9
56	0 2 2 2 4
57	3 5

Key: 50|3=503

12. Make a line plot of the data below.

11 9 13 9 11 10 12 16

13 8 13 11 8 13 15 16 13 11

Variables and Rules

Under the Sea A *formula* is an algebraic "recipe" for finding the value of one variable based on the values of one or more other variables. Letters are usually used to represent the variables. For example, a scuba diver can use this formula to figure out how long she can stay under water.

$$T = 120V \div d$$

In the formula, T represents the time in minutes the diver can stay under water, V represents the volume of air in the diver's tank in cubic meters, and d represents the water's depth in meters.

Think About It Suppose a diver has 2 cubic meters of compressed air in her tank and is 4 meters underwater. How long can the diver stay under water?

Family Letter

Dear Student and Family Members,

Earlier in the year, you worked with many patterns, finding rules to describe them. Here's a pattern from Chapter 1.

| Term 1 | Term 2 | Term 3 | Term 4 |

Each new term has one more square added on, made up of three more toothpicks. A rule that relates the term number to the number of toothpicks is:

$$\text{number of toothpicks} = 3 \cdot (\text{term number}) + 1.$$

Can you use the rule to find how many toothpicks would be in Term 5?

In this chapter, you will learn about shorter and more useful ways of writing rules to describe patterns—an important and fundamental part of algebra. Quantities that change are called *variables*. For example, the term number above is a variable because it has different values at different terms. If t represents the number of toothpicks, and n represents the term number, then the algebraic rule is $t = 3 \cdot n + 1$.

You will also look at rules that apply to everyday situations. For example, suppose a phone company charges $3.00 per month plus $0.07 per minute for each phone call. One way to understand the situation is by making a table.

Month	June	July	August	September	October
Minutes	120	90	95	150	80
Cost	$11.40	$9.30	$9.65	$13.50	?

You can also write a rule to calculate the phone bill for any month using m to represent the number of minutes and c to represent the cost. Using the rule $c = \$0.07 \cdot m + \$3.00$, can you find the cost of October's phone bill?

What can you do at home?

During the next few weeks, be on the lookout for rules in real life, such as finding the cost of m movie tickets or c CDs. Ask your student to write rules using variables for each situation.

Patterns and Variables

In Chapter 1 you explored several sequences. A *sequence* is an ordered list of items such as numbers, symbols, or figures. Here are two sequences:

4, 7, 10, 13, 16, . . .

Each item in a sequence is called a *term.* So, in the second sequence above, Term 1 is a diamond, Term 2 is a heart, and Term 3 is a diamond.

Explore

You looked at this sequence of toothpick figures in Chapter 1:

□	□ □	□ □ □	□ □ □ □
Term 1	**Term 2**	**Term 3**	**Term 4**

Write a rule that describes how to create each term of this sequence from the previous term.

In this sequence, there are many *variables*—things that change, or vary. For example, the *number of squares* increases by 1 with each term. You can write a number sequence to show the values of this variable:

1, 2, 3, 4, . . .

Look for a pattern in the way each variable given below changes from term to term. First describe the pattern in words. Then write a number sequence to show the values of the variable.

• the number of toothpicks in a term

• the number of toothpicks along the top of each term

Look for at least three more variables in the toothpick sequence above. Describe how each variable changes from term to term, and give the first four terms of the matching number sequence.

Try to find a variable in the toothpick sequence that changes according to each of these number sequences:

2, 3, 4, 5, . . .   2, 4, 6, 8, . . .   1, 3, 6, 10, . . .

Just the facts

A numeric sequence created by adding the same number to each term to get the next term is called an *arithmetic sequence.* Here are two examples:

4, 9, 14, 19, 24, . . .

$\frac{3}{4}$, $1\frac{1}{2}$, $2\frac{1}{4}$, 3, . . .

Investigation ▶ 1 Sequences, Rules, and Variables

In the Explore, you looked at variables in a sequence of toothpicks, including the number of squares, the number of toothpicks, and the number of toothpicks along the top of each term. The term number is also a variable.

Making a table is a good way to compare the values of variables. This table shows the term number and the number of toothpicks for the first five terms:

Term Number	1	2	3	4	5
Number of Toothpicks	4	7	10	13	16

You can sometimes write a rule to show how two variables are related. In Chapter 1, you wrote a rule relating the number of toothpicks to the term number. Here is one possible rule:

$$\text{number of toothpicks} = 3 \cdot \text{term number} + 1$$

In the next problem set, you will make tables and find rules for some of the other variables in the toothpick sequence.

Just the facts

A numeric sequence created by multiplying each term by the same number to get the next term is called a geometric sequence. Here are two geometric sequences:

1, 4, 16, 64, 256, . . .

72, 36, 18, 9, 4. 5, . . .

Problem Set A

1. In Parts a–d, make a table showing the values of the given variable for the first four terms of the sequence. The first table has been started for you.

Term 1   Term 2   Term 3   Term 4

a. number of squares

Term Number	1	2	3	4
Number of Squares	1	2		

b. number of vertical toothpicks

c. number of horizontal toothpicks

d. number of rectangles (Hint: In each term, count the squares, the rectangles made from two squares, the rectangles made from three squares, and so on.)

2. For each table from Problem 1, try to find a relationship between the term number and the other variable. Then write a rule to describe the relationship.

    a. number of squares =

    b. number of vertical toothpicks =

    c. number of horizontal toothpicks =

    d. Challenge number of rectangles =

3. To check your rules, you can test them for a particular term number. Although this won't tell you for certain that a rule is correct, it's a good way to find mistakes. For each part of Problem 2, use your rule to predict the value of the variable for Term 5. Then draw Term 5 and check your predictions.

4. Explain how you know that the rules you wrote in Parts a–c of Problem 2 will work for every term.

In algebra, letters are often used to represent variables. For example, consider this rule:

$$\text{number of toothpicks} = 3 \cdot \text{term number} + 1$$

If you use the letter *n* to represent the term number and the letter *t* to represent the number of toothpicks, you can write the rule like this:

$$t = 3 \cdot n + 1$$

This rule is much shorter and easier to write than the original rule.

When a number is multiplied by a variable, the multiplication symbol is often left out. So, you can write the rule above in an even shorter form:

$$t = 3n + 1$$

You can use any letter to represent a variable, as long as you say what the letter represents. For example, you could let *w* represent the term number and *z* represent the number of toothpicks, and write the rule as $z = 3w + 1$.

A single rule can usually be written in many ways. Here are six ways to write the rule for the number of toothpicks in a term:

$$t = n \cdot 3 + 1 \qquad t = (n * 3) + 1 \qquad t = 1 + 3n$$

$$t = 1 + (3 \cdot n) \qquad t = 1 + n \cdot 3 \qquad t = 3 \times n + 1$$

None of the rules above need parentheses, because order of operations tells you to multiply before you add. However, it is not incorrect to include them. Some rules do need parentheses, so be careful when you write your rules.

Remember

It is not enough to show that a rule works in a few specific cases. Try to explain why it works based on how the terms are built.

Remember

Order of operations:

• Evaluate expressions inside parentheses and above and below fraction bars.

• Do multiplications and divisions from left to right.

• Do additions and subtractions from left to right.

Problem Set B

1. Rewrite your rules from Problem Set A in a shorter form by using *n* for the term number and a different letter for the other variable. Make sure to state what variable each letter represents.

2. Consider this toothpick sequence:

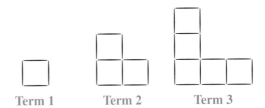

Term 1   Term 2   Term 3

   a. Choose a variable other than the term number.

   b. Create a table showing the value of your variable for each term.

Term Number, *n*	1	2	3	4
Your Variable				

   c. Try to find a rule that connects the term number and your variable. Write the rule as simply as you can, using *n* to represent the term number and a different letter to represent your variable.

3. Complete Parts a–c of Problem 2 for this sequence of toothpicks:

   Term 1   Term 2   Term 3   Term 4

4. Complete Parts a–c of Problem 2 for this sequence of dots:

   Term 1   Term 2   Term 3   Term 4

5. Consider the rule $t = 4 \cdot n + 2$, where *n* represents the term number and *t* represents the number of toothpicks in a sequence.

   a. Write the first four numbers in the sequence.

   b. Draw a toothpick sequence that fits the rule.

6. Consider the rule $d = 3 \cdot (n + 1)$, where *n* represents the term number and *d* represents the number of dots in a sequence.

   a. Write the first four numbers in the sequence.

   b. Draw a dot sequence that fits the rule.

Remember

When a number is multiplied by a quantity in parentheses, the multiplication symbol is often left out. So, $3 \cdot (n + 1)$ can be written $3(n + 1)$.

1. Draw a toothpick or dot sequence. Make sure your sequence changes in a predictable way.

2. Name two variables in your sequence.

3. For each variable you named, try to write a rule relating the term number to the variable. Use letters to represent the variables, and tell what each letter represents.

Investigation 2 ▶ Are These Rules the Same?

Sometimes two people can look at the same pattern and write rules that look very different. This may have happened when you and your classmates wrote rules in the last investigation.

Consider this toothpick sequence:

Term 1   Term 2   Term 3   Term 4

Remember

The multiplication symbol is often left out when a number is multiplied by a variable. So, 2*n* is the same as 2 · *n*.

Rosita and Conor wrote rules for the number of toothpicks in each term. Both students used *n* to represent the term number and *t* to represent the number of toothpicks.

Rosita's rule: $t = 3 + 2 \cdot (n - 1)$ Conor's rule: $t = 1 + 2n$

Think & Discuss

Use the two rules to find the number of toothpicks in Term 10. Check your results by drawing Term 10 and counting toothpicks.

Show that both rules give the same result for Term 20 and for Term 100.

Do you think the rules will give the same result for every term?

One way to show that the two rules will give the same result for every term is to explain why both rules must work for any term in the sequence.

Rosita explains why her rule works.

Conor explains why his rule works.

Rosita's and Conor's rules both correctly describe the toothpick sequence, so they *will* give the same result for every term. Two rules that look different, but that describe the same relationship, are said to be *equivalent*.

Problem Set C

1. Consider this sequence:

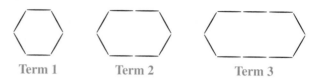

Term 1   Term 2   Term 3

Althea and Marcus wrote equivalent rules for this sequence. Both students used n to represent the term number and t to represent the number of toothpicks.

Althea's rule: $t = 2 \cdot n + 4$ Marcus' rule: $t = 2 \cdot (n + 2)$

Althea used diagrams to explain why her rule is correct.

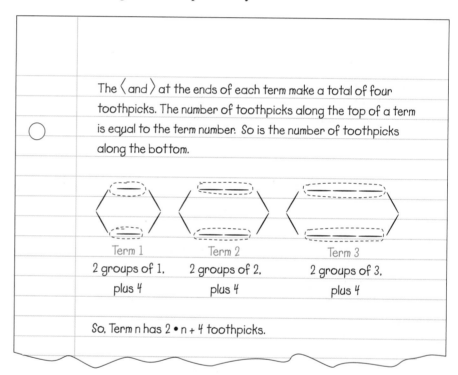

The $\langle$ and $\rangle$ at the ends of each term make a total of four toothpicks. The number of toothpicks along the top of a term is equal to the term number. So is the number of toothpicks along the bottom.

Term 1   Term 2   Term 3

2 groups of 1,   2 groups of 2,   2 groups of 3,
plus 4   plus 4   plus 4

So, Term n has 2 • n + 4 toothpicks.

Use diagrams to help explain why Marcus' rule is correct.

2. Consider this sequence:

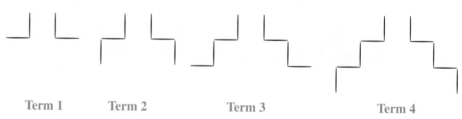

Term 1   Term 2   Term 3   Term 4

Caroline and Jing wrote equivalent rules for this sequence. Both girls used n to represent the term number and t to represent the number of toothpicks.

Caroline's rule: $t = 2n + 2$ Jing's rule: $t = 2 \cdot (n + 1)$

The Mayan temple of Kukulcan on the Yucatan Peninsula, Mexico

a. Copy and complete the table to show that both rules work for the first five terms of the sequence.

Term Number, n	1	2	3	4	5
Number of Toothpicks, t					
$2n + 2$					
$2 \cdot (n + 1)$					

b. Use words and diagrams to explain why Caroline's rule is correct.

c. Use words and diagrams to explain why Jing's rule is correct.

3. Consider this sequence:

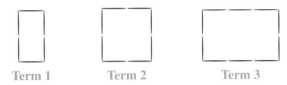

Term 1   Term 2   Term 3

a. Write two equivalent rules for the number of toothpicks in each term.

b. Use words and diagrams to explain why each rule is correct.

4. Consider this sequence:

a. Write two equivalent rules for the number of dots in each term.

Term 1   Term 2   Term 3

b. Use words and diagrams to explain why each rule is correct.

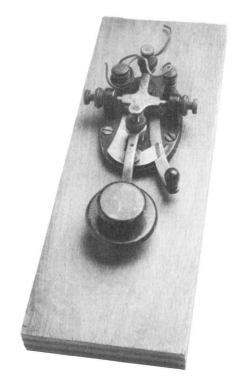

5. Conor, Jahmal, and Rosita each wrote a rule for this sequence:

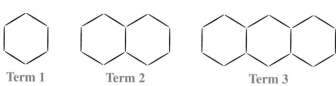

Term 1   Term 2   Term 3

Conor's rule: $t = 5n + 1$, where t is the number of toothpicks and n is the term number

Jahmal's rule: $t = 3 \cdot n + 3$, where t is the number of toothpicks and n is the term number

Rosita's rule: $t = 6h - (h - 1)$, where t is the number of toothpicks and h is the number of hexagons

Determine whether each rule correctly describes the sequence. If it does, explain how you know it is correct. If it does not, explain why it isn't correct.

6. Consider strips of T-shapes like this one, in which a strip can have any number of T-shapes.

a. Find a rule that connects the number of toothpicks in a strip to the number of T-shapes.

b. Explain why your rule is correct. Use diagrams if they help you to explain.

Share & Summarize

1. How can you show that two rules for a sequence are equivalent?

2. Suppose two different rules give the same number of dots for Term 4 of a sequence. Can you conclude that the two rules are equivalent? Give an example to support your answer.

Investigation 3 ▶ What's My Rule?

In Chapter 1, you played the game *What's My Rule?* Here's how you play:

- One player, the rule-maker, thinks of a secret rule for calculating an output number from a given input number. For example:

  To find the output, add 3 to the input and multiply by 4.

- The other players take turns giving the rule-maker input numbers. For each input, the rule-maker calculates the output and says the result out loud.

- By comparing each input to its output, the players try to guess the secret rule. The first player to guess the rule correctly wins.

In *What's My Rule?* the input and output are variables. In this investigation, you will play *What's My Rule?* using letters to represent these variables. For example, if you let *i* represent the input and *o* represent the output, you can write the rule above as

$$o = (i + 3) \cdot 4$$

Remember

The multiplication symbol is often left out when a number is multiplied by a variable. So, 3b is the same as $3 \cdot b$.

Think & Discuss

Rosita, Jahmal, and Althea are playing *What's My Rule?* Rosita's secret rule is $a = 3b + 4$, where *a* is the output and *b* is the input.

- Jahmal guesses that the rule is $a = 4 + 3b$, where *a* is the output and *b* is the input.

- Althea guesses that the rule is $x = 3y + 4$, where *x* is the output and *y* is the input.

Rosita is not sure whether the rules Jahmal and Althea wrote are the same as her secret rule. Tell whether each rule is correct, and explain how you know.

Problem Set D

Play *What's My Rule?* with your group, using these added rules:

- The rule-maker should write the secret rule with symbols, using letters for the variables.

- The other players should make a table to keep track of the inputs and outputs.

- When a player guesses the rule, he or she should write it with symbols, using letters for the variables.

Take turns being the rule-maker so everyone has a chance. As you play, you may have to decide whether a guessed rule is equivalent to the secret rule even though it looks different.

After your group has played several games, write a paragraph describing what you learned while playing. In your paragraph, you might discuss the following:

- Strategies you used to help you guess the rule

- A description of what makes a rule easy to guess and what makes a rule difficult to guess

- Strategies you used to decide whether two rules were equivalent even when they looked different

Problem Set E

These tables were made during games of *What's My Rule?* Two rules are given for each table. Determine whether each rule could be correct, and explain how you know.

1.

q	1	2	3	10
p	7	11	15	43

$p = q + q + q + 4$

$p = 4 \cdot q + 3$

2.

s	1	2	5	10
t	10	20	50	100

$t = 4 \cdot s + 6 \cdot s$

$t = 10s$

3.

k	0	2	5	10
j	1	17	101	402

$j = 5 \cdot k^2 + 1$

$j = 5 \cdot k \cdot k + 1$

In the next problem set, you will try to figure out the rules for some *What's My Rule?* games.

Remember

An exponent tells how many times a number is multiplied together. For example, $4^2 = 4 \cdot 4$ and $t^4 = t \cdot t \cdot t \cdot t$.

Problem Set F

These tables were made during games of *What's My Rule?* In each table, the values in the top row are the inputs and the values in the bottom row are the outputs.

Write a rule for each table, using the given letters to represent the variables.

1.

a	2	5	3	6	1
b	9	21	13	25	5

2.

y	4	5	6	1	$\frac{3}{5}$
z	18	23	28	3	1

3.

w	$\frac{12}{7}$	11	19	4	7
g	$\frac{2}{7}$	$1\frac{5}{6}$	$3\frac{1}{6}$	$\frac{2}{3}$	$1\frac{1}{6}$

4.

q	10	5.5	1	2	3
p	6	3.75	1.5	2	2.5

5.

c	100	42	17	1	0.3
d	10,000	1,764	289	1	0.09

6.

s	1	3.1	10	5	6.5
t	3	11.61	102	27	44.25

Share & Summarize

Abby and Ji-Young were playing *What's My Rule?* Ji-Young's secret rule was "To get the output, multiply the input by itself and subtract 1." Abby guessed that the rule was "Subtract 1 from the input, and multiply the result by itself."

1. Write both rules with symbols. Use *m* to represent the output and *n* to represent the input.

2. Is Abby's rule equivalent to Ji-Young's rule? Explain.

Practice & Apply

1. Consider this sequence of toothpick "houses":

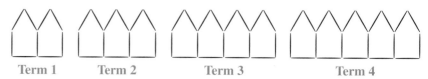

Term 1   Term 2   Term 3   Term 4

   a. Write a number sequence for the number of houses in each term.

   b. Write a rule that connects the number of houses to the term number. Use letters to represent the variables, and tell what each letter represents.

2. Consider this dot sequence:

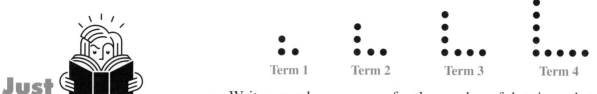

Term 1   Term 2   Term 3   Term 4

   a. Write a number sequence for the number of dots in each term.

   b. Write a rule that connects the number of dots to the term number. Use letters to represent the variables, and tell what each letter represents.

In Exercises 3 and 4, give the first four numbers in the sequence. Then draw a sequence of toothpicks or dots that fits the rule.

3. $t = 4 \cdot (k + 1)$, where t represents the number of toothpicks and k represents the term number

4. $d = 3 \cdot p - 2$, where d represents the number of dots and p represents the term number

Just the facts

Braille is a system of writing used by the blind. Each Braille character is made from 1 to 6 raised dots. The system was invented in 1824 by Louis Braille while he was a student at the National Institute for Blind Youth in Paris.

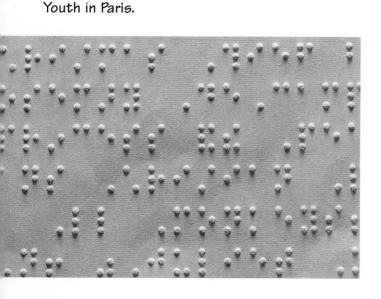

impactmath.com/self_check_quiz

5. Consider this sequence:

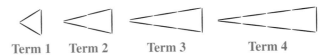

Term 1   Term 2   Term 3   Term 4

 a. Write two equivalent rules for the number of toothpicks in each term.

 b. Use words and diagrams to explain why each of your rules is correct.

6. Three students wrote rules for this dot sequence. Determine whether each rule correctly describes the pattern. If it does, explain how you know it is correct. If it does not, explain why it is not correct.

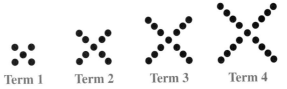

Term 1   Term 2   Term 3   Term 4

 a. Ilsa's rule: $d = 5 \cdot m$, where d is the number of dots and m is the term number

 b. Mattie's rule: $d = 1 + 4k$, where d is the number of dots and k is the term number

 c. Mateo's rule: $d = 4 \cdot (j - 1) + 5$, where d is the number of dots and j is the term number

7. When tiling a walkway, a particular contractor surrounds each pair of blue tiles with white tiles as shown at right. Four copies of this design are put together below.

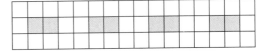

 a. Copy and complete the table to show the number of white tiles needed for each given number of blue tiles.

Blue Tiles	2	4	6	8	20	100
White Tiles	10					

 b. Find a rule that describes the connection between the number of white tiles and the number of blue tiles. Use letters for the variables in your rule, and tell what each letter represents.

 c. Explain how you know your rule is correct. Use diagrams if you need to.

8. Consider this sequence of cubes:

Term 1 　　　　　 Term 2 　　　　　 Term 3

a. Find a rule for the number of cubes in each term. You may want to make a table first. Use letters for the variables in your rule, and tell what each letter represents.

b. Explain how you know your rule is correct. Use diagrams if you need to.

The tables below were made during games of *What's My Rule?* The values in the top row are inputs, and the values in the bottom row are outputs. Write a rule for each table, using the given letters to represent the variables.

9.

f	11	4	1	7	2
g	43	22	13	31	16

10.

j	12	9	2	16	23
k	6	5.25	3.5	7	8.75

11. In a game of *What's My Rule?* Hiam's secret rule was $a = b^2 \cdot 3$, where b is the input and a is the output. Complete the table to show the outputs for the given inputs.

b	8	0	15	7	4
a					

The tables in Exercises 12 and 13 were made during games of *What's My Rule?* Two rules are given for each table. Determine whether each rule could be correct, and explain how you know.

12.

s	2	5	11	7	10
t	5	14	32	20	29

$s = (t - 1) * 3$

$t = 2s + (s - 1)$

13.

p	2	5	1	4	3
m	10	127	3	66	29

$m = 2 + p \cdot p \cdot p$

$m = p^3 + 2$

Connect & Extend

14. Here are the first and fifth terms of a toothpick sequence:

Term 1   Term 5

a. What might Terms 2, 3, and 4 look like?

b. Write a rule that connects the term number and the number of toothpicks in your sequence. Use letters to represent the variables.

15. Each term of this sequence is made from a 1-inch straw cut into equal-sized pieces:

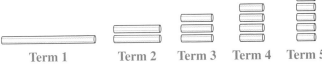

Term 1   Term 2   Term 3   Term 4   Term 5

a. How many pieces of straw will be in Term 10? What fraction of an inch will the length of each piece be?

b. Write a rule that connects the number of straw pieces to the term number.

c. Write a rule that connects the length of each straw piece in a term to the term number.

16. The Briggs made a patio using square bricks. They worked in stages, completing one stage before moving on to the next.

Stage 2   Stage 3   Stage 4   Stage 5

a. Write a number sequence showing the number of bricks in the patio in each stage, starting with Stage 2.

b. Write a rule that connects the number of bricks in the patio to the stage number. Use letters to represent the variables, and tell what each letter represents.

c. Use your rule to figure out how many bricks were in the patio in Stage 1.

d. Use your rule to predict how many bricks will be in the patio at Stage 14.

Remember

The *perimeter* of a figure is the distance around it.

17. **Geometry** Gage and Tara wrote formulas for the perimeter of a rectangle. Both students used P for the perimeter, L for the length, and W for the width.

Length

Width

Gage's formula: $P = 2 \cdot L + 2 \cdot W$

Tara's formula: $P = 2 \cdot (L + W)$

a. Tell whether each formula is correct. If it is correct, draw diagrams showing why it is correct. If it is not correct, explain what is wrong.

b. Gage said, "I wonder whether I could just write my formula as $P = 2 \cdot L \cdot W$." Does this formula give the correct perimeter for a rectangle? Explain how you know.

18. Ben made this sequence from watermelon seeds:

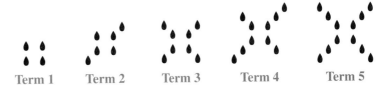

Term 1   Term 2   Term 3   Term 4   Term 5

a. Complete the table to show the number of seeds in each term.

Term Number	1	2	3	4	5
Number of Seeds					

b. On a set of axes like the one below, graph the data in your table.

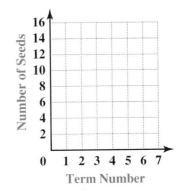

c. Describe the pattern of points in your graph.

d. Use your graph to predict the number of seeds in Terms 6 and 7. Check your prediction by drawing these terms and counting the seeds.

19. Jacob made the graph below to show the number of stars in each term of this sequence:

Term 1   Term 2   Term 3   Term 4   Term 5

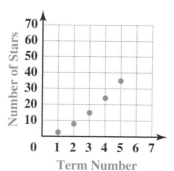

a. Describe the pattern of points in Jacob's graph.

b. Use the graph to predict the number of stars in Terms 6 and 7. Do you think your predictions are accurate? Explain.

c. Write a rule for finding the number of stars in any term of the sequence. Use your rule to check your predictions from Part b.

20. This table shows values of the variables m and n:

m	6	0	11	7	3
n	78	0	143	91	39

a. Complete this rule for the relationship between m and n:

$$n = \underline{\hspace{3cm}}$$

b. Complete this rule for the relationship between m and n:

$$m = \underline{\hspace{3cm}}$$

c. Explain how the two rules you wrote could describe the same relationship.

In your
own
words

Describe how you can determine whether two rules for a sequence are equivalent.

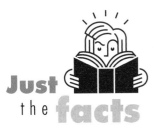
Mixed Review

21. Sports Danilo and his brother Adan are having a walking race. Since Adan is younger, Danilo gives him a head start. The table shows the number of minutes after the start of the race that each boy has reached the given distance. The boys are walking at a steady pace, and all the blocks are about the same length.

Distance (blocks)	1	4	6	10
Adan's Time (min)	3	12	18	30
Danilo's Time (min)	22	28	32	40

a. Write a rule for finding the number of minutes M that it takes Adan to reach block N.

b. Write a rule for finding the number of minutes M, after the start of the race, that it takes Danilo to reach block N.

c. If both boys stop walking an hour after the race began, will Danilo catch up to Adan? Explain.

Find each product without using a calculator.

22. $44 * 781$

23. $4.4 \cdot 0.781$

24. $440 * 781{,}000$

25. 0.44×7.81

26. $0.044 \cdot 0.0781$

27. 440×78.1

Fill in the blanks.

28. ___ % of $152 = 114$

29. 0.5% of $200 =$ ___

30. $33\frac{1}{3}\%$ of ___ $= 15$

31. ___ % of $20 = 60$

32. 45% of $45 =$ ___

33. 80% of ___ $= 320$

34. Statistics Jaleesa received the following scores on her 20-point spelling quizzes:

   20   15   18   19   20   0   14   17   16

a. Find the mean, median, and mode of her scores.

b. Which measure of center do you think best represents Jaleesa's typical quiz score? Give reasons for your choice.

c. Jaleesa's teacher has agreed to drop each student's lowest score. Drop the lowest score, and compute the new mean, median, and mode of Jaleesa's scores.

You have seen that using letters to stand for variables lets you write rules more simply. In Lesson 7.1, the rules you worked with came from sequences. In this lesson, you will look at rules for everyday situations.

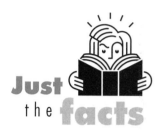

Just the facts

Long-distance service between London and Paris was introduced in 1891.

Explore

Last summer, Jing's best friend moved to another state. Since then, Jing has been making lots of long-distance calls. Jing's mother has asked her to use her allowance to help pay the phone bill.

• Jing's family uses TalkCo long-distance service. The service charges $3.00 per month plus $0.07 per minute for each phone call. Use this information to complete the table below.

TalkCo Charges

	Jun	Jul	Aug	Sep	Oct
Minutes	120	90			80
Cost			$9.65	$13.50	

• Write a rule Jing's family could use to calculate their monthly long-distance bill. Use *m* to stand for the number of minutes they talked and *c* to stand for the cost in dollars.

• Complete this table to show what Jing's family's monthly charges would have been if they had used Chatterbox long-distance service instead.

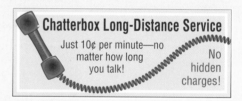

Chatterbox Long-Distance Service

Just 10¢ per minute—no matter how long you talk!   No hidden charges!

Chatterbox Charges

	Jun	Jul	Aug	Sep	Oct
Minutes	120	90	95	150	80
Cost					

• Write a rule Chatterbox customers could use to calculate their monthly bills. Use *m* to stand for the minutes and *c* for the cost.

• Do you think Jing's family should switch to Chatterbox? Explain.

Investigation Rules in Context

Remember

A *variable* is a quantity that can change.

Just the facts

The sale of sports utility vehicles grew from 960,852 in 1988 to 2,796,310 in 1998—a 191% increase!

Just like sequences and patterns, real situations often involve variables you can represent with letters. In the telephone-service problem, the variables are the number of long-distance minutes *m* and the cost *c*. In this lesson, you will write lots of rules to represent real situations you might encounter.

Problem Set A

Haley's car rental company rents sedans and sports utility vehicles (SUVs).

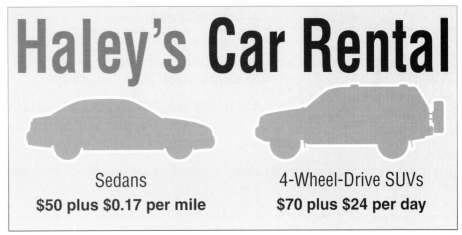

Haley's Car Rental

Sedans
$50 plus $0.17 per mile

4-Wheel-Drive SUVs
$70 plus $24 per day

1. The ad shows Haley's car rental rates.

   a. If a customer rents an SUV for 14 days and drives it 2,000 miles, what will the cost be?

   b. If a customer rents a sedan for 14 days and drives it 2,000 miles, what will the cost be?

   c. What are the variables in this situation?

2. Consider the cost of renting a sedan.

   a. Explain in words how to calculate the cost of renting a sedan.

   b. Create a table that shows the cost of renting a sedan and driving it for various numbers of miles.

   c. Write a rule for calculating the cost of renting a sedan. Use *m* to represent the number of miles driven and *c* to represent the cost.

   d. A customer rented a sedan and drove it 567.6 miles. Use your rule to determine how much the customer should be charged.

3. Consider the cost of renting an SUV.

    a. Explain in words how to calculate the cost of renting an SUV. How is this rule different from the rule for the sedan?

    b. Create a table that shows the cost of renting an SUV for various numbers of days.

    c. Write a rule for calculating the cost of renting an SUV. Use d to represent the number of days and c to represent the cost.

    d. A customer rented an SUV for 27 days. Use your rule to determine how much the customer should be charged.

4. Each customer described below wants to rent a vehicle. Use the rules you wrote in Problems 2 and 3 to decide which type of vehicle would be less expensive. Explain your answers.

    a. Mr. Houlihan is taking a business trip to Denver, Colorado. He will be driving 650 miles and will be gone for 3 days.

    b. Ms. Basil is traveling to Orlando, Florida, for 10 days. She will be driving 1,350 miles round-trip.

    c. Mr. and Mrs. Iso are planning to visit their grandchildren in Portland, Maine. They will be driving 125 miles and will be gone for 2 weeks.

5. Haley's has changed its rate for renting a sedan. It now charges $.85 per mile, with no fixed amount.

    a. Write a new rule for calculating the rental cost for a sedan. Use c to represent the cost and m to represent the number of miles driven.

    b. Does this change make it more or less expensive to rent a sedan? Explain.

Share & Summarize

In Chapter 1, you worked with real-life rules written in words. In this investigation, you wrote and used symbolic rules, with letters for the variables. What are some advantages of using rules written with letters and symbols?

Investigation ▶ Crossing a Bridge

While walking at night, a group of eight hikers—six adults and two children—arrives at a rickety wooden bridge. A sign says the bridge can hold a maximum of 200 pounds. The group estimates that this means

- one child can cross alone, *or*
- one adult can cross alone, *or*
- two children can cross together.

Anyone crossing the bridge will need to use a flashlight. Unfortunately, the group has only one flashlight.

Try It Out

1. Find a way to get all the hikers across the bridge in the fewest number of trips. Count one trip each time one or two people walk across the bridge.

   a. Describe your plan using words, drawings, or both.

   b. How many trips will it take for everyone in the group to get across the bridge?

2. A second group of hikers approaches the bridge. This group has 10 adults, 2 children, and one flashlight.

   a. Can everyone in this group get across the bridge? If so, describe how.

   b. What is the least number of trips it will take for this group to cross the bridge?

3. In Parts a–c, find the least number of trips it would take for the group to get across the bridge. Assume each group has only a single flashlight.

   a. 8 adults and 2 children

   b. 1 adult and 2 children

   c. 100 adults and 2 children

4. Suppose a group has two children, *a* adults, and one flashlight. Write a rule that relates these two variables: the least number of trips *t* needed to get everyone across, and the number of adults.

5. Could a group with 15 adults, one child, and one flashlight cross the bridge? Explain.

Try It Again

Now you will explore how the number of children in a group affects the number of trips needed for the group to cross the bridge. You will start by thinking about groups with no adults at all.

6. Tell how many trips it would take to get each of these groups across the bridge. (Hint: First figure out a method for getting everyone across, and then think about how many trips it would take.)

   a. 2 children   **b.** 3 children

   c. 6 children   **d.** 10 children

7. Look for a pattern relating the number of children to the number of trips.

   a. Write rule that relates the number of trips *t* to the number of children in the group *c*. Assume the group has no adults, more than one child, and just one flashlight.

   b. Part a states that the group must have more than one child. How many trips would it take a single child to get across the bridge? Does your rule give the correct result for one child?

8. Tell how many trips it would take to get each of these groups across the bridge:

   a. 6 adults and 2 children   **b.** 6 adults and 3 children

   c. 6 adults and 6 children   **d.** 8 adults and 10 children

9. **Challenge** Write a rule that relates the number of trips to the number of children and adults in the group. Assume the group has at least two children and exactly one flashlight.

What Did You Learn?

Another bridge has a different weight restriction. For this bridge,

- one adult can cross alone, *or*

- one child can cross alone, *or*

- one adult and one child can cross together, *or*

- two children can cross together.

10. With these new rules, a group might be able to use a different method to get across the bridge.

    a. Find a way to get a group of eight adults and two children across the bridge in the fewest trips. Describe your method using words, pictures, or both.

    b. How many trips does it take for everyone to get across?

    c. How does the number of trips needed to get the group across this bridge compare to the number of trips needed for the first bridge? (See your answer for Part a of Problem 3.)

11. A group of 15 adults and 1 child could not get across the first bridge. Could this group cross the second bridge? If so, explain how, and tell how many trips it would take.

12. Tell how many trips it would take each of these groups to get across the second bridge:

    a. 3 adults and 1 child

    b. 5 adults and 1 child

    c. 100 adults and 1 child

13. Write a rule that relates the number of trips needed to cross the second bridge to the number of adults in the group. Assume the group has one child and one flashlight.

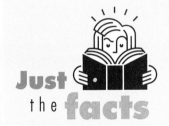

Just the facts

The world's longest overwater bridge is the Lake Pontchartrain Causeway, which connects Mandeville and Metairie, Louisiana. The bridge is almost 24 miles long.

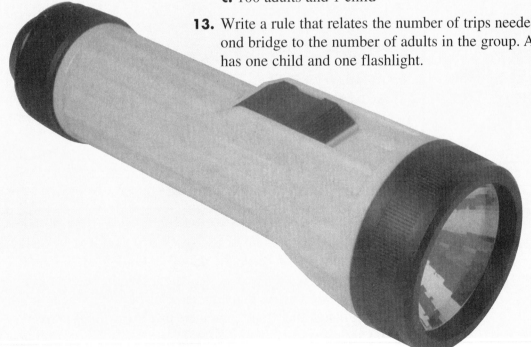

Investigation 2 Translating Words into Symbols

Writing a rule for a real-life situation can be difficult, even when the situation is fairly simple. It's easy to make a mistake if you don't pay close attention to the details.

Just the facts

The body and legs of a tarantula are covered with hairs. Each hair has a tiny barb on the end. When a tarantula is being attacked, it can rub its hairs on its attacker, causing itching and even temporary blindness.

Think & Discuss

A spider has eight legs. If S represents the number of spiders and L represents the number of legs, which of the following rules is correct? How do you know?

$$S = 8 \cdot L \qquad\qquad L = 8 \cdot S$$

In the spider situation, it is easy to confuse the two rules. The Example shows how Luke thought about the situation.

EXAMPLE

Creating a table and looking for a pattern can make finding a rule a little easier. Notice that, after Luke wrote his rule, he checked it by testing a value he knew the answer for. It is a good idea to test a value whenever you write a rule. Although this won't guarantee that your rule is correct, it's a helpful way to uncover mistakes.

Problem Set B

1. To tile his bathroom, Mr. Drury needs twice as many blue tiles as white tiles.

   a. What are the two variables in this situation?

   b. Make a table of values for the two variables.

   c. Look for a pattern in your table. Describe how to calculate the values of one variable from the values of the other.

   d. Write a rule for the relationship between the two variables. Use letters to represent the variables, and tell what each letter represents. Be sure to check your rule by testing a value.

In Problem 2–6, complete Parts a–d of Problem 1.

2. In packages of Cool Breeze mints, there are four green mints for every pink mint.

3. A pet store always carries six times as many fish as hamsters.

4. In a factory, each assembly worker earns one seventh as much money as his or her manager.

5. A community theater charges $3 less for a child's ticket than for an adult's ticket.

6. In a toothpick sequence, the total number of toothpicks in a term is 4 more than twice the number of vertical toothpicks in the term.

Problem Set C

Below are more situations involving rules. Make a table whenever you feel it will help you understand the problem. Also, be sure to test all your rules.

1. Nick and his friends collect the prizes hidden in boxes of Flako cereal. Joel has twice as many prizes as Nick. Rafael has 3 more prizes than Nick. Andrea has half as many prizes as Nick.

   a. If Nick has 6 prizes, how many prizes do the other friends have?

   b. Write a rule for the relationship between the number of prizes Joel has j and the number of prizes Nick has n.

   c. Write a rule for the relationship between the number of prizes Rafael has r and the number Nick has n.

   d. Write a rule for the relationship between the number of prizes Andrea has a and the number Nick has n.

2. Suppose Gish has 2 more prizes than Joel.

   a. Write a rule for the relationship between the number of prizes Gish has g and the number Joel has j.

   b. If Nick has 19 prizes, how many prizes does Joel have? How many does Gish have?

   c. Describe in words the relationship between the number of prizes Gish has and the number Nick has.

   d. Write a rule for the relationship between the number of prizes Gish has g and the number Nick has n.

3. Spiro Papadopoulos weighs s pounds.

   • His wife, Toula, is 20 pounds lighter than Spiro.

   • His son, Michael, weighs half as much as Spiro.

   • His daughter, Christa, weighs 3 pounds more than his son.

   a. If Spiro weighs $155\frac{1}{2}$ pounds, how much do each of the others weigh?

   b. Write a rule for calculating the weight of each member of Spiro's family based on Spiro's weight s. Tell what variable each letter represents.

Investigation 3 ▶ Equivalent Rules

Every rule can be written in a variety of ways. As you discovered when working with toothpick sequences, when two rules look different, it is sometimes difficult to tell whether they represent the same relationship.

Think & Discuss

Each of these rectangles is five times as long as it is wide:

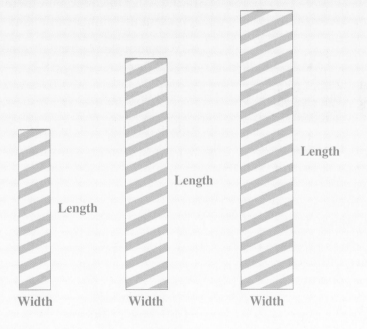

Four students wrote rules to describe the relationship between the length and width of these rectangles. In each rule, W represents the width and L represents the length.

Caroline's rule: $W = 5 \cdot L$

Jahmal's rule: $L = 5W$

Jing's rule: $W = L \div 5$

Rosita's rule: $\frac{L}{W} = 5$

Which of these rules are correct? How do you know?

Remember

The multiplication symbol is often left out when a number is multiplied by a variable. So, 5W is the same as 5 · W.

You can sometimes use what you know about the relationship between multiplication and division, or between addition and subtraction, to rewrite a rule in a different way. For example, you know that for every multiplication equation, there are two related division equations:

Multiplication Equation	Division Equations	
$3 \cdot 5 = 15$	$15 \div 5 = 3$	$15 \div 3 = 5$

You can use this idea to rewrite a rule involving multiplication, such as $2 \cdot q = p$:

Multiplication Equation	Division Equations	
$2 \cdot q = p$	$p \div q = 2$	$p \div 2 = q$

All three of the rules above represent the same relationship—that is, they are all *equivalent*.

In the next problem set, you will consider different rules that represent the same situation, and you will rewrite rules in different ways.

Problem Set D

1. In the Aster banquet hall, there are four chairs at every table. Let c represent the number of chairs and t represent the number of tables.

   a. Which of these rules correctly describes this situation?

   $$c \cdot 4 = t \qquad t = c \div 4 \qquad c = t \div 4$$
   $$c = 4t \qquad c + 4 = t \qquad t + 4 = c$$

   b. Explain how you know each of the other rules is incorrect.

2. A hardware store sells nuts and bolts in three package sizes.

32 Nuts
30 Bolts

20 Nuts
18 Bolts

62 Nuts
60 Bolts

Althea and Luke wrote rules to describe the relationship between the number of nuts n and the number of bolts b:

Althea's rule: $n = b + 2$ Luke's rule: $n - 2 = b$

  a. Explain in words what each rule says.

  b. Do the rules represent the same relationship? In other words, are the two rules equivalent? Explain.

3. The rule $p = 7e$ expresses the relationship between the number of envelopes e and the number of sheets of paper p in a package of stationery.

  a. What does the rule tell you about relationship between paper and envelopes in the package?

  b. Rewrite this rule in at least two different ways.

4. The rule $\frac{1}{24} \cdot h = d$ expresses the relationship between the number of days d and the number of hours h.

  a. What does the rule tell you about the relationship between days and hours?

  b. Rewrite this rule in at least two different ways.

5. A jar contains black and green jelly beans. There are $\frac{2}{5}$ as many black jelly beans as green jelly beans.

  a. Draw two jars that fit this description.

  b. Write a rule that relates the number of black jelly beans b to the number of green jelly beans g.

  c. Rewrite your rule in at least two different ways.

6. Three students wrote rules connecting the number of tires (including the spare) to the number of cars in a parking lot. If all the rules below are correct, tell what variable each letter must stand for.

Lee's rule: $s = 5 \cdot t$

Gavin's rule: $5 \times b = a$

Chitra's rule: $c = w \div 5$

So far in this chapter, you have looked at and written rules that fit specific situations. Now you will describe situations that match given rules.

Problem Set E

1. Think about this rule: $t = 3 \cdot c$.

Describe three different situations this rule could represent. Tell what the variables stand for in each situation.

2. Sandra and Tim collect old cameras. In the rules below, s represents the number of cameras Sandra has and t represents the number of cameras Tim has. Explain what each rule tells you.

a. $s = t + 1$

b. $s = 2 \times t$

c. $s = 2t + 1$

d. $t = s + 3$

Share & Summarize

Think about these rules:

$$\tfrac{1}{5} \cdot f = b \qquad g = a - 7$$

1. Describe a situation each rule could represent. Tell what variable each letter stands for.

2. Rewrite each rule in a different way so that it still describes the situation.

On Your Own Exercises

Just the facts

The first windsurfing board was patented in 1968. The board was developed by Hoyle Schweitzer (a surfer) and Jim Drake (a sailor), who wanted to combine surfing with sailing.

1. When he grades a test, Mr. Bonilla awards 2 points for each correct answer, plus 6 points just for taking the test.

  a. Copy and complete the table to show the score a student would receive for the given numbers of correct answers.

Correct Answers	30	35	40	45
Test Score				

  b. Write a rule you could use to calculate a test score if you know the number of questions a student answered correctly. Use c to represent the number of correct answers and s to represent the score.

2. Sports The Get Wet beach shop rents sailboards for windsurfing. The cost to rent a sailboard is $10 plus $5 per hour. Get Wet also teaches people how to windsurf for $15 per lesson.

  a. What are the variables in this situation?

  b. Complete the table to show the cost of renting a sailboard for various numbers of hours.

Hours	3		4.5	5	
Cost		$30.00			$36.25

  c. Write a rule you could use to calculate the cost of renting a sailboard for any number of hours. Use c to represent the cost and h to represent the number of hours.

  d. Create a table to show the costs for different numbers of lessons.

  e. Use letters and symbols to write a rule for calculating the cost for any number of lessons. Tell what variable each letter represents.

  f. Get Wet is running an early-summer special. If you take two lessons, you get 50% off the cost of your next rental. How much would it cost to take two lessons and then rent a sailboard for 2 hours? Show how you found your answer.

3. Measurement There are approximately 1.6 kilometers in a mile.

　a. Create a table comparing miles to kilometers.

　b. Write a rule to represent the relationship between the number of miles and the number of kilometers. Tell what each letter in your rule represents.

　c. The driving distance from New York to Boston is about 200 miles. Use your rule to calculate how many kilometers this is.

4. Economics Tickets for a school play cost $3.75 each. Write a rule connecting the total cost in dollars *c* and the number of tickets bought *t*. Make a table if you need to, and check your rule by testing a value.

5. The cooking time for a turkey is 18 minutes for every pound plus an extra 20 minutes.

　a. How long will it take to cook a 12-pound turkey?

　b. Write a rule for this situation, using *m* for the number of minutes and *p* for the number of pounds. Make a table if you need to, and check your rule by testing a value.

6. Sports Ana and her friends are training for a marathon. Today Ana ran *a* miles. In Parts a–d, write a rule expressing the relationship between the number of miles the person ran and the number of miles Ana ran. Be sure to state what the letters in your rules represent. Make a table if you need to, and check your rule by testing a value.

　a. Juan ran twice as far as Ana.

　b. Kai ran 3 miles more than Ana.

　c. Toshio ran $\frac{2}{3}$ as far as Ana.

　d. Challenge Melissa ran three times as far as Toshio. (Remember, your rule should relate Melissa's distance to Ana's distance.)

　e. If Toshio ran 2 miles, how far did Ana, Juan, Kai, and Melissa run?

Just the facts

On April 14, 2002, Khalid Khannouchi of Morocco completed the Flora London Marathon in 2 hours 5 minutes 38 seconds, setting a new world record.

7. Sports In football, a team receives 6 points for a touchdown, 1 point for making the kick after a touchdown, and 3 points for a field goal.

a. If a team gets three touchdowns, makes two of the kicks after the touchdowns, and scores two field goals, what is the team's total score?

b. Write a rule for a team's total score S if the team gets t touchdowns, makes p kicks after touchdowns, and scores g field goals. Be sure to check your rule by testing it for a specific case.

8. Three students wrote rules for the relationship between the number of eyes and the number of noses in a group of people. Each student used e to represent the number of eyes and n to represent the number of noses. Which of the rules are correct? Explain how you decided.

Miguel's rule: $2 \cdot e = n$

Althea's rule: $n \times 2 = e$

Hannah's rule: $n = e \div 2$

9. Measurement The rule $L = W + 1$ expresses the relationship between the length L and width W of a room. Both dimensions are in feet.

a. What does this rule tell you about the dimensions of the room?

b. Rewrite this rule in at least two different ways. Explain how you know your rules are correct.

10. The rule $s = 3{,}600h$ gives the relationship between the number of hours and the number of seconds.

a. What does this rule tell you about the relationship between hours and seconds?

b. Rewrite this rule in at least two different ways.

11. The rule $m = \frac{3}{7} \times b$ describes the relationship between the number of mountain bikes m in a cycling shop and the total number of bikes b.

a. What does this rule tell you about the relationship between the number of mountain bikes and the total number of bikes?

b. Rewrite this rule in at least two different ways.

In Exercises 12 and 13, describe a situation the rule could represent. Tell what variables the letters stand for.

12. $n = m - 8$

13. $p = 3r + 4$

Just the facts

Penny farthing bicycles, introduced in England in the early 1870s, had front wheels with diameters up to 5 feet.

14. Physics Suppose you throw a ball straight up into the air and catch it again. The relationship between the time the ball is in the air t and the approximate speed s at which you threw the ball, in meters per second, is given by this rule:

$$s = 5 \cdot t$$

a. If the ball was in the air for 4.5 seconds, about how fast did you throw it?

b. If you throw the ball at a speed of about 25 meters per second, how long will it stay in the air?

15. Science This rule is used to calculate the amount of a particular medicine a child should be given:

$$\text{child's dose} = \text{adult's dose} \times \frac{\text{child's age}}{\text{child's age} + 12}$$

a. Rewrite this rule using c to represent the child's dose, d to represent the adult's dose, and y to represent the child's age.

b. If the adult's dose is 50 milliliters, how much medicine should a 3-year-old child be given?

16. In a children's story, peacocks and rabbits lived in a king's garden. A peacock has two legs, and a rabbit has four legs.

a. Complete the table to show the total number of legs in the garden for the given numbers of peacocks and rabbits.

Peacocks	2	4	6	8	10
Rabbits	3	6	9	12	15
Legs					

b. Describe how you calculated the total number of legs for each group of animals.

c. Use letters and symbols to write a rule to calculate the total number of legs in the garden if you know the number of rabbits and the number of peacocks. Tell what variable each letter represents.

17. Lina has q quarters in her pocket. Alan has d dimes in his pocket.

    a. Write a rule to express the value of Lina's quarters *in dollars.* Use v to represent the value.

    b. Write a rule to express the value of Alan's dimes *in dollars.* Use v to represent the value.

    c. Alan combines his dimes with Lina's quarters. Write a rule to express the total value of the coins.

    d. Together, Lina and Alan have $5.20. Find three possibilities for the values of d and q.

18. **Physical Science** Water is made up of hydrogen atoms and oxygen atoms. A sample of water contains twice as many hydrogen atoms as oxygen atoms.

A molecule of water

    a. Write a rule to express the relationship between the number of hydrogen atoms and the number of oxygen atoms in a sample of water. Tell what the letters in your rule represent.

    b. A particular sample of water contains 476 hydrogen atoms. How many oxygen atoms does it contain?

19. **Science** A pharmacist is filling bottles with capsules. Let n represent the number of bottles she has.

    a. She takes half of the bottles and puts 40 capsules in each. Write a rule for the total number of capsules p in these bottles. (Your rule should include the letter n.)

    b. She puts 50 capsules in each of the remaining bottles. Write a rule for the total number of capsules q in these bottles. (Your rule should include the letter n.)

    c. Use your answers to Parts a and b to help you write a rule for the total number of capsules t the pharmacist has put into all the bottles.

    d. If the pharmacist started with 4 bottles, how many capsules did she use? If she started with 100 bottles, how many capsules did she use?

20. Technology In this exercise, you will investigate some keys on your calculator.

   a. Describe in words what the ⌨x^2 key does. If you are not sure, enter a number, press the key, then press ⌨ENTER. Try this a few times until you are certain you know what the key does.

   b. Suppose you enter a number into your calculator and press ⌨x^2 ⌨ENTER. Write a rule to describe the relationship between the number you enter n and the result r.

   c. Your calculator has the command x^{-1}, accessed by pressing ⌨2nd $[x^{-1}]$. Use this command a few times with different numbers until you think you know what it does. Describe in words what this command does.

   d. Suppose you enter a number into your calculator and press ⌨2nd $[x^{-1}]$ ⌨ENTER. Write a rule to describe the relationship between the number you enter m and the result r.

   e. Challenge Suppose you enter a number, press ⌨x^2, and then press ⌨2nd $[x^{-1}]$ ⌨ENTER. Write a rule to describe the relationship between the number you enter p and the result r. Check that your rule gives the same result as the calculator for at least three values of p.

21. Physical Science To convert a temperature from degrees Fahrenheit (°F) to degrees Celsius (°C), use this rule: *Subtract 32, multiply the result by 5, and then divide by 9.*

   a. Use symbols to write the rule for converting degrees Fahrenheit f to degrees Celsius c. (Hint: Make sure you think about order of operations.)

   b. To convert a temperature from degrees Celsius to degrees Fahrenheit, just "undo" the operations in the reverse order. In other words, use this rule: *Multiply the Celsius temperature by 9, divide the result by 5, and then add 32.*

   Use this information to write a rule for converting degrees Celsius to degrees Fahrenheit.

   c. Use the rule you wrote in Part a to convert 212°F to degrees Celsius. Then convert your answer back to degrees Fahrenheit using the rule you wrote in Part b. Do you end with 212°F?

Write each decimal as a fraction in lowest terms.

22. 0.05   **23.** $0.\overline{6}$ **24.** 0.6

25. 0.0075   **26.** 0.4545   **27.** 0.99

Write each fraction as a decimal.

28. $\dfrac{1}{8}$ **29.** $\dfrac{3}{5}$ **30.** $\dfrac{1}{11}$

31. $\dfrac{10}{12}$ **32.** $\dfrac{228}{475}$ **33.** $\dfrac{8}{9}$

34. Stewart answered 14 out of 18 questions correctly on his math test. What percent did he answer correctly? Round your answer to the nearest whole percent.

35. **Economics** Mr. Rosen's car cost $13,500 when he bought it. Since then, the value of the car has decreased by 45%. What is Mr. Rosen's car worth now?

36. **Statistics** Althea asked 20 of her classmates to keep track of the number of movies they watched on video or at the theater over a two-month period. She made a stem-and-leaf plot of the results.

Number of Movies Watched

Stem	Leaf
0	0 2 3 5 6 8 9 9
1	2 2 4 5 6
2	0 4 4 4 7
3	2
4	
5	5

Key: $2|4=24$

a. What was the least number of movies watched? What was the greatest number?

b. Describe the overall distribution of data values.

c. Find the mode, median, and mean of the data.

7.3 Explaining Number Relationships

You have used letters and symbols to describe sequences and real-life situations. In this lesson, you will use symbols to help describe and explain some number relationships.

Think & Discuss

Try this number trick with your class. Each student should start with his or her own number.

• Think of a number.

• Add 5.

• Double the result.

• Subtract twice the number you started with.

What number do you end with? How does your result compare with other students' results?

Now try this number trick:

• Think of a number.

• Add 5.

• Multiply the result by 5.

• Subtract 25.

• Divide the result by 5.

What number do you end with? How does your result compare with other students' results?

Investigation ▶ 1 Think of a Number

In this investigation, you will try some more number tricks. You will then use variables to help you understand how the tricks work.

Problem Set A

1. The first column of this table gives the steps for a number trick:

Step	Result			
Think of a number.				
Add 2.				
Multiply by 3.				
Subtract 3 times the number you started with.				

   a. Choose four different starting numbers, and record the result of each step in your table. One of your starting numbers should be a fraction and one should be a decimal.

   b. Describe your results.

2. This table gives the steps for another number trick:

Step	Result			
Think of a number.				
Multiply by 10.				
Add 10.				
Subtract 9 times the number you started with.				
Subtract 10.				

   a. Choose four different starting numbers, and record the result of each step in the table. One of your starting numbers should be a fraction and one should be a decimal.

   b. Describe your results.

Because you can change the starting number each time you try a number trick, the starting number is a variable. Using a letter to represent this variable can help you understand how the trick works. The Example explains the first trick in the Think & Discuss on page 450.

EXAMPLE

- Think of a number.

  Use the letter n to represent the starting number.

- Add 5.

  Adding 5 to n gives $n + 5$.

- Double the result.

  Doubling the result of the previous step gives $2 \cdot (n + 5)$.

  Multiplying a number by 2 is the same as adding the number to itself. You can use this idea to rewrite $2 \cdot (n + 5)$:

  $$2 \cdot (n + 5) = (n + 5) + (n + 5)$$
  $$= n + 5 + n + 5$$

  When you add a string of numbers, you can always rearrange them:

  $$n + 5 + n + 5 = n + n + 5 + 5$$
  $$= 2 \cdot n + 10$$

  So, the result of this step is $2 \cdot n + 10$.

- Subtract twice the number you started with.

  The number you started with is n. Twice this number is $2 \cdot n$. Subtract this from the result of the last step:

  $$2 \cdot n + 10 - 2 \cdot n$$

  $2 \cdot n$ represents a number, and if you subtract a number from itself, the result is 0. In other words, $2 \cdot n - 2 \cdot n = 0$. You can use this fact to simplify the expression above:

  $$2 \cdot n + 10 - 2 \cdot n = 2 \cdot n - 2 \cdot n + 10$$
  $$= 0 + 10$$
  $$= 10$$

So, no matter what number you start with, the result will be 10!

Just the facts

Coins with heads on both sides, and cards with back designs on both sides, are common magician's accessories.

The table summarizes the steps in the Example.

Step in Words	Step in Symbols
Think of a number.	n
Add 5.	$n + 5$
Double the result.	$2 \cdot (n + 5) = n + 5 + n + 5$ $\qquad\qquad\quad = 2 \cdot n + 10$
Subtract twice the number you started with.	$2 \cdot n + 10 - 2 \cdot n = 10$

To understand the steps in the Example, you need to think about and work with expressions as if they were numbers. For instance, you need to think about doubling $n + 5$. It takes practice to become comfortable thinking in this way.

Think & Discuss

You can think of multiplication as repeated addition. For example, $3 \cdot 5$ can be thought of as $5 + 5 + 5$.

Rewrite each expression below by thinking about the multiplication as repeated addition. Then rearrange and rewrite the result to make it as simple as you can.

$$3 \cdot (n + 2) \qquad 5 \cdot (n - 1) \qquad 4 \cdot (3 + n) \qquad 2 \cdot (10 + n)$$

Compare the original expressions above and the final expressions you wrote. Find a pattern or shortcut you could use for multiplying an expression, such as $(n + 4)$ or $(n - 3)$, by a number without writing out all the steps.

Use your shortcut to find $3 \cdot (6 + n)$. Check your answer by writing out all the steps.

In Problem Set B, you will use symbols to try to explain how some of the number tricks you have tried in this investigation work.

Problem Set B

1. The second number trick on page 450 always gives back the number you start with. You can use symbols to understand why this happens.

   a. Copy and complete the table to show each step in symbols.

Step in Words	Step in Symbols
Think of a number.	n
Add 5.	
Multiply by 5.	
Subtract 25.	
Divide by 5.	

   b. In the second step, you add 5 to your number. By the end of the trick, the 5 has "disappeared." What happens to the 5?

2. Here is a trick you worked with in Problem Set A. For this trick, you end up with 6 no matter what number you start with.

   a. Complete the table to show each step in symbols.

Step in Words	Step in Symbols
Think of a number.	n
Add 2.	
Multiply by 3.	
Subtract 3 times the number you started with.	

   b. In the first step, you start with n. What happens to n by the end?

3. Here is the second trick from Problem Set A. For this trick, you always end up with your original number. Complete the table.

Step in Words	Step in Symbols
Think of a number.	n
Multiply by 10.	
Add 10.	
Subtract 9 times the number you started with.	
Subtract 10.	

Share & Summarize

Explain how using letters and symbols can help you understand how a number trick works.

Investigation 2 > Consecutive Numbers

Consecutive numbers are numbers that follow one after the other.

- 199, 200, and 201 are consecutive whole numbers.
- 5, 7, 9, and 11 are consecutive odd numbers.
- 52, 54, 56, 58, and 60 are consecutive even numbers.

Think & Discuss

If *n* is some whole number, how would you write the next three consecutive whole numbers?

If *n* is some whole number, what whole number comes just before *n*? What whole number comes before that?

If *e* is an even number, what are the next two consecutive even numbers? What two even numbers come just before *e*?

Problem Set C

Now you will think about sums of three consecutive whole numbers.

1. Write down three consecutive whole numbers and find their sum. Repeat this with at least four more sets of numbers. Record your results in a table.

Three Consecutive Numbers	Sum

2. Look for a pattern in your table. See if you can find a connection between the sum and the middle consecutive number. Write a rule in words that you could use to find the sum of three consecutive whole numbers without actually adding the numbers.

3. Test your rule on three more sets of consecutive numbers. Does it work for all the numbers you tried?

4. To show that your rule will *always* work, you need to show that it works for *any* three numbers, not just a few specific examples. You can do this by using letters to represent the variables. Use *m* to stand for any whole number. How would you represent the number just before *m* and the number just after *m*?

5. Now write the sum of three consecutive numbers—that is, the sum of *m*, the number just before *m*, and the number just after *m*. Then rearrange and rewrite the sum to make it as simple as possible.

6. Does your result show that your rule always works? Explain.

Since it is impossible to test a rule for *every* number, you can't prove that a rule is always true by giving lots of examples. However, if you use a letter to represent "any number," as you did in Problem Set C, you can often use symbols to show that a rule is always true.

Problem Set D

1. Rosita thinks she has found another rule involving three consecutive whole numbers. She says that if you add the first number and the last number, the result will be twice the middle number.

   a. Test Rosita's rule for at least three sets of numbers. Does it work for all the numbers you tried? If not, find a correct rule relating the sum of the first and last numbers to the middle number.

   b. Using *m* to stand for the middle number, explain why the correct rule (either Rosita's rule or the rule you wrote) always works.

2. Marcus wonders whether the sum of four consecutive whole numbers is always 4 times the second number.

   a. Test Marcus' rule for at least three sets of numbers. Do you think it is correct? If not, find a correct rule relating the sum of four consecutive numbers to the second number.

   b. Using *n* to represent the second number, show that the correct rule (either Marcus' rule or the rule you wrote) always works.

3. Consider the sum of three consecutive even numbers—such as 6, 8, 10 or 42, 44, 46.

   a. Find the sum of at least three sets of three consecutive even numbers.

   b. Write a rule in words that you could use to find the sum of three consecutive even numbers without actually adding the numbers.

   c. Use symbols to help explain why your rule is always true.

Share & Summarize

Suppose you want to convince someone that a relationship involving consecutive numbers is always true. Explain why it is better to use letters and symbols than to test lots of examples.

On Your Own Exercises

Practice & Apply

In Exercises 1–3, the first column of the table gives the steps for a number trick. Complete Parts a–c for each table.

a. Choose at least three different starting numbers, and record the result of each step in the table.

b. Describe your results.

c. Use symbols to explain how the trick works.

1.

Step	Result		
Think of a number.			
Add 6.			
Multiply by 2.			
Add your original number.			
Subtract 3 times your original number.			

2.

Step	Result		
Think of a number.			
Divide by 2.			
Add 1.			
Multiply by 4.			
Subtract your original number.			
Subtract 3.			

3.

Step	Result		
Think of a number.			
Add 2.			
Multiply by 5.			
Subtract 10.			
Subtract 2 times your original number.			
Divide by 6.			

4. Jahmal said that for any four consecutive whole numbers, the sum of the first and last numbers is equal to the sum of the two middle numbers.

a. Test Jahmal's statement for at least three sets of numbers. Is it true for all the numbers you tried?

b. Using f to represent the first number, use symbols to help explain why Jahmal's statement is true for any four consecutive numbers.

5. Consider the sum of five consecutive whole numbers.

    a. Find the sum of at least four sets of five consecutive whole numbers.

    b. Write a rule in words relating the sum of five consecutive whole numbers to the middle number.

    c. Use symbols to help show that your rule must always be true.

Connect & Extend

6. Make up a number trick with at least three steps that always gives a result of 0. Use symbols to show why your trick works.

7. In the addition puzzles below, each letter stands for a single digit. The same letter stands for the same digit, and different letters stand for different digits.

    a. Look at this subtraction problem:

        Z must be equal to 3, since $8 - 5 = 3$.

$$\begin{array}{r} 7\ 2\ 8 \\ -\ 2\ 4\ 5 \\ \hline X\ Y\ Z \end{array} \qquad \begin{array}{r} 7\ 2\ 8 \\ -\ 2\ 4\ 5 \\ \hline X\ Y\ 3 \end{array}$$

    Figure out what digits X and Y must represent.

    b. Solve this puzzle. In other words, figure out what digits A, B, and C represent.

$$\begin{array}{r} A\ 1\ 2 \\ +\ C\ B\ A \\ \hline 9\ C\ 7 \end{array}$$

    c. Find at least three different solutions to this puzzle:

$$\begin{array}{r} U \\ +\ U \\ \hline H\ E \end{array}$$

    d. In Part c, are there any digits U cannot be? Explain your answer.

    e. In Part c, are there any digits H cannot be? Explain your answer.

    f. Solve this puzzle:

$$\begin{array}{r} B\ B\ E \\ +\ H\ E \\ \hline H\ E\ E\ E \end{array}$$

In your own words

Describe how using letters and symbols can help you

• show how number tricks work

• explain why relationships involving consecutive numbers are always true

8. Paul is 3 times as old as Ann will be next year.

    a. Suppose *A* stands for Ann's age. How old will Ann be in a year?

    b. Write a rule you can use to find Paul's age *P* if you know Ann's age.

    c. If Ann is now 10 years old, how old is Paul? Explain how you found your answer.

    d. If Ann is 15, how old is Paul?

    e. If you know that Paul is 24, explain how you can find Ann's age.

9. Jonas is twice as old as Eva was last year.

   a. If J represents Eva's age this year, how old was she last year?

   b. Write a rule you can use to find Jonas' age C if you know Eva's age.

   c. Suppose Eva is 15 *this* year. How old is Jonas this year?

   d. Suppose Eva was 12 *last* year. How old is Jonas this year?

   e. Suppose Jonas is 30 this year. How old is Eva this year?

10. Natalie is 3 times as old as Terry was last year.

   a. Write a rule for this situation, using letters and symbols. Tell what the letters in your rule stand for.

   b. If Terry is now 7, how old is Natalie?

Number Sense Fill in each ⬤ with >, <, or =.

11. $33\frac{1}{3}\%$ ⬤ $\frac{1}{3}$

12. $\frac{7}{8}$ ⬤ 85%

13. 0.398 ⬤ $\frac{2}{5}$

14. $^-5$ ⬤ $^-1$

15. $0.\overline{5}$ ⬤ $\frac{5}{9}$

16. $\frac{31}{40}$ ⬤ 75%

17. $\frac{347}{899}$ ⬤ $\frac{347}{900}$

18. $\frac{6}{7}$ ⬤ $\frac{7}{8}$

19. 80% ⬤ $\frac{45}{60}$

20. 0.01 ⬤ 0.1%

Measurement Fill in the blanks.

21. 356 cm = _____ m

22. 356 cm = _____ mm

23. 44 m = _____ mm

24. 5 mm = _____ m

25. 5 mm = _____ cm

26. 89,000 mm = _____ m

Geometry Estimate the measure of each angle.

27.

28.

29.

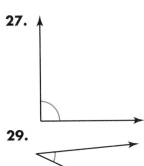

30.

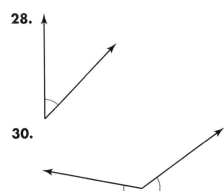

Chapter Summary

In this chapter, you worked with *symbolic rules,* in which letters are used to represent variables. You started by writing rules to relate variables in sequences of dots and toothpicks. You found ways to explain why a rule works for every term in a sequence and to show that two different rules are equivalent. Then you played the game *What's My Rule?* and expressed input/output rules by using symbols.

You saw that real-life situations often involve variables, and you wrote symbolic rules to represent many such situations. You found that the same relationship can often be expressed in several ways, and you practiced rewriting rules to form equivalent rules.

Finally, you saw how using letters to represent variables can help you show how number tricks work and explain why certain relationships involving consecutive numbers are always true.

Strategies and Applications

The questions in this section will help you review and apply the important ideas and strategies developed in this chapter.

Writing and interpreting rules for sequences and input/output tables

1. Consider this toothpick pattern:

Term 1   Term 2   Term 3

  a. Choose a variable other than the term number.

  b. Create a table showing the value of your variable for each term.

  c. Try to find a rule that connects the term number and your variable. Write the rule as simply as you can, using *n* to represent the term number and another letter to represent your variable.

2. Stacey drew a dot sequence. The rule $d = 5 \cdot n - 2$, where n represents the term number and d represents the number of dots in a term, describes her sequence.

a. Find the number of dots in each of the first four terms of Stacey's sequence.

b. Draw a dot sequence that fits the rule.

3. Write two different rules to express the relationship between a and b.

a	0	1	2	3	4
b	3	6	9	12	15

Showing that two rules for a sequence are equivalent

4. Consider this toothpick sequence:

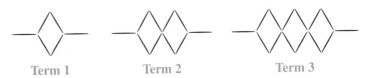

Term 1   Term 2   Term 3

a. Write two equivalent rules for the number of toothpicks in a term.

b. Use words and diagrams to explain why each of your rules is correct.

5. Suppose two different rules give the same number of dots for Term 2 of a sequence. Can you conclude that the rules are equivalent? Give an example to support your answer.

Writing and interpreting rules for real-life situations

6. At the entrance to Fairmont Park, two stores rent in-line skates. Rolling Along charges $4.50, plus $2.50 per hour. The Skate Shop has a flat rate of $4 per hour.

a. How much would it cost to rent skates from Rolling Along for 2 hours? How much would it cost to rent skates from the Skate Shop for 2 hours?

b. Write a rule for the cost c of renting skates from Rolling Along for h hours.

c. Write a rule for the cost c of renting skates from the Skate Shop for h hours.

d. Suppose you want to rent in-line skates to use in Fairmont Park. Assuming the stores have skates of the same quality, how would you decide where to rent your skates?

7. Consider this rule:

$$t = 3.5 + d$$

a. Describe a situation this rule could represent. Tell what variables the letters stand for.

b. Write the rule in two different ways. Explain how you know your rules are correct.

Using variables to explain number relationships

8. The first column of this table gives the steps for a number trick:

Step	Result		
Think of a number.			
Add 3.			
Multiply by 4.			
Subtract 12.			
Subtract your original number.			
Divide by 3.			

a. Choose at least three different starting numbers, and record the result of each step in the table.

b. Describe your results.

c. Use symbols to explain how the trick works.

9. Consider sums of three consecutive whole numbers.

a. Write down three consecutive whole numbers and find their sum. Repeat this with at least three more sets of numbers. Record your results in a table.

Three Consecutive Numbers	Sum

b. Look for a connection between the first number and the sum. Write a rule in words that you could use to find the sum of three consecutive whole numbers based only on the value of the first number.

c. Using f to stand for the first number, explain why your rule always works.

Demonstrating Skills

Find the value of t when n is 7.

10. $t = 3n - 13$ **11.** $t = 2.5 + 1.5 \cdot n$ **12.** $t = 5 \cdot (n - 4)$

Find the value of y when x is $\frac{2}{5}$.

13. $y = 20 \cdot x$ **14.** $y = 4 \cdot x - x$ **15.** $y = x \div \frac{7}{10}$

Write a rule for the relationship between x and y.

16.

x	0	1	2	3	4
y	1	2	5	10	17

17.

x	9	0.75	1	7	5.5
y	34	1	2	26	20

18. Write two rules that are equivalent to $v = 157 \cdot q$.

19. Write two rules that are equivalent to $m = n + 9$.

20. The number of cantaloupes Miguel has growing in his garden is $\frac{1}{3}$ the number of pumpkins he has. Write a rule to express this relationship. Be sure to tell what the letters in your rule represent.

21. A movie theater charges $7.50 for an adult's ticket and $3.75 for a child's ticket. Write a rule for calculating the total cost t if a adults and c children see a movie.

22. Members of the student council sold tickets to an after-school concert. Carlos sold 3 times as many tickets as Angie. Yoshi sold 5 fewer tickets than Angie. Mel sold 12 more tickets than Carlos.

    a. Write a rule for the relationship between the number of tickets Carlos sold c and the number of tickets Angie sold a.

    b. Write a rule for the relationship between the number of tickets Yoshi sold y and the number of tickets Angie sold a.

    c. Write a rule for the relationship between the number of tickets Mel sold m and the number of tickets Angie sold a.

CHAPTER 8

Geometry and Measurement

Olympic Proportions The 2000 Olympic Games in Sydney, Australia, featured 35 sports. The table below lists the dimensions, perimeter, and area of the rectangular courts, fields, mats, and pools where some of these sports were played.

Sport (meters)	Length (meters)	Width (meters)	Perimeter (meters)	Area (square meters)
Football (soccer) field	100	70	340	7,000
Field hockey field	91.4	55	292.8	5,027
Swimming pool	50	25	150	1,250
Handball court	40	20	120	800
Water polo pool	30	20	100	600
Volleyball court	18	9	54	162
Badminton court (singles)	13.4	5.18	37.16	69.41
Fencing piste	14	1.5	31	21

Think About It How do the dimensions of your classroom compare to the dimensions of the volleyball court?

Family Letter

Dear Student and Family Members,

It's time for a little change of pace. Our next chapter in mathematics is about measurement in geometry. In this chapter, you will learn about angles, angle measurement, and perimeters and areas of squares, rectangles, and irregular shapes. Finally, you will learn about the famous Pythagorean Theorem, $a^2 + b^2 = c^2$, that is used with right triangles.

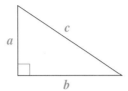

Perimeter is a measurement of the distance around a figure. When a figure is a circle, the distance around it is called the *circumference*. To find the circumference of a circle, you will discover that you can multiply the diameter by pi. Pi (π) is approximately equal to 3.14.

Perimeter →   ← Circumference

Vocabulary
There are many vocabulary terms in this chapter. Some of them may be familiar to you.

acute angle	**parallelogram**
area	**perfect square**
chord	**perimeter**
circumference	**perpendicular**
diameter	**radius**
hypotenuse	**right angle**
inverse operations	**right triangle**
leg	**square root**
obtuse angle	

What can you do at home?

Geometry is everywhere! Ask your student for examples of perimeter and area in his or her daily life.

8.1 Angles

In Chapter 1, you investigated angles. You learned that an angle is defined as two rays with a common endpoint, called the *vertex*.

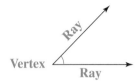

Remember

You can think of an angle as a rotation. A 360° angle is a rotation around a complete circle. A 180° angle is a rotation $\frac{1}{2}$ of the way around a circle. A 90° angle is a rotation $\frac{1}{4}$ of the way around a circle.

Angles are measured in degrees. In Chapter 1, you used 90°, 180°, and 360° angles as benchmarks to help estimate the measures of other angles.

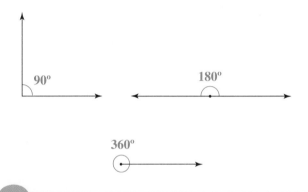

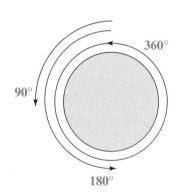

Think & Discuss

Each diagram is constructed from angles of the same size. Estimate the measure of each marked angle, and explain how you found it.

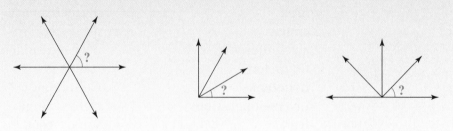

Investigation 1 ▶ Measuring Angles

A *protractor* is a tool for measuring angles. A protractor has two sets of degree labels around the edge of a half circle (or sometimes a full circle). The line that goes through 0° is called the *reference line.*

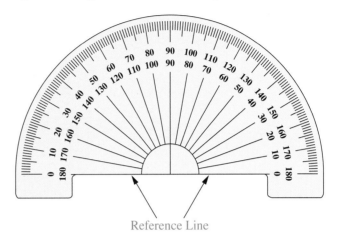

Reference Line

Just the facts

In in-line skating, 360s (full turns) and 540s (one and a half turns) are two of the more difficult stunts.

To measure an angle, follow these steps:

• Place the bottom center of the protractor at the vertex of the angle.

• Line up the reference line with one ray of the angle. Make sure the other ray can be seen through the protractor. (You may need to extend this ray so that you can see where it meets the tick marks along the edge of the protractor.)

• Read the angle measurement.

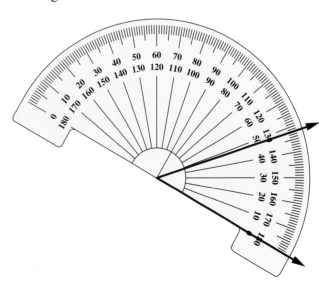

MATERIALS

- protractor
- copies of the angles

Just the facts

A *mariner's astrolabe* was a navigational instrument of the 15th and 16th centuries. Sailors used it to measure the angle of elevation of the sun or other star. This measurement could help them determine their ship's latitude.

Think & Discuss

The angle below measures about 48°. Or is it about 132°? How do you know which number to use?

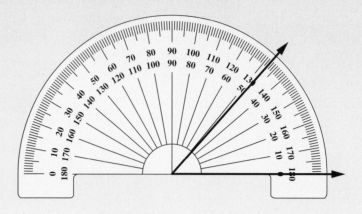

Is the measure of the angle below a little more than 90° or a little less than 90°? How do you know?

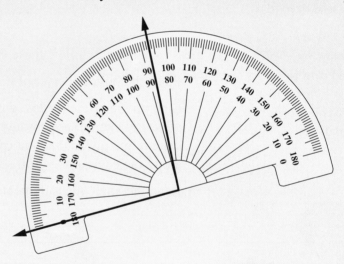

Measure these two angles. How do the measures compare?

Find the measure of the angle below. Describe the method you used.

468 CHAPTER 8 Geometry and Measurement

You have seen that when you measure an angle with a protractor, you must determine which of two measurements is correct. One way to decide is to compare the angle to a 90° angle. Because 90° is such an important benchmark, angles are sometimes classified by how their measures compare to 90°.

VOCABULARY
acute angle
obtuse angle
perpendicular
right angle

Acute angles measure less than 90°.

Obtuse angles measure more than 90° and less than 180°.

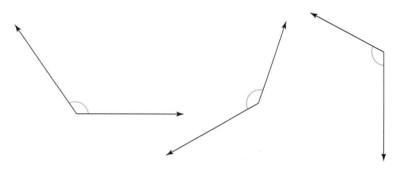

Right angles measure exactly 90°. Right angles are often marked with a small square at the vertex.

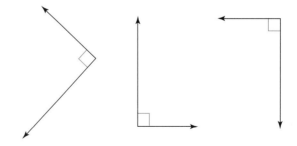

Two lines or segments that form a right angle are said to be **perpendicular.**

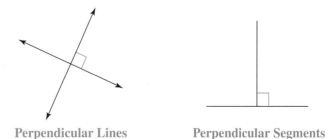

Perpendicular Lines   Perpendicular Segments

Problem Set A

Tell whether each angle is acute or obtuse. Then find its measure.

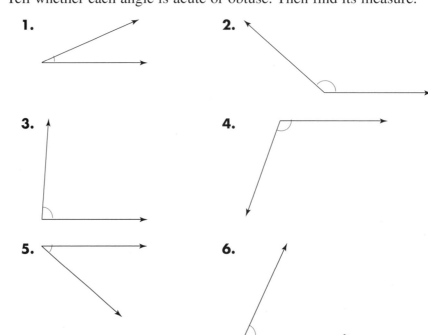

1.

2.

3.

4.

5.

6.

You have seen that a protractor is a useful tool for measuring angles. You can also use a protractor to draw angles with given measures.

EXAMPLE

To create a 25° angle, start by drawing a line segment. This segment will be one side of the angle.

Line up the reference line of the protractor with the segment, with the center of the protractor at one endpoint of the segment. This endpoint will be the vertex of the angle.

Draw a mark next to the 25° label on the protractor. (Be sure to choose the correct 25° label!)

Remove the protractor, and draw a segment from the vertex (the endpoint that was at the center of the protractor) through the mark.

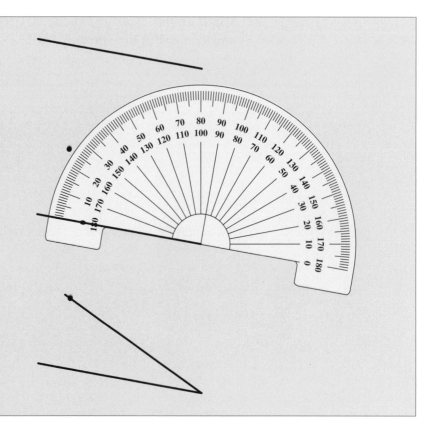

- protractor
- ruler

Just the facts

A *theodolite* is an angle-measuring instrument used in navigation, meteorology, and surveying.

Problem Set B

1. Draw a 160° angle. Include a curved angle mark to show which angle is 160°.

2. Draw a 210° angle. Include an angle mark to show which angle is 210°.

3. Draw two perpendicular segments.

4. Draw a triangle in which one angle measures 50° and the other angles have measures greater than 50°. Label each angle with its measure.

5. Draw a triangle with one obtuse angle. Label each angle with its measure.

6. Draw a triangle with two 60° angles.

7. Measure the sides of the triangle you drew in Problem 6. What do you notice?

8. Draw a square. Make sure all the sides are the same length and all the angles measure 90°.

9. Draw a polygon with any number of sides and one angle that has a measure greater than 180°. Mark that angle.

Share & Summarize

1. When you measure an angle with a protractor, how do you know which of the two possible numbers to choose?

2. The protractor on page 470 has a scale up to 180°. Describe how you would use such a protractor to draw an angle with a measure greater than 180°. Give an example if it helps you to explain your thinking.

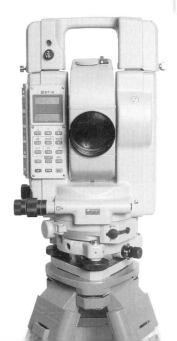

Investigation 2 Investigating Angle Relationships

You can refer to the angles in a drawing more easily if you label them with numbers or letters.

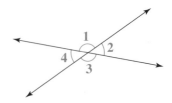

In the drawing above, the measure of Angle 1 is 135°. You can write this in symbols as $m\angle 1 = 135°$. The "*m*" stands for "measure," and $\angle$ is the symbol for "angle."

MATERIALS
• protractor
• ruler

Problem Set C

Two lines that cross each other—like the lines in the drawing above—are said to *intersect*.

1. Measure angles 1, 2, 3, and 4 in the drawing above.

2. Which angles have the same measure?

3. Use a ruler to draw another pair of intersecting lines. Measure each of the four angles formed, and label each with its measure.

4. Draw one more pair of intersecting lines, and label each angle with its measure.

5. What patterns do you see relating the measure of the angles formed by two intersecting lines?

When two lines intersect, two angles that are not directly next to each other are called *vertical angles*. In the drawing below, $\angle a$ and $\angle c$ are vertical angles, and $\angle b$ and $\angle d$ are vertical angles.

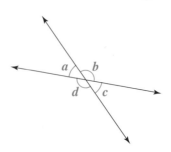

Below, Conor explains the relationship he discovered in Problem Set C.

Think & Discuss

In the cartoon, Conor showed that $m\angle 1 = m\angle 3$. Explain why $m\angle 2 = m\angle 4$.

The *interior angles* of a polygon are the angles inside the polygon. In this quadrilateral, the interior angles are marked.

In Chapter 1, you discovered that the sum of the measures of the interior angles of any triangle is 180°. In the next problem set, you will look for similar rules about the angle sums of other polygons.

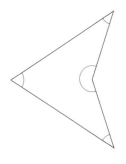

MATERIALS
- ruler
- protractor

Problem Set D

1. Use a ruler and pencil to draw a quadrilateral. Measure each interior angle, and then find the sum of the four angles.

2. Now draw a pentagon. Measure each interior angle, and find the sum of the five angles.

3. Finally, draw a hexagon. Measure each interior angle, and find the sum of the six angles.

Remember

A *concave polygon* has an interior angle with measure greater than 180°. Concave polygons look "dented."

Think & Discuss

For each problem in Problem Set D, you and your classmates probably all drew different polygons. Compare the angle sums you found with the sums found by your classmates. What patterns do you see?

Describe a rule you could use to predict the sum of the interior-angle measures of a polygon when you know only the number of angles.

Use your rule to predict the interior-angle sums for each concave polygon below. Check your predictions by measuring the angles. Be sure to measure the *interior* angles.

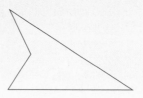
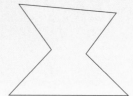

By now, you have probably concluded that the sum of the angle measures in a polygon depends only on the number of angles (or the number of sides). You may have also discovered a rule for predicting the angle sum of any polygon when you know the number of angles.

Hannah and Jahmal wondered whether they could use what they know about the angle sum for triangles to think about the angle sums for other polygons.

Remember

A *diagonal* is a segment that connects two vertices of a polygon but is not a side of the polygon.

We can divide any quadrilateral into two triangles by drawing a diagonal.

180° + 180° = 360°

The angle sum for each triangle is 180°, so the angle sum for a quadrilateral must be twice that. Every quadrilateral must have an angle sum of 360°.

I wonder if our strategy will work for other polygons, like these.

In the next problem set, you will investigate whether Hannah and Jahmal's strategy applies to other polygons. You will also see how their strategy leads to a rule for calculating angle sums.

MATERIALS
ruler

Problem Set E

1. First consider pentagons.

    a. Draw two pentagons. Make one of the pentagons concave. Divide each pentagon into triangles by drawing diagonals from one of the vertices.

    b. Into how many triangles did you divide each pentagon?

    c. Use your answer to Part b to find the sum of the interior angles in a pentagon.

2. Now consider hexagons.

    a. Draw two hexagons. Make one of the hexagons concave. Divide each hexagon into triangles by drawing diagonals from one of the vertices.

    b. Into how many triangles did you divide each hexagon?

    c. Use your answer to Part b to find the sum of the interior angles in a hexagon.

Just the facts

Crystallographers—people who study the geometric properties and internal structures of crystals—use reflecting goniometers to measure the angles between the faces of a crystal.

3. Now think about octagons, which are 8-sided polygons.

   a. Without making a drawing, predict how many triangles you would divide an octagon into if you drew all the diagonals from one of the vertices. Explain how you made your prediction.

   b. Draw an octagon, and check your prediction.

   c. Use your answer to find the interior-angle sum for an octagon.

4. Suppose you drew a 15-sided polygon and divided it into triangles by drawing diagonals from one of the vertices.

   a. How many triangles would you make?

   b. Use your answer to find the interior-angle sum for a 15-sided polygon.

5. Suppose a polygon has *n* angles.

   a. How many triangles would you make if you divided the polygon into triangles by drawing diagonals from one of the vertices?

   b. Use your answer to write a rule for calculating the sum of the angle measures *s* in a polygon with *n* angles.

6. A particular quadrilateral has three 90° angles.

   a. What is the measure of the fourth angle? How do you know?

   b. What kind of quadrilateral is it?

Just the facts

A 15-sided polygon is called a *pentadecagon.*

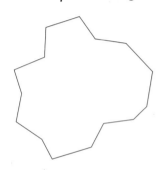

Share & Summarize

1. Marcus said the sum of the angle measures for a quadrilateral must be 720° because a quadrilateral can be split into four triangles by drawing both diagonals.

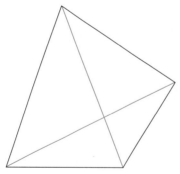

Explain what is wrong with Marcus' argument.

2. What is the sum of the angle measures of a nonagon (a 9-sided polygon)?

On Your Own Exercises

Practice & **Apply**

Find the measure of each angle.

1.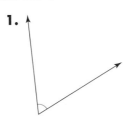

2.

3.

4.

Without measuring, find the missing angle measures.

5.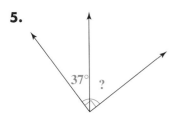

6.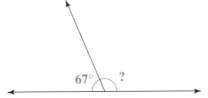

7.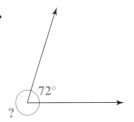

Draw an angle with the given measure.

8. 17°   **9.** 75°   **10.** 164°   **11.** 290°

Draw the figure described. Label every angle in the figure with its measure.

12. a quadrilateral with two 60° angles

13. a pentagon with two 90° angles

14. a quadrilateral with one 200° angle

Without measuring, find the measure of each lettered angle.

15.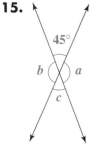

16.

17.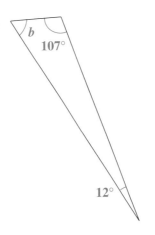

18.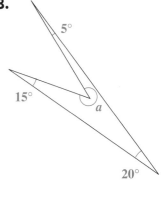

19. In this polygon, ∠*a* and ∠*b* have the same measure. What is it?

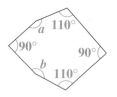

20. What is the measure of each angle of a regular pentagon?

21. What is the measure of each angle of a regular hexagon?

22. **Sports** The drawings below show angles formed by a soccer player and the goalposts. The greater the angle, the better chance the player has of scoring a goal. For example, the player has a better chance of scoring from Position A than from Position B.

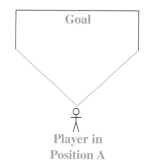

In Parts a and b, it may help to trace the diagrams and draw and measure angles.

a. Seven soccer players are practicing their kicks. They are lined up in a straight line in front of the goalposts. Which player has the best (the greatest) kicking angle?

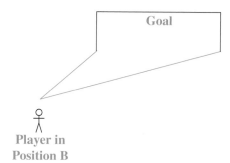

b. Now the players are lined up as shown. Which player has the best kicking angle?

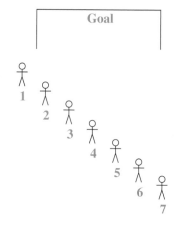

23. The *diameter* of a circle is a segment that passes through the center of the circle and has both its endpoints on the circle.

Diameter

The four triangles below have all three vertices on a circle and the diameter as one side.

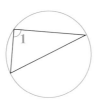  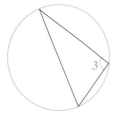 

Measure each numbered angle. What do the measures have in common?

24. You discovered a rule about the sums of the interior angles of polygons. Polygons also have *exterior* angles, which can be found by extending their sides. In the drawings below, the exterior angles are marked.

In Parts a–e, find the measure of each exterior angle. Then find the sum of the measures.

a.

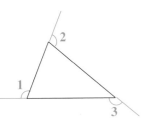

b.

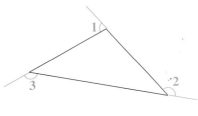

c.

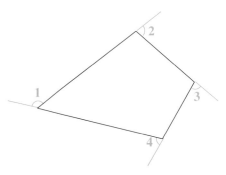

d.

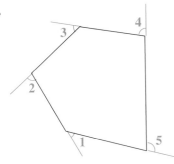

e.

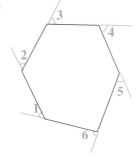

f. Describe any pattern you find in the measures in Parts a–e.

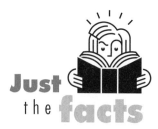
25. Sports The angle at which a pool ball hits the side of a table has the same measure as the angle at which it bounces off the side. This is shown in the drawing at right. The marked angles have the same measure, and the arrow shows the ball's path.

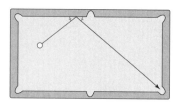

In Parts a–c, trace the drawing. Then use your protractor to find the path the ball will take when it bounces off the side. Tell whether the ball will go into a pocket or hit another side. (Draw just one bounce.)

a.

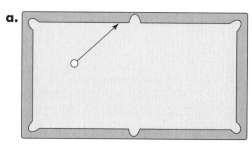

b.

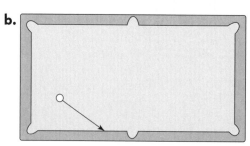

c.

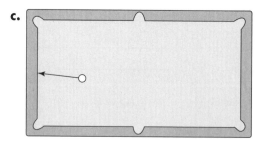

d. Challenge Trace this drawing. Draw a path for which the ball will bounce off a side and land in the lower-right pocket.

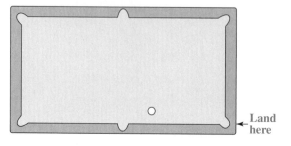

Land here

Mixed Review

Find the least common multiple of each pair of numbers.

26. 14 and 21

27. 17 and 51

28. 54 and 24

29. 13 and 5

30. 100 and 75

31. 32 and 36

Evaluate each expression.

32. $\frac{1}{3} + \frac{3}{4} \cdot \frac{4}{9}$

33. $\frac{5}{14} \div \frac{6}{7} - \frac{2}{7}$

34. $2\frac{1}{5} \div \frac{11}{15} \cdot \frac{7}{10}$

35. $3\frac{5}{8} - 1\frac{1}{2} + \frac{3}{4}$

36. $\frac{21}{34} \cdot \left(\frac{7}{12} + \frac{1}{8} \right)$

37. $\left(\frac{5}{6} \div \frac{5}{12} \right) + \left(1\frac{5}{6} \cdot 6 \right)$

38. **Technology** The graphs give information about two video games. Use the graphs to determine whether each statement is true or false, and explain how you decided.

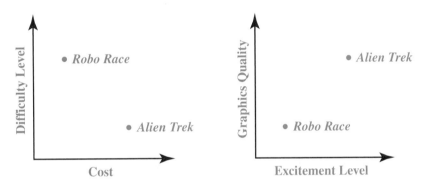

a. The more exciting game is less expensive.

b. The more difficult game has lower-quality graphics.

c. The less difficult game is more exciting.

d. The less expensive game has better graphics.

e. The game with better graphics is less exciting.

8.2 Measuring Around

VOCABULARY
perimeter

The **perimeter** of a two-dimensional shape is the distance around the shape. The perimeter of the shape at right is 10.8 cm.

2 cm   2 cm

2 cm   2 cm

2.8 cm

Think & Discuss

Describe as many methods as you can for measuring the perimeter of the floor of your classroom.

Which of your methods do you think will give the most accurate measurement?

Which of your methods do you think is the most practical?

Investigation 1 ▶ Finding Perimeter

To find the perimeter of a polygon, you simply add the lengths of its sides.

MATERIALS
metric ruler

Problem Set A

This is the floor plan of the second floor of Millbury Middle School. On the drawing, each centimeter equals 2 meters.

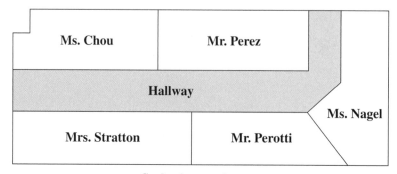

Ms. Chou   Mr. Perez

Hallway

Ms. Nagel

Mrs. Stratton   Mr. Perotti

Scale: 1 cm = 2 m

1. Without measuring, tell whose classroom you think has the greatest perimeter, and explain why you think so.

2. Look at the floor plan for Ms. Nagel's room.

   a. What type of polygon is the floor of Ms. Nagel's room?

   b. Find the perimeter of Ms. Nagel's floor plan to the nearest tenth of a centimeter. Then calculate the perimeter of the actual floor in meters.

3. Find the perimeter of Mrs. Stratton's floor plan to the nearest tenth of a centimeter. Then calculate the perimeter of the actual floor in meters.

4. Look at the floor plan for Mr. Perez's room.

   a. Is Mr. Perez's floor a rectangle? How do you know?

   b. Describe how to find the perimeter of Mr. Perez's floor plan by making only two measurements.

   c. Measure the perimeter of Mr. Perez's floor plan to the nearest tenth of a centimeter. Then calculate the perimeter of the actual floor in meters.

5. To find the perimeter of Ms. Chou's floor plan, Althea made the measurements labeled below. She claims these are the only measurements she needs to make.

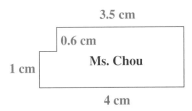

   3.5 cm
   0.6 cm
   Ms. Chou
   1 cm
   4 cm

   a. Is Althea correct? If so, explain how to find the perimeter of Ms. Chou's room using only these measurements. If not, tell what other measurements you would need.

   b. Find the perimeter of Ms. Chou's floor plan to the nearest tenth of a centimeter. Then calculate the perimeter of the actual floor in meters.

6. Which teacher's classroom floor has the greatest perimeter? What is the perimeter?

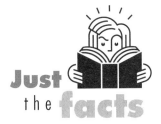

Just the facts

Many colleges and universities offer classes over the Internet, allowing students to earn college credit—and even a college degree—without setting foot in a classroom!

In Problem Set A, you probably realized you could find the perimeter of a rectangle without measuring every side. This is because the opposite sides of a rectangle are the same length. If you measure the length and the width of a rectangle, you can find the perimeter using either of two rules:

Add the length and the width, and double the result.

Double the length, double the width, and add the results.

If you use P to represent the perimeter and L and W to represent the length and width, you can write these rules in symbols.

Geometric rules expressed using symbols, like those above, are often called *formulas.*

Perimeter of a Rectangle

$$P = 2 \cdot (L + W) \qquad\qquad P = 2L + 2W$$

In these formulas, P represents the perimeter and L and W represent the length and width.

Problem Set B

1. Use one of the perimeter formulas to find the perimeter of a rectangle with length 5.7 meters and width 2.9 meters.

2. The floor of a rectangular room has a perimeter of 42 feet. What are three possibilities for the dimensions of the floor?

3. A square floor has a perimeter of 32.4 meters. How long are the sides of the floor?

4. Write a formula for the perimeter of a square, using P to represent the perimeter and s to represent the length of a side. Explain why your formula works.

This floor plan is of the auditorium at Marshville Middle School.

Scale: 1 cm = 5 m

Since part of the floor is curved, it is difficult to find the perimeter using just a ruler. You could use a measuring tape or a piece of string to find the length of the curved part. Another method is to use a polygon to *approximate* the shape of the floor.

Luke drew a pentagon to approximate the shape of the floor:

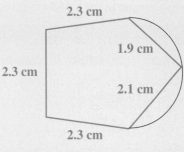

2.3 cm

1.9 cm

2.3 cm

2.1 cm

2.3 cm

Scale: 1 cm = 5 m

Then he found the pentagon's perimeter:

$$2.3 + 2.3 + 2.3 + 1.9 + 2.1 = 10.9 \text{ cm}$$

MATERIALS

- copies of the auditorium floor
- string
- ruler

Problem Set C

1. You can get a closer approximation than Luke's by using a polygon with more sides.

   a. Try using a hexagon—a polygon with six sides—to approximate the shape of the floor plan. What perimeter estimate do you get using a hexagon?

   b. Is the actual perimeter greater than or less than your estimate? Explain.

   c. Now try a heptagon—a polygon with seven sides—to approximate the shape of the floor plan. What perimeter estimate do you get using a heptagon?

   d. Is the actual perimeter greater than or less than your estimate? Explain.

   e. Using a polygon with more than seven sides, make another estimate. What is your estimate?

2. Wrap a piece of string around the floor plan. Try to keep the string as close to the sides of the floor plan as possible. Then mark the string to indicate the length of the perimeter, and measure the string's length up to the mark. What is your perimeter estimate?

3. Which of your estimates do you think is most accurate? Explain.

Investigation 2 Approximating π

In the last investigation, you found perimeters of polygons, and you estimated perimeters of a shape with curved sides. In this investigation, you will focus on circles.

VOCABULARY
chord
circumference
diameter
radius

The perimeter of a circle is called its **circumference.** Although you can estimate the circumference of a circle by using string or by approximating with polygons, there is a formula for finding the exact circumference. Before you begin thinking about circumference, you need to learn some useful words for describing circles.

A **chord** is a segment connecting two points on a circle. The **diameter** is a chord that passes through the center of the circle. *Diameter* also refers to the distance across a circle through its center. The **radius** is a segment from the center to a point on the circle. *Radius* also refers to the distance from the center to a point on the circle. The plural of *radius* is *radii*.

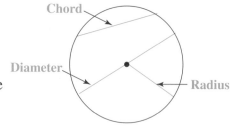

Think & Discuss

Are all the chords of a circle the same length? If not, which are the longest?

Are all the diameters of a circle the same length? Are all the radii the same length?

Write a rule for the relationship between the radius of a circle r and its diameter d.

This quote from the novel *Contact* by Carl Sagan mentions a relationship between the circumference and diameter of any circle:

*In whatever galaxy you happen to find yourself,
you take the circumference, divide it by its diameter,
measure closely enough, and uncover a miracle.*

In the next problem set, you will examine the relationship that Sagan is describing.

MATERIALS

- 5 objects with circular faces (for example, a CD, a coffee can, a roll of tape, a plate, and a nickel)
- string or measuring tape
- ruler
- scissors

Problem Set D

For this problem set, your group will need five objects with circular faces.

1. Follow these steps for each object:

   - Use string or a measuring tape to approximate the circumference of the object.

   - Trace the circular face of the object. Cut out the tracing, and fold it in half to form a crease along the circle's diameter. Measure the diameter.

   Record your measurements in a table like this one.

Object	Circumference, *C*	Diameter, *d*

2. Do you see a relationship between the circumference and the diameter of each circle? If so, describe it.

3. The quotation from *Contact* mentions dividing the circumference by the diameter. Add a column to your table showing the quotient $C \div d$ for each object. Describe any patterns you see.

4. On the board, record the values of $C \div d$ that your group found. How do your results compare with those of other groups?

5. Does the $C \div d$ value depend on the size of the circle? Explain.

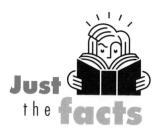
No matter what size a circle is, the circumference divided by the diameter is always the same value. You probably discovered that this quotient is a little more than 3. The exact value is a decimal number whose digits never end or repeat. This value has been given the special name "pi" and is represented by the Greek letter π.

$\pi = C \div d$, where C is the circumference of a circle and d is the diameter

Since the digits of π never end or repeat, it is impossible to write its exact numeric value. The number 3.14 is often used as an approximation of π. You can press the $\boxed{\pi}$ key on your calculator to get a closer approximation.

3.141592654

If you start with the division equation $\pi = C \div d$, you can write the related multiplication equation: $C = \pi \cdot d$. This is the formula for computing the circumference C of a circle when you know its diameter d.

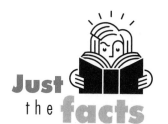

Circumference of a Circle

$$C = \pi \cdot d$$

In this formula, C is the circumference and d is the diameter. Since the diameter of a circle is twice the radius r, you can also write the formula in these ways:

$$C = \pi \cdot 2 \cdot r \qquad C = 2 \cdot \pi \cdot r$$

Since the radius of this circle is 2.5 cm, the diameter is 5 cm. So,

$$C = \pi \cdot d$$
$$= \pi \cdot 5$$

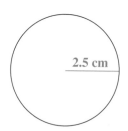

2.5 cm

The exact circumference of the circle is $5 \cdot \pi$ cm. Although you can't write the circumference as an exact numeric value, you can use the $\boxed{\pi}$ key on your calculator to find an approximation.

$$C = \pi \cdot 5 \text{ cm} \approx 15.71 \text{ cm}$$

The symbol $\approx$ means "is approximately equal to."

Problem Set E

For this problem set, use your calculator's $\boxed{\pi}$ key to approximate π. (If your calculator does not have a $\boxed{\pi}$ key, use 3.14 as an approximation for π.) Give your answers to the nearest hundredth.

1. Find the circumference of a circle with diameter 9 centimeters.

2. A circular pool has a circumference of about 16 meters. What is the pool's diameter?

3. Find the circumference of this circle:

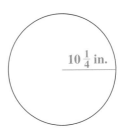

$10\frac{1}{4}$ in.

4. The radius of Earth at the equator is about 4,000 miles.

   a. Suppose you could wrap a string around Earth's equator. How long would the string have to be to reach all the way around? (Assume the equator is a perfect circle.)

4,000 mi

   b. Now suppose you could raise the string 1 mile above Earth's surface. How much string would you have to add to your piece from Part a to go all the way around?

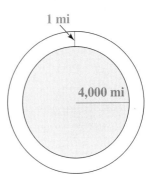

1 mi

4,000 mi

Share & Summarize

Explain what π is in your own words. Be sure to discuss

- how it is related to circles

- its approximate value

On Your Own Exercises

Sports In Exercises 1–3, use this diagram of a baseball field.

1. Consider the baseball diamond in this diagram.

  a. Find the perimeter of the diamond to the nearest $\frac{1}{4}$ inch.

  b. An actual baseball diamond is a square with sides 90 feet long. What is the perimeter of an actual baseball diamond?

  c. The perimeter of an actual baseball diamond is about how many times the perimeter of the baseball diamond in the diagram?

2. Rosita approximated the perimeter of the infield using a quadrilateral. She found a perimeter of about $5\frac{1}{8}$ inches.

  a. Trace the shape of the infield. Use a polygon with more than four sides to find a better approximation of the infield's perimeter. Make all measurements to the nearest $\frac{1}{8}$ of an inch.

  b. How does your approximation compare to Rosita's?

3. Suppose the manager tells a player to run five laps around the entire baseball field (including the outfield), staying as close to the outer edge as possible.

  a. Measure the perimeter of the field in the diagram at the top of this page to the nearest $\frac{1}{8}$ of an inch.

  b. If 1 inch on the diagram represents approximately 100 feet on the actual field, about how many miles will the player run in his five laps around the field?

Remember
1 mile = 5,280 feet

 impactmath.com/self_check_quiz

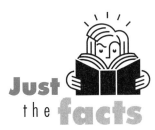
4. This is the floor plan of the Harperstown Library. What is the perimeter of the floor?

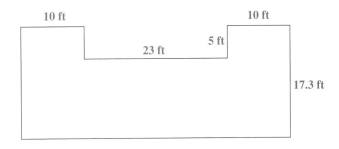

5. Give the dimensions of five rectangles that have a perimeter of 50 feet.

6. Find the circumference of a circle with diameter 7 meters. Round your answer to the nearest hundredth.

7. Find the circumference of a circle with radius 4.25 inches. Round your answer to the nearest hundredth.

8. The circumference of a tire is 150 inches. What is the tire's radius? Round your answer to the nearest hundredth.

9. Challenge The radius of the wheel on Jahmal's bike is 2 feet.

   a. If he rides 18.9 feet, how many full turns will the wheel make?

   b. If the wheel on Jahmal's bike turned 115 times, how many feet did Jahmal ride? About how many miles is this?

   c. If Jahmal rides 20 miles, how many times will his wheel turn?

10. Two shapes are *nested* when one is completely inside the other.

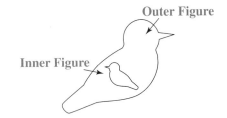

   a. Draw two nested shapes so that the outer shape has a greater perimeter than the inner shape. Give the perimeters of both shapes.

   b. Draw two nested shapes so that the inner shape has a greater perimeter than the outer shape. Give the perimeters of both shapes.

   c. Draw two nested shapes so that the outer shape has the same perimeter as the inner shape. Give the perimeters of both shapes.

   d. Look at your shapes from Parts a–c. In each case, which shape has more space inside: the inner shape or the outer shape? How do you know?

11. Fine Arts The artist M. C. Escher often incorporated mathematics into his artwork. Many of his well-known works are tessellations. A *tessellation* is a design made of identical shapes that fit together without gaps or overlaps.

One way to make a shape that will tessellate is to cut a rectangle into two pieces and slide one piece to the other side.

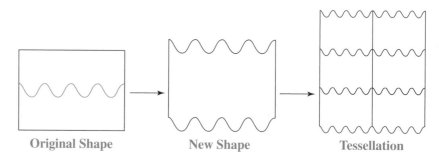

Original Shape   New Shape   Tessellation

a. Find the perimeter of the original shape above.

b. Trace the new shape, and estimate its perimeter by using a polygon approximation or a piece of string.

c. When the new shape is formed from the original, the space inside the shape—the area—stays the same, but the perimeter changes. Explain why this happens.

12. Sports This is a diagram of the outer lane of the track at Albright Middle School. The lane is made of two straight segments and two semicircles (half circles). If a student runs one lap around the track in this lane, how many yards will she run?

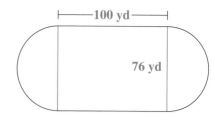

100 yd

76 yd

13. Caroline wrapped a piece of string around the circumference of a circle with a diameter of 23 inches. She cut the string to the length of the circumference and then formed a rectangle with the string. Give the approximate dimensions of three rectangles she could make.

14. A circle with radius 6.5 inches is cut into four wedges and rearranged to form another shape.

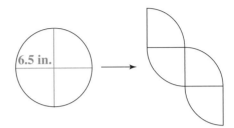

Does the perimeter change? How do you know? If it does change, by how much does it increase or decrease?

Find the greatest common factor of each pair of numbers

15. 14 and 21   **16.** 17 and 51   **17.** 54 and 24

18. 13 and 5   **19.** 100 and 75   **20.** 32 and 36

Write a rule for each relationship between p and q.

21.

p	2	3	4	5	6	7
q	0.5	0.75	1	1.25	1.5	1.75

22.

p	14	6	8.35	8	12	9
q	9	1	3.35	3	7	4

23.

p	2	$\frac{2}{3}$	7	3	10	1
q	4	0	19	7	28	1

24. Statistics Jing kept a record of the length (in minutes) of each phone call she made last week. Here are the results:

7   37   3   24   29   54   12   18   25

15   19   22   32   35   18   21   15   22

a. Make a histogram of the phone-call times. Use 10-minute intervals on the horizontal axis.

b. Describe the distribution of phone-call times.

In your **own words**

Describe what perimeter is and how to find it for various shapes. Give an example of a situation in which finding a shape's perimeter would be useful.

VOCABULARY
area

You know that the perimeter of a two-dimensional shape is the distance around the shape. The **area** of a two-dimensional shape is the amount of space inside the shape.

MATERIALS
* copies of the two shapes
* scissors

Explore

Consider these shapes:

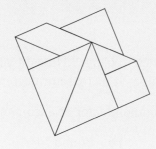

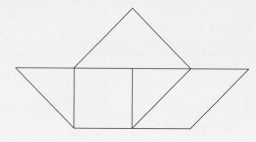

Shape 1   Shape 2

Which shape do you think is larger? That is, which shape do you think has the greater area?

Cut out Shape 1 along the lines, and rearrange the pieces to make a square. Do the same for Shape 2.

Of the two squares you made, which has the greater area? How can you tell?

Do the original shapes have the same areas as the squares? Why or why not?

When determining which shape has the greater area, is it easier to compare the original shapes or the squares? Why?

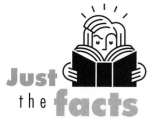

Just the facts

Shape 1 is made with tangram pieces. A tangram is a Chinese puzzle consisting of a square cut into five triangles, a square, and a parallelogram that can be put together to form various shapes.

Squares are the basic unit used for measuring areas. In this lesson, you will look closely at areas of squares and at a special operation associated with the areas of squares.

Investigation 1 Counting Square Units

Area is measured in *square units*, such as square inches and square centimeters. A *square inch* is the area inside a square with sides 1 inch long. A *square centimeter* is the area inside a square with sides 1 centimeter long.

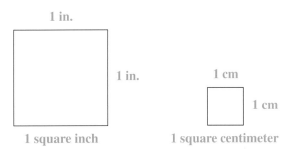

1 in.

1 in.

1 square inch

1 cm

1 cm

1 square centimeter

The area of a shape is the number of square units that fit inside it.

Area = 8 square centimeters

MATERIALS
1-inch tiles

Problem Set A

1. Use your tiles to create two rectangles with perimeters of 12 inches but different areas. Sketch your rectangles, and label them with their areas.

2. Now create two rectangles with areas of 12 square inches but different perimeters. Sketch your rectangles, and label them with their perimeters.

3. Now use your tiles to create this shape:

   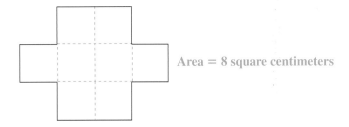

   a. Find the perimeter and area of the shape. (Don't forget to give the units.)

   b. Move one tile to create a shape with a smaller perimeter. Sketch the new shape, and give its perimeter. How does the new shape's area compare to the original area?

   c. Reconstruct the original shape. Move one tile to create a shape with a greater perimeter. Sketch the new shape, and give its perimeter. How does the new area compare to the original area?

4. Use your tiles to create two new shapes so that the shape with the smaller area has the greater perimeter. Sketch your shapes, and label them with their perimeters and areas.

Problem Set B

The shapes below are drawn on dot grids. Find the area of each shape. Consider the horizontal or vertical distance between two dots to be 1 unit.

1.

2.

3.

4.

In Problems 5–8, draw the shape by connecting dots on a sheet of 1-inch dot paper.

5. a square with area 4 square inches

6. a rectangle with area 2 square inches

7. a shape with an area of at least 15 square inches and a perimeter of no more than 25 inches

8. Challenge a square with an area of 2 square inches

Find the area of each shape.

9.
7 in.

7 in.

10.
$\frac{1}{2}$ mi

$\frac{1}{2}$ mi

11.
50 cm

70 cm

12.
2 in.

$\frac{1}{4}$ in.

13. If you know the length and width of a rectangle, how can you find the rectangle's area without counting squares?

Finding the area of a shape by counting squares is not always easy or convenient. Fortunately, there are shortcuts for some shapes.

To find the area of a rectangle, just multiply the length by the width.

Area of a Rectangle

$$A = L \cdot W$$

In this formula, A represents the area of a rectangle, and L and W represent the length and width.

Think & Discuss

On dot or grid paper, draw a rectangle with side lengths 5 units and $7\frac{1}{2}$ units.

Use the formula above to find the area of your rectangle. Check that your answer is correct by counting the squares.

Problem Set C

1. On your newspaper page, draw rectangles around the major items (photographs and art, advertisements, articles, and headlines).

   a. Measure the sides of each rectangle to the nearest tenth of a centimeter.

   b. Calculate the area of each rectangle.

   c. Calculate the area of the entire page.

2. What percent of your newspaper page is used for

   a. photographs and art?

   b. advertisements?

   c. articles?

   d. headlines?

Investigation 2 ▶ Squaring

Recall that an exponent tells you how many times a number is multiplied by itself. You can write the product of a number times itself using the exponent 2:

$$5 \cdot 5 = 5^2$$

Multiplying a number by itself is called *squaring* the number. The expression 5^2 can be read "5 squared."

MATERIALS
dot paper

Think & Discuss

Evaluate 5^2. Then, on a sheet of dot paper, draw a square with an area equal to that many square units.

How long is each side of the square?

Why do you think 5^2 is read "5 squared"?

You can use the $\boxed{x^2}$ key on your calculator to square a number. To calculate 5^2, press these keys:

$\boxed{5}$ $\boxed{x^2}$ $\boxed{\overset{\text{ENTER}}{=}}$

The exponent 2 is often used to abbreviate square units of measurement. For example, *square centimeter* can be abbreviated cm^2, and *square inch* can be abbreviated $in.^2$.

Problem Set D

Fill in the blanks. Here is an example:

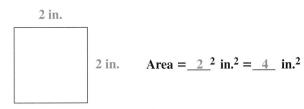

2 in.

2 in.   Area = _2_ ² in.² = _4_ in.²

1.   13 ft

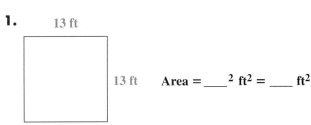

13 ft   Area = ___ ² ft² = ___ ft²

2. 1.25 cm

1.25 cm   Area = ___ ² cm² = ___ cm²

3. $\frac{7}{4}$ in.

$\frac{7}{4}$ in.   Area = ___ ² in.² = ___ in.²

4. Write a formula for finding the area *A* of a square if you know the side length *s*. Use an exponent in your formula.

Find the area of a square with the given side length.

5. 1 in.   **6.** $\frac{1}{3}$ in.   **7.** 19 cm

Find the side length of a square with the given area.

8. 144 ft²   **9.** 10,000 in.²   **10.** 53.29 cm²

In the next problem set, you will look at the squares that can be made with square tiles.

Problem Set E

1. Find every square that can be made from 100 tiles or fewer. Give the side length and area of each square.

2. Is it possible to make a square with 20 tiles? If so, explain how. If not, explain why not.

3. Is it possible to make a square with 625 tiles? If so, explain how. If not, explain why not.

4. **Challenge** Miguel tried to make a square with area 8 in^2 using tiles. After several tries, he said, "I don't think I can make this square using my tiles. But I know I can make it on dot paper."

   On dot paper, draw a square with area 8 in^2.

5. How can you tell whether a given number of tiles can be made into a square without actually making the square?

A number is a **perfect square** if it is equal to a whole number multiplied by itself. In other words, a perfect square is the result of *squaring* a whole number.

Whole Number Squared	1^2	2^2	3^2	4^4	5^2
Perfect Square	1	4	9	16	25

Geometrically, a perfect square is the area of a square with whole-number side lengths. In Problem Set E, the perfect squares were the numbers of tiles that could be formed into squares.

Problem Set F

1. Find three perfect squares greater than 1,000.

2. Is 50 a perfect square? Why or why not?

Tell whether each number is a perfect square, and explain how you know.

3. 3,249   4. 9,196.81   5. 12,225   6. 184,041

7. Find two perfect squares whose sum is also a perfect square.

8. Find two perfect squares whose sum is not a perfect square.

1. How is the idea of squaring a number related to the area of a square?

2. Can *any* number be squared? Why or why not?

3. Can *any* number be a perfect square? Why or why not?

Investigation ▶ 3 More about Squaring

Squaring is an operation, just like addition, subtraction, division, and multiplication. In Chapter 1, you learned about *order of operations*—a rule that specifies the order in which the operations in an expression should be performed. Below, the rule has been extended to include squares and other exponents.

Order of Operations

• Evaluate expressions inside parentheses and above and below fraction bars.

• Evaluate all exponents, including squares.

• Do multiplications and divisions from left to right.

• Do additions and subtractions from left to right.

Think & Discuss

Evaluate each expression.

$$2 \cdot 11^2 \qquad\qquad (2 \cdot 11)^2$$

Explain how the order in which you performed the operations is different for the two expressions.

Problem Set G

1. Does $(3 + 5)^2$ have the same value as $3^2 + 5^2$? Explain.

2. Does $(5 \cdot 3)^2$ have the same value as $5^2 \cdot 3^2$? Explain.

3. Is $(5 \cdot x)^2$ equivalent to $5^2 \cdot x^2$? Explain.

4. This equation is *not* true:

$$2 \times 5 + 2^2 = 11^2 - 23$$

   a. Show that the equation above is not true by finding the value of each side.

   b. **Challenge** Place one pair of parentheses in the equation to make it true. Show that your equation is true by finding the value of each side.

5. Consider the four digits of the year you were born. Write at least three expressions using these four digits and any combination of parentheses, squaring, addition, subtraction, multiplication, and division. Use each digit only once in an expression. Evaluate each expression.

In the next problem set, you will compare squaring to doubling.

Problem Set H

In this problem set, you and a partner will play the game *Square to a Million.* The object of the game is to get a number as close to 1 million as possible, without going over, using only the operation of squaring.

Here are the rules for the game:

- Player 1 enters a number greater than 1 into a calculator.

- Starting with Player 2, players take turns choosing to continue or to end the game. In either case, the player states his or her decision and then presses x^2 $\boxed{\text{ENTER}}$.

   —*If the player chooses to continue the game* and the result is greater than or equal to 1 million, the player loses the round. If the result is less than 1 million, it is the other player's turn.

   —*If the player chooses to end the game* and the result is greater than or equal to 1 million, the player wins. If it is less than 1 million, the player loses.

Play six games with your partner, switching roles for each round.

1. On your turn, how did you decide whether to continue or to end the game?

2. What is the greatest whole number whose square is less than 1 million?

3. What is the greatest whole number you could start with, press x^2 twice, and get a number less than 1 million?

4. What is the greatest whole number you could start with, press x^2 three times, and get a number less than 1 million?

5. What is the greatest whole number you could start with, press x^2 four times, and get a number less than 1 million?

6. What would happen if you started the game with a positive number less than or equal to 1?

7. Imagine you are playing the game *Double to a Million,* in which you double the number in the calculator instead of squaring it. If you start with the given number, how many times will you have to double until you produce a number greater than or equal to 1 million?

    a. 50   **b.** 5   **c.** 1   **d.** 0.5

8. For each part of Problem 7, describe what would happen if you repeatedly squared the result instead of doubling it.

Share & Summarize

1. Write an expression that involves parentheses, squaring, and at least two other operations. Explain how to use order of operations to evaluate your expression.

2. Copy the table, and fill in the missing information. The first row has been completed for you.

Number	**Double It** Is the result greater than, less than, or equal to the original number?	**Square It** Is the result greater than, less than, or equal to the original number?	**Which gives the greater result: squaring or doubling?**
Between 0 and 1	greater than	less than	doubling
1			
Between 1 and 2			
2			
Greater than 2			

Investigation Taking Square Roots

V O C A B U L A R Y
inverse operations

Two operations that "undo" each other are called **inverse operations.** Addition and subtraction are inverse operations.

Add 12 to 15 to get 27. To undo the addition, subtract 12 from 27 to get 15.

Subtract 12 from 27 to get 15. To undo the subtraction, add 12 to 15 to get 27.

Similarly, multiplication and division are inverse operations.

Multiply 7 by 5 to get 35. To undo the multiplication, divide 35 by 5 to get 7.

Divide 35 by 5 to get 7. To undo the division, multiply 7 by 5 to get 35.

In this lesson, you will explore the operation that undoes squaring.

Think & Discuss

Luke squared some numbers on his calculator. His results are shown below. In each case, find the number he started with.

    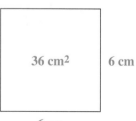

V O C A B U L A R Y
square root

In each part of the Think & Discuss, you found the number you need to square to get a given number. The number you found is the **square root** of the original number. For example, the square root of 36 is 6.

The square root is shown using a *radical sign,* $\sqrt{}$. You can think of $\sqrt{36}$ in any of these ways:

- the number you multiply by itself to get 36: $6 \cdot 6 = 36$
- the number you square to get 36: $6^2 = 36$
- the side length of a square with area 36:

36 cm²   6 cm

6 cm

Remember

Every positive number has both a positive and a negative square root. For example, the square roots of 36 are 6 and ⁻6. In this lesson, you will focus on positive square roots.

Squaring and taking the square root are inverse operations.

Square 6 to get 36. To undo the squaring, take the square root of 36 to get 6.

6   36   **Take the square root of 36 to get 6. To undo taking the square root, square 6 to get 36.**

Problem Set

Fill in the blanks. Here is an example:

81 in.^{2} Side length = $\sqrt{81}$ in. = __9__ in.

1.

49 ft^2 Side length = $\sqrt{\underline{}}$ ft = ____ ft

2.

2.25 cm^2 Side length = $\sqrt{\underline{}}$ cm = ____ cm

3.

$\frac{121}{4}$ yd^2 Side length = $\sqrt{\underline{}}$ yd = ____ yd

4. Jing drew a square with area 10 in.^{2} on a sheet of dot paper. She knew that the sides of the square must be $\sqrt{10}$ in. long. About how long is this? How do you know?

Area = 10 in.^{2}

Not all whole numbers have whole-number square roots. In fact, whole numbers that are not perfect squares have square roots that are decimals that never end or repeat. So, you can only estimate the decimal equivalents of numbers such as $\sqrt{10}$ and $\sqrt{41}$.

Luke estimated the decimal equivalent of $\sqrt{10}$.

Problem Set J

Find the two whole numbers each given square root is between. Do not use your calculator.

1. $\sqrt{2}$

2. $\sqrt{75}$

3. $\sqrt{20}$

4. To the nearest tenth, between which two numbers is $\sqrt{26}$? Do not use your calculator.

5. In the Example, Luke estimated $\sqrt{10}$ to one decimal place. Use Luke's method to estimate $\sqrt{10}$ to two decimal places. Explain each step in your work.

Your calculator has a command for finding square roots accessed by pressing ⌧[√]. For example, to find $\sqrt{4}$, press ⌧ [√] 4 ⌤ .

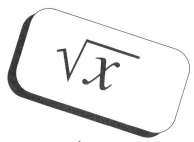

Problem Set K

Use your calculator to approximate each square root to the nearest hundredth. Compare your results with your answers in Problem Set J.

1. $\sqrt{2}$ **2.** $\sqrt{75}$ **3.** $\sqrt{20}$ **4.** $\sqrt{26}$

5. In this problem, you will look at the result your calculator gives for $\sqrt{1,000}$.

  a. Use your calculator to approximate $\sqrt{1,000}$. Write down the exact result shown in the display. Do not clear the screen.

  b. Square the number on your calculator by pressing ⌧ ⌤ . What result does your calculator give?

  c. Clear your calculator screen. Then enter your result from Part a and press ⌧ ⌤ . What result does your calculator give?

  d. Why do you think your results from Parts b and c are different?

6. Althea entered a number into her calculator. She then pressed ⌧ ⌤ repeatedly until the calculator showed 43,046,721. What could her original number have been? List all the possibilities.

7. Suppose you enter a positive number less than 10 into your calculator and press ⌧ ⌤ 5 times. Then you start with the final result and take the square root 5 times. What will happen? Explain why.

Share & Summarize

1. Describe the relationship between squaring a number and taking its square root.

2. Describe a method for approximating a square root *without* using a calculator.

On Your Own Exercises

1. On dot paper or grid paper, draw a rectangle with an area of 20 square units, whole-number side lengths, and the greatest possible perimeter. What is the perimeter of your rectangle?

2. On dot paper or grid paper, draw a rectangle with an area of 20 square units, whole-number side lengths, and the least possible perimeter. What is the perimeter of your rectangle?

These shapes are drawn on centimeter dot grids. Find the area of each shape.

3.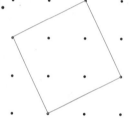

4.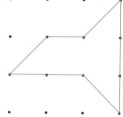

5.

6. Find the area of a rectangle with length 7.5 feet and width 5.7 feet.

7. Find the length of a rectangle with width 11 centimeters and the given area.

  a. 165 square centimeters   **b.** 60.5 square centimeters

8. Find the length of a rectangle with area 484 square inches and the given width.

  a. 10 inches   **b.** 22 inches

9. A square garden has area 289 square feet. How long is each side of the garden?

10. If one rectangle has a greater perimeter than another, must it also have a greater area? Explain your answer.

Square each number.

11. 14   **12.** 21.5   **13.** $\frac{9}{10}$ **14.** 0.3

15. List five perfect squares between 100 and 500.

Tell whether each number is a perfect square, and explain how you know.

16. 40   **17.** 81   **18.** 125   **19.** 256

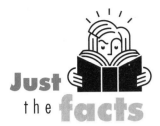
20. If a square has area 30.25 square feet, how long is each side?

Find the value of each expression.

21. $5 \cdot 3^2 - 2$ **22.** $2 \cdot (5^2 - 10)$ **23.** $3^2 - 2^2$ **24.** $7 + \frac{6^2}{3}$

25. Does $(1 + 3)^2$ have the same value as $1^2 + 3^2$? Explain.

26. Does $(4 - 2)^2$ have the same value as $4 - 2^2$? Explain.

27. Does $(11 \cdot 7)^2$ have the same value as $11^2 \cdot 7^2$? Explain.

28. Challenge Place one pair of parentheses in the equation below to make it true. Show that it is true by computing the value of each side.

$$22 - 7 - 5\ ^2 \cdot 2 = 2 \cdot 3\ ^2 - 4$$

29. Suppose you are playing *Square to a Million*. You chose the starting number 5, and your partner squared it. Now it is your turn. Should you continue or end the game? Explain.

30. Suppose you are playing *Square to a Million*. Your partner chose the starting number 1,001. Should you continue or end the game? Explain.

In Exercises 31–35, suppose you enter the number into a calculator and press $\boxed{x^2}$ three times. Without doing any calculations, tell whether the result will be *less than*, *greater than*, or *equal to* the original number.

31. 0.75   **32.** $\frac{2}{3}$ **33.** 1   **34.** 1.5   **35.** 5

36. Luke squared a number and got 28,900. What number did he square?

37. Jing squared a number and got $\frac{121}{25}$. What number did she square?

38. Find the side length of a square playground that has an area of 1,521 square yards.

39. Without using your calculator, determine which two whole numbers $\sqrt{72}$ is between.

40. Without using your calculator, determine which two whole numbers $\sqrt{3}$ is between.

41. $\sqrt{53}$ is between 7 and 8. Without using your calculator, find five other whole numbers whose square roots are between 7 and 8.

42. Without using your calculator, determine whether $\sqrt{39}$ is closer to 6 or to 7. Explain how you know.

43. Without using a calculator, approximate $\sqrt{75}$ to the nearest tenth.

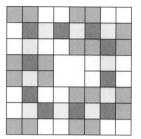

44. Ms. Johnson built this tile patio around a square fountain. The tiles measure 1 foot on each side. The patio is constructed of white, light green, dark green, and blue tiles.

a. What is the total perimeter of the patio? (Add the inner and outer perimeters.)

b. What is the area of the patio?

c. Express the portion of the patio that each color makes up as a fraction and as a percent.

45. Hannah wants to build a fenced-in play area for her rabbit. She has 30 feet of fencing. Give the dimensions and area of the largest rectangular play area she can fence in.

46. Each of these rectangles has whole-number side lengths and an area of 25 square units:

Below is the only rectangle with whole-number side lengths and an area of 5 square units:

a. How many different rectangles are there with whole-number side lengths and an area of 36 square units? Give the dimensions of each rectangle.

b. Consider every whole-number area from 2 square units to 30 square units. For which of these areas is there only one rectangle with whole-number side lengths?

c. What do the areas you found in Part b have in common?

d. For which area from 2 square units to 30 square units can you make the greatest number of rectangles with whole-number side lengths? Give the dimensions of each rectangle you can make with this area.

The length of a soccer field can vary from 100 yards to 130 yards. The width can vary from 50 yards to 100 yards. So, the least possible area is 100 · 50, or 5,000, square yards, and the greatest possible area is 130 · 100, or 13,000, square yards.

47. Althea squared a number, and the result was the same as the number she started with. What number might she have squared? Give all the possibilities.

48. Conor squared a number, and the result was 10 times the number he started with. What was his starting number?

49. Marcus squared a number, and the result was less than the number he started with. Give two possible numbers that Marcus might have started with.

50. In this exercise, you will explore what happens to the area of a square when you double its side lengths.

    a. Draw and label four squares of different sizes, and calculate their areas.

    b. For each square you drew, draw a square with sides twice as long. Calculate the areas of the four new squares.

    c. When you doubled the side lengths of your squares, did the areas double as well? If not, how did the areas change? Why do you think this happened?

    d. If you double the side lengths of a rectangle that is not a square, do you think the same pattern would hold? Why or why not?

    e. If you triple the side lengths of a square, what do you think will happen to the area? Test your hypothesis on two or three squares.

51. Melissa receives $1.00 a week as an allowance. Her older brother Owen gets $1.50 each week. Her younger sister Simone receives $0.75 each week. All three children have asked their parents for larger allowances. Their parents have given them these choices:

    • Option I: Add $0.50 to your current weekly allowance.

    • Option II: Square your current weekly allowance.

    Which option should each child choose? Why?

52. Miguel squared a number and got 390,625. Without using your calculator, find the possible ones digits of his original number. Explain.

53. Rosita squared a number and got 15,376. Without using your calculator, find the possible ones digits of her original number. Explain.

54. Caroline squared a number and got 284,089. Without using your calculator, find the possible ones digits of her original number. Explain.

In y o u r
own
words

Explain how squaring and taking square roots is different for numbers greater than 1 than for numbers less than 1.

55. Does $\sqrt{36 - 25}$ have the same value as $\sqrt{36} - \sqrt{25}$? Explain.

56. Does $\sqrt{36 \cdot 25}$ have the same value as $\sqrt{36} \cdot \sqrt{25}$? Explain.

57. Does $\dfrac{\sqrt{36}}{\sqrt{25}}$ have the same value as $\sqrt{\dfrac{36}{25}}$? Explain.

58. Economics The school store is having a two-day sale. If you shop Thursday, the sale price of any item is the square root of the original price. If you shop Friday, the sale price of any item is half of the original price.

In Parts a–e, tell which day you should shop to get the item at the lowest price.

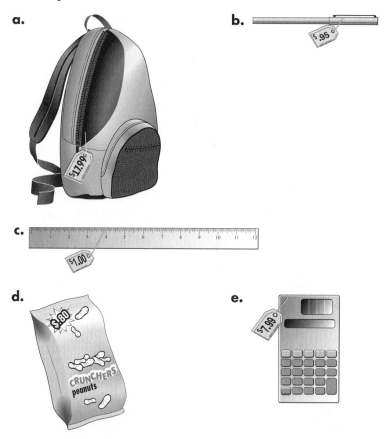

a.

b. $5.95

c. $1.00

d. $.80

e. $7.99

f. In general, for what prices do you save more by taking the square root? For what prices do you save more by taking half?

Find the prime factorization of each number.

59. 432　　　　　**60.** 224　　　　　**61.** 1,053

62. 935　　　　　**63.** 198　　　　　**64.** 736

65. In Art's Art Supply store, each box of colored pencils contains 12 pencils. Write a rule for the number of pencils *p* in *b* boxes.

66. Economics T. J. sells magazine subscriptions over the phone. He earns $50 per day, plus $1.25 for each subscription he sells. Write a rule for T. J.'s daily earnings *d* if he sells *s* subscriptions.

67. In the Spring Valley hot-air balloon race, there were 7 striped balloons entered for every 6 solid-color balloons. Write a rule for the relationship between the number of striped balloons *p* and the number of solid-color balloons *s*.

68. Economics A store is having a "25% off everything" sale. Write a rule for calculating the sale price *s* of an item originally priced at *d* dollars.

Find the measure of each angle.

69.

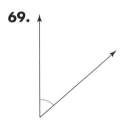

70.

71.

72.

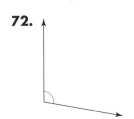

8.4 Calculating Areas

In the last lesson, you learned that the area of a shape is the number of square units that fit inside it.

MATERIALS
- 1-inch grid paper
- $\frac{1}{2}$-inch grid paper

Explore

Place one hand on a sheet of 1-inch grid paper, with your fingers held together. Trace around your hand.

- Estimate the area of your hand tracing in square inches by counting grid squares.

Now trace your hand onto a sheet of $\frac{1}{2}$-inch grid paper.

- Estimate the number of squares inside the tracing.

- On $\frac{1}{2}$-inch grid paper, each small square has side length $\frac{1}{2}$ inch. What is the area of each small square in square inches?

- Use the previous two answers to estimate the area of your hand in square inches.

Which estimate do you think is more accurate: the estimate based on the 1-inch grid, or the estimate based on the $\frac{1}{2}$-inch grid? Why?

Just the facts

The area of the palm of your hand is about 1% of the area of your skin. Doctors use this approximation to estimate the percent of a person's skin that is affected by a burn or other problem. It is known as the "rule of palms."

When you want to estimate the area of an odd shape such as your hand, counting grid squares is a fairly good method, although it does take time. For many other shapes, you can use formulas to find the area quickly. You already know formulas for areas of squares and rectangles. In this lesson, you will explore formulas for areas of parallelograms, triangles, and circles.

Investigation Areas of Parallelograms

VOCABULARY
parallelogram

A **parallelogram** is a quadrilateral with opposite sides that are the same length. The term *parallelogram* refers to the fact that the opposite sides are *parallel*—that is, they never meet no matter how far they are extended.

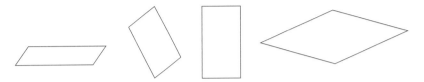

In this investigation, you will use what you know about finding areas of rectangles to develop a formula for the area of a parallelogram.

MATERIALS
- copies of the parallelograms
- metric ruler

Problem Set A

1. Find the area of each parallelogram below, and explain the method you used.

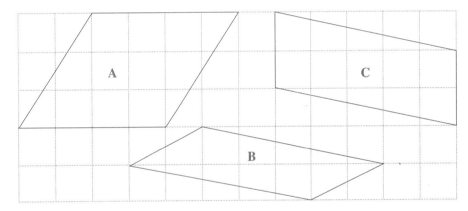

2. Measure the lengths of the sides of each parallelogram to the nearest tenth of a centimeter.

3. Is the area of a parallelogram equal to the product of the lengths of its sides?

VOCABULARY
base of a parallelogram
height of a parallelogram

The **base of a parallelogram** can be any of its sides. The **height of a parallelogram** is the distance from the side opposite the base to the base. The height is always measured along a segment perpendicular to the base (or to the line containing the base).

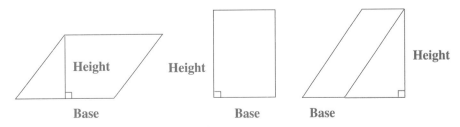

In Problem Set B, you will explore how the base and height of a parallelogram are related to its area.

- copies of the parallelograms
- metric ruler
- scissors
- tape
- protractor

Problem Set B

1. Complete Parts a–c for each parallelogram in Problem Set A.

   a. Choose a side of the parallelogram as the base. Draw a segment perpendicular to the base that extends to the side opposite the base. The segment should be completely inside the parallelogram. For Parallelogram A, you might draw the segment shown in green.

   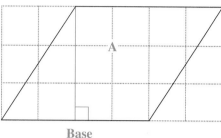

   b. Find the lengths of the base and the height. (The height is the length of the segment you drew in Part a.) Record these measurements in a table like the one below.

	Parallelogram		Rectangle	
	Base	Height	Length	Width
A				
B				
C				

   c. Divide the parallelogram into two pieces by cutting along the segment you drew in Part a. Then reassemble the pieces to form a rectangle. Record the length and width of the rectangle in your table.

2. How do the base and height of each parallelogram compare with the length and width of the rectangle formed from the parallelogram?

3. How does the area of each parallelogram compare with the area of the rectangle formed from the parallelogram?

4. How can you find the area of a parallelogram if you know the length of a base and the corresponding height? Use what you have discovered in this problem set to explain why your method works.

5. Find the area of this parallelogram without forming it into a rectangle. Explain each step of your work.

You can find the area of a parallelogram by multiplying the length of the base by the height. This can be stated using a formula.

Area of a Parallelogram

$$A = b \cdot h$$

In this formula, A represents the area, b represents the base, and h represents the height.

Problem Set C

Find the area of each parallelogram to the nearest hundredth of a square unit.

1.

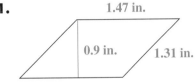

1.47 in.

0.9 in.   1.31 in.

2.

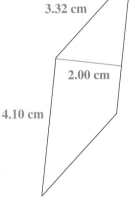

3.32 cm

2.00 cm

4.10 cm

3.

5.27 cm

1.73 cm

3.21 cm

4. The area of the parallelogram below is 12.93 cm². Find the value of b to the nearest hundredth.

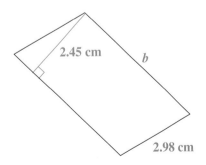

2.45 cm

b

2.98 cm

Share & Summarize

How is finding the area of a parallelogram similar to finding the area of a rectangle? How is it different?

Investigation 2 ▶ Areas of Triangles

You have looked at areas of rectangles and parallelograms. Now you will turn your attention to triangles.

MATERIALS
copies of the triangle

Problem Set D

Find the area of each triangle, and explain the method you used.

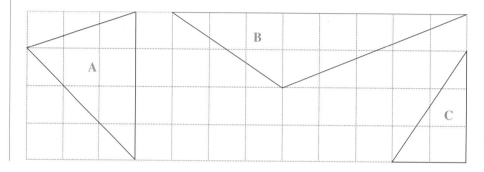

VOCABULARY
base of a triangle
height of a triangle

The **base of a triangle** can be any of its sides. The **height of a triangle** is the distance from the base to the vertex opposite the base. The height is always measured along a segment perpendicular to the base (or the line containing the base).

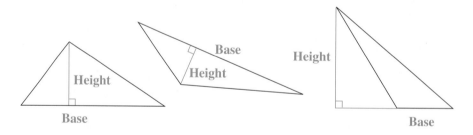

In Problem Set D, you may have used a variety of methods for finding the areas of the triangles. In the next problem set, you will see how you can find the area of a triangle by relating it to a parallelogram.

MATERIALS
• copies of the triangles
• scissors
• tape
• protractor
• ruler

Problem Set E

Cut out two copies of each triangle from Problem Set D.

1. Complete Parts a–d for each triangle.

   a. Make as many different parallelograms as you can by putting together the two copies of the triangle. (Do not tape them together.) Make a sketch of each parallelogram.

   b. How does the area of the triangle compare to the area of each parallelogram?

c. Tape the two copies of the triangle together to form one of the parallelograms you sketched in Part a. Choose one side of the parallelogram as the base, and draw a segment perpendicular to the base extending to the opposite side.

d. Do the base and height of the parallelogram correspond to a base and height of the triangle?

2. Think about what you learned in Problem 1 about the relationship between triangles and parallelograms. How can you find the area of a triangle if you know the length of a base and the corresponding height?

Find the area of each triangle to the nearest hundredth.

3.

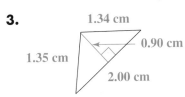

4.

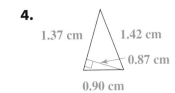

5.

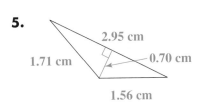

Just the facts

Triangles are rigid shapes. If you build a triangle out of a strong material, it will not collapse or change shape when you press on its sides or vertices. Because of this property, triangles are used frequently as supports for buildings, bridges, and other structures.

In Problem Set E, you probably discovered that the area of a triangle is half the length of the base times the height. You can state this using a formula.

Area of a Triangle

$$A = \tfrac{1}{2} \cdot b \cdot h$$

In this formula, A represents the area, b represents the base, and h represents the height.

Problem Set F

Three students found the area of this triangle. Caroline used the 12-centimeter side as the base, Rosita used the 11-centimeter side, and Marcus used the 10-centimeter side.

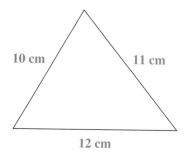

10 cm   11 cm

12 cm

1. Assuming the students did the calculations correctly, do you think they found the same area or different areas? Explain.

2. Complete Parts a–c to find the area using the 12-centimeter side as the base.

   a. Draw a segment perpendicular to the base from the vertex opposite the base. Use your protractor to make sure the base and the segment form a right angle.

   b. Measure the height to the nearest tenth of a centimeter.

   c. Use the base and height measurements to calculate the area of the triangle.

3. Repeat Parts a–c of Problem 2 using the 11-centimeter side as the base.

4. Repeat Parts a–c of Problem 2 using the 10-centimeter side as the base.

5. Compare your results for Problems 2, 3, and 4. Did the area you calculated depend on the base you used? Explain.

Problem Set G

△ABD and △ABE were created by shearing △ABC. *Shearing* a triangle means "sliding" one of its vertices along a line parallel to the opposite side. In this case, △ABC was sheared by sliding vertex *C* to vertex *D* and then to vertex *E*.

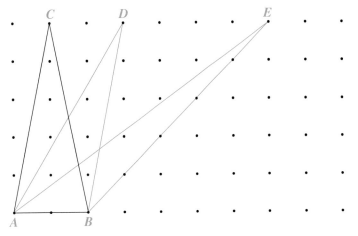

1. How are △ABC, △ABD, and △ABE alike? How are they different?

2. Draw two more triangles by shearing △ABC.

3. Does shearing △ABC change its area? Explain.

4. Does shearing △ABC change its perimeter? Explain.

Share & Summarize

Describe how finding the area of a triangle is related to finding the area of a parallelogram.

Investigation 3 ► Areas of Circles

Finding the area of a figure with curved sides often requires counting grid squares or using another estimation method. However, there is a surprisingly simple formula for calculating the area of a circle.

Problem Set H

These circles are drawn on 1-centimeter grid paper:

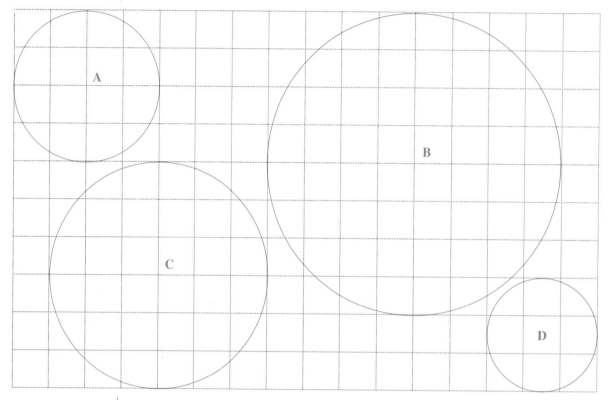

1. Copy the table. Find the radius of each circle, and record your results in the "Radius" column.

Circle	Radius, r (cm)	Estimated Area, A (cm^2)	$A \div r$	$A \div r^2$
A				
B				
C				
D				

2. Estimate the area of each circle by counting grid squares. Record your estimates in your table.

3. For each circle, divide the area by the radius, and record the results.

4. For each circle, divide the area by the radius squared, and record the results.

5. Look at the last two columns of the table. Do the values in either column show an obvious pattern? If so, does it remind you of other patterns you have seen in this chapter?

6. Jahmal estimated that the area of a circle with radius 10 cm is about 40 cm^2.

    a. Explain why Jahmal's estimate is not reasonable.

    b. What is a reasonable estimate for the area of a circle with radius 10 cm?

    c. What is a reasonable estimate for the radius of a circle with area 40 cm^2?

Remember

π is a decimal number with digits that never end or repeat. It can be approximated as 3.14.

In Lesson 8.2, you learned about the number π and how it is related to the circumference of a circle. You found that if C is the circumference of a circle and d is the diameter, the following is true:

$$\pi = C \div d$$

Amazingly, the number π is also related to the area of a circle. If A is the area of any circle and r is the radius, the following is true:

$$\pi = A \div r^2$$

You can use this fact to develop the formula for the area of a circle.

Area of a Circle

$$A = \pi \cdot r^2$$

In this formula, A is the area and r is the radius.

Problem Set I

For this problem set, use the $\boxed{\pi}$ key on your calculator to approximate π. (If your calculator doesn't have a $\boxed{\pi}$ key, use 3.14 to approximate π.)

1. What is the area of a circle with radius 15 in.? Give your answer to the nearest hundredth of a square inch.

2. What is the area of a circle with radius 10.15 cm? Give your answer to the nearest hundredth of a square centimeter.

3. Which has the greater area: a circle with radius 7.2 cm, or a circle with diameter 12.75 cm? Explain your answer.

4. What is the radius of a circle with area 100 in.^{2}? Give your answer to the nearest hundredth of an inch.

5. A pizza parlor makes pizzas in two shapes. The circular pizza has a diameter of 10 inches. The rectangular pizza measures 16 inches by 10 inches. A circular cheese pizza costs $8, and a rectangular cheese pizza costs $14. Which shape gives you more pizza for your money? Explain how you found your answer.

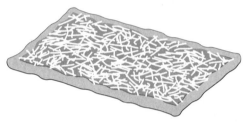

6. This is a diagram of the inner lane of the track at Walker Middle School. The lane is made of two straight segments and two semicircles (half circles). The area inside the track is covered with grass. To the nearest tenth of a square yard, what is the area of the grass inside the track? Explain how you found your answer.

— 160 yd —

120 yd

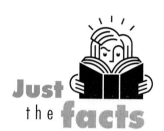

Share & Summarize

1. Give the formula for the area of a circle. Tell what the letters in the formula represent.

2. How can you calculate the area of a circle if you know only its diameter?

3. How can you calculate the area of a circle if you know only its circumference?

Using a Spreadsheet to Maximize Area

Miguel's dog Max loves to play outside. When he is in the backyard, he is usually tied to a 10-foot leash in the center of the yard.

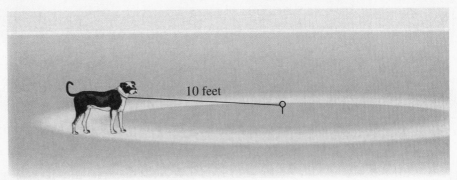

10 feet

Just the **facts**

Dogs resembling modern-day grey-hounds lived in Egypt as early as 4500 B.C. Hieroglyphs of dogs can be found on ancient Egyptian tombs.

Miguel has earned $400 mowing lawns, and he wants to use the money to build a pen for Max. The fencing Miguel has chosen costs $4.50 per foot. In this lab investigation, you will use a spreadsheet to figure out the largest rectangular pen Miguel can afford to build.

Think about the Problem

1. If you know the length and width of a rectangular pen, how can you figure out the area and perimeter?

2. How can you figure out how much the pen will cost?

Create a Spreadsheet

You will create a spreadsheet that automatically calculates the area, perimeter, and cost of the pen when you enter the length and the width. First enter column headings for length, width, perimeter, area, and cost.

	A	**B**	**C**	**D**	**E**	**F**	**G**
1	Length	Width	Perimeter	Area	Cost		
2							

You want your spreadsheet to calculate perimeter, area, and cost when you enter length and width values in Cells A2 and B2. For this to work, you will enter *formulas* that tell the spreadsheet how to do the calculations.

The perimeter is 2 times the length value in Cell A2 plus 2 times the width value in Cell B2. So, in Cell C2, enter this formula:

$$=2*A2+2*B2$$

The = sign lets the spreadsheet know that the entry is a formula to evaluate. So, the formula you entered in Cell C2 tells the spreadsheet to add 2 times the value in Cell A2 to 2 times the value in Cell B2.

	A	B	C	D	E	F	G
1	Length	Width	Perimeter	Area	Cost		
2			=2*A2+2*B2				
3							
4							
5							

3. In Cell D2, you need to enter a formula to calculate the area. What formula should you enter? (Your formula must begin with an = sign and should contain the cell names A2 and B2.)

4. In Cell E2, you need to enter a formula to calculate the cost. What formula should you enter?

Enter your area and cost formulas into the spreadsheet. Test the formulas by entering the length value 11 in Cell A2 and the width value 7 in Cell B2.

5. Does your spreadsheet give the correct perimeter, area, and cost values for a pen that is 11 feet by 7 feet? How do you know? If your formulas are not correct, adjust them.

6. Can Miguel afford to build an 11-ft-by-7-ft pen?

7. Change the length and width values in Cells A2 and B2 to the length and width of your choice. What happens to the values in the other columns?

Now you will copy your formulas down to Row 25 of the spreadsheet. This will allow you to enter different length and width values in each row. Here are the steps for copying the perimeter formula:

• Highlight Cells C2 through C25.

• Select the Fill Down command.

	A	B	C	D	E	F	G
1	Length	Width	Perimeter	Area	Cost		
2			=2*A2+2*B2				
3							
4							
5							
6							
7							
8							
9							
10							

8. Select Cell C3. What formula appears in this cell? Now select Cell C18. What formula appears here? What does the Fill Down command do?

Copy the formulas for area and cost down to Row 25. Now you can look at perimeter, area, and cost for several different pens at the same time.

Find the Largest Pen

Test different length and width values to try to find the largest pen Miguel can afford. Enter a different pair of values in each row.

9. What are the dimensions and area of the largest rectangular pen Miguel can afford?

10. How do you know that the pen you found in Question 9 is the largest possible?

11. Will Max have more room to move in his new pen than he now has on his leash? Explain.

What Did You Learn?

12. Create a spreadsheet that calculates the circumference, area, and cost of circular pens with different radii. Use 3.14 as an approximation for π.

	A	B	C	D
1	Radius	Circumference	Area	Cost
2				

13. Find the approximate radius and area of the largest circular pen Miguel could build with the fencing.

14. How does the area of the largest circular pen compare to the area of the largest rectangular pen?

Just the facts

Dogs similar to modern-day mastiffs were used by ice-age hunters to help catch large game such as the woolly mammoth.

On Your Own Exercises

Practice & Apply

1. Choose an object in your home with a nonrectangular surface that will fit on a piece of grid paper. (Some ideas: a can of soup, your shoe, an iron.) Trace the surface onto the grid paper, and estimate its area.

2. These parallelograms are drawn on a centimeter grid:

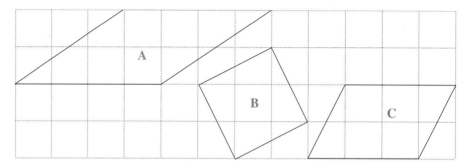

   a. Find the area of each parallelogram.

   b. Sketch a rectangle that has the same area as Parallelogram A.

   c. Sketch a rectangle that has the same area as Parallelogram C.

3. Find the area of this parallelogram:

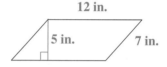

4. Can you use the area formula for a parallelogram, $A = b \cdot h$, to find the area of a rectangle? If so, where are the base and height on the rectangle? If not, why not?

5. A parallelogram has an area of 42.6 cm². The height of the parallelogram is 8 cm. What is the length of the base?

6. These triangles are drawn on a centimeter grid:

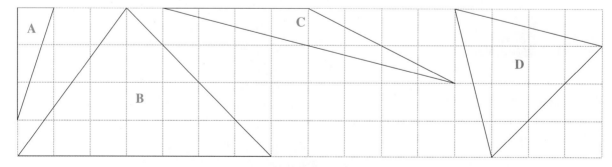

   a. Find the area of each triangle.

   b. For each triangle, sketch a parallelogram with twice the area of the triangle.

impactmath.com/self_check_quiz

Find the area of each triangle. Round your answers to the nearest hundredth.

7.

3.29 cm

1.72 cm

3.78 cm

8.

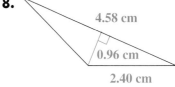

4.58 cm

0.96 cm

2.40 cm

9. Consider this triangle.

a. Which of the given measurements would you use to find the area of the triangle? Why?

2.17 cm   3.46 cm

4.08 cm

b. What is the triangle's area? Round your answer to the nearest hundredth.

10. The green parallelogram was created by shearing Parallelogram Z. *Shearing* a parallelogram means "sliding" one of its sides along a line parallel to the opposite side.

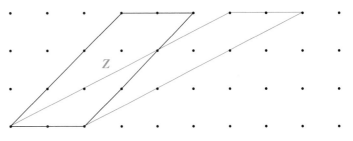

a. Create two more parallelograms by shearing Parallelogram Z. (In each case, "slide" the top of the parallelogram.)

b. Does shearing a parallelogram change its area? Explain.

c. Now shear Parallelogram Z to create a parallelogram with the smallest possible perimeter. What does this new parallelogram look like?

d. Meela sheared Parallelogram Z to create this figure. She says she has drawn the sheared parallelogram with the greatest possible perimeter. Do you agree with her? Explain.

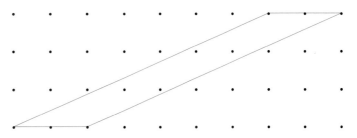

In Exercises 11–13, use the ⬚π⬚ key on your calculator to approximate π. (If your calculator does not have a ⬚π⬚ key, use 3.14 as an approximation for π.) Round your answers to the nearest hundredth.

11. Calculate the area of a circle with radius 8.5 inches.

12. Calculate the area of a circle with diameter 15 feet.

13. Calculate the area of a circle with circumference 90 feet.

14. A dog is tied to a 15-foot leash in the center of a yard.

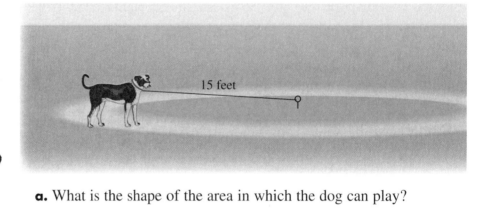

15 feet

a. What is the shape of the area in which the dog can play?

b. To the nearest square foot, what is the area of the space in which the dog can play?

c. Suppose that, instead of being tied in the center of the yard, the dog is tied to the corner of the house. To the nearest square foot, what is the area of the space in which the dog can play? (The sides of the house are longer than the leash.)

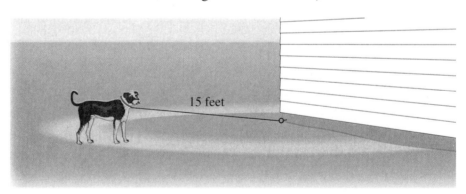

15 feet

Connect & Extend

15. This parallelogram has an area of 20.03 cm². Find the values of a, b, c, and d to the nearest hundredth of a centimeter.

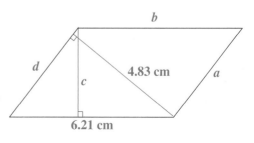

b

d

4.83 cm

c

a

6.21 cm

16. In this exercise, you will draw parallelograms.

   a. Draw three different parallelograms with base length 15 cm and height 7 cm.

   b. Which of your parallelograms has the least perimeter? Which has the greatest perimeter?

   c. Could you draw a parallelogram with the same base and height and an even smaller perimeter? If so, draw it. If not, explain why not.

17. A deck of cards has been pushed as shown. Notice that the sides of the deck are shaped like parallelograms.

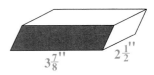

The deck contains 52 cards. Each card is $\frac{1}{48}$ of an inch thick, $3\frac{7}{8}$ inches long, and $2\frac{1}{2}$ inches wide. Find the area of the shaded parallelogram.

18. Below is a floor plan for a museum, divided into four parallelograms and a rectangle. Find the area of the floor to the nearest hundredth of a square meter.

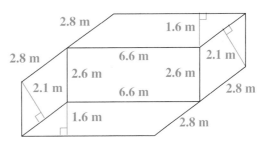

19. The area of this triangle is 782 square centimeters. Find a and b to the nearest tenth of a centimeter.

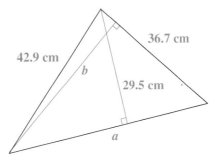

20. In an *equilateral triangle,* all three sides are the same length. Suppose the area of an equilateral triangle is 27.7 cm² and the height is 6.9 cm. How long are each of the triangle's sides?

21. A *trapezoid* is a quadrilateral with exactly one pair of parallel sides, called *bases.* The *height* is the length of a perpendicular segment from one base to the other.

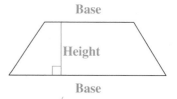

a. Find the area of this trapezoid:

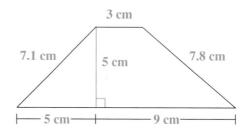

b. Which of the given measurements did you use to find the area?

c. **Challenge** Write a formula for the area of a trapezoid *A* if the lengths of the bases are *B* and *b* and the height is *h*. Explain how you found your answer. (Hint: Divide the trapezoid into two triangles, or form a parallelogram from two copies of the trapezoid.)

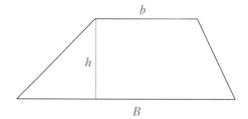

Remember

In a *regular polygon*, all sides are the same length and all angles have the same measure.

22. Any regular polygon can be divided into identical triangles. This hexagon is divided into six identical triangles.

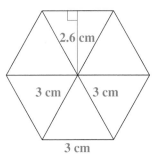

a. Find the area of each triangle and the area of the hexagon to the nearest tenth of a square centimeter. Explain how you found the areas.

b. This formula can be used to find the area of a regular polygon:

$$A = \tfrac{1}{2} \times \text{polygon perimeter} \times \text{height of one triangle}$$

Show that this formula gives you the correct area for the hexagon above.

c. Why do you think the formula works?

d. A stop sign is in the shape of a regular octagon. This sketch of an octagon has been divided into eight identical triangles.

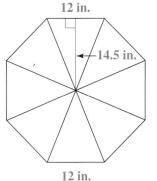

Use the formula from Part b to find the area of a stop sign.

e. Brett found the area of the stop sign by surrounding it with a square.

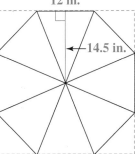

How long are the sides of the square? How long are the perpendicular sides of the small triangles in the corners of the square?

f. Explain how Brett might have calculated the area. Show that this method gives the same area you found in Part d.

23. The Smallville town council plans to build a circular fountain surrounded by a square concrete walkway. The fountain has a diameter of 4 yards. The walkway has an outer perimeter of 28 yards.

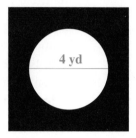

Find the area of the walkway to the nearest tenth of a square yard.

24. The *surface area* of a three-dimensional figure is the sum of the areas of its faces. For example, this cube is made up of six faces, each with area 9 in.^{2}. So, its total surface area is $9 \cdot 6$, or 54, in.^{2}.

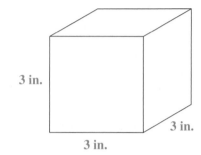

a. Find the surface area of this rectangular box:

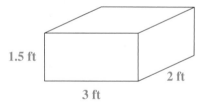

b. Challenge To find the surface area of a cylinder, you can imagine it as three separate pieces: the circular top and bottom, and the rectangle wrapped around them. Find the surface area of this cylinder. (Hint: You need to figure out what the length of the rectangle is. To do this, think about how this length is related to the circles.)

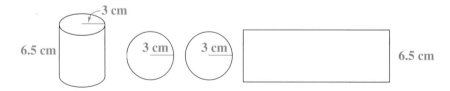

Mixed Review

Geometry Find the perimeter of each figure.

25.

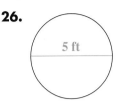

3.65 cm

2.19 cm

26.

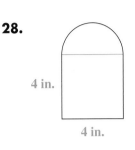

5 ft

27.

2 cm

4 cm

1 cm   2 cm

2 cm

5 cm

28.

4 in.

4 in.

Find each sum or difference.

29. $3\frac{1}{2} - 1\frac{5}{8}$

30. $1\frac{7}{12} + 4\frac{2}{3}$

31. $12\frac{6}{7} + 5\frac{5}{6}$

32. $37.42 - 9.04$

33. $553.89 + 332.7$

34. $2{,}545 - 1{,}365.787$

35. Hannah and Rosita each wrote rules for the number of toothpicks in each term of this sequence:

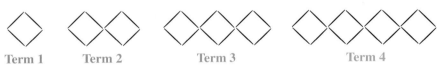

Term 1   Term 2   Term 3   Term 4

Both girls used t to represent the number of toothpicks and n to represent the term number. Use words or diagrams to explain why each rule is correct.

a. Hannah's rule: $t = 2 \cdot n + 2 \cdot n$

b. Rosita's rule: $t = 4 + 4 \cdot (n - 1)$

36. **Preview** Rachel set a thermometer in a beaker of liquid. She recorded the liquid's temperature every 5 minutes. From the data she collected, she wrote an equation to represent the temperature of the liquid T in degrees Celsius after m minutes.

$$T = 72 \times 0.97^{m}$$

Use Rachel's equation to approximate the liquid's temperature after 10 minutes.

The Pythagorean Theorem

VOCABULARY
right triangle

In this lesson, you will learn a famous mathematical fact known as the *Pythagorean Theorem*. The Pythagorean Theorem expresses a remarkable relationship between **right triangles**—triangles that have one 90° angle— and squares.

Before you investigate the theorem, you will practice finding the area of a square drawn on dot paper.

MATERIALS
copy of the square

Explore

Find the exact area of this square. Describe the method you use.

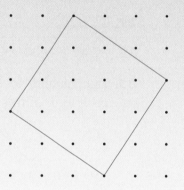

Investigation Right Triangles and Squares

VOCABULARY
hypotenuse
leg

Every right triangle has one right angle. The side opposite the right angle is called the **hypotenuse.** The other two sides are called the **legs.**

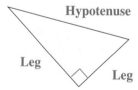

The relationship expressed in the Pythagorean Theorem involves the areas of squares built on the sides of a right triangle. In the next problem set, you will try to discover that relationship.

MATERIALS
- copies of the figures
- dot paper

Just **the facts**

The Pythagorean Theorem is named for Greek mathematician Pythagoras, who lived from approximately 580 to 500 B.C.

An Ionic column in Olympia, Greece

Problem Set A

Problems 1–4 show right triangles with squares drawn on their sides. Find the exact area of each square, and record your results in a copy of the table.

Problem	Area of Square on Side a (units^2)	Area of Square on Side b (units^2)	Area of Square on Side c (units^2)
1			
2			
3			
4			

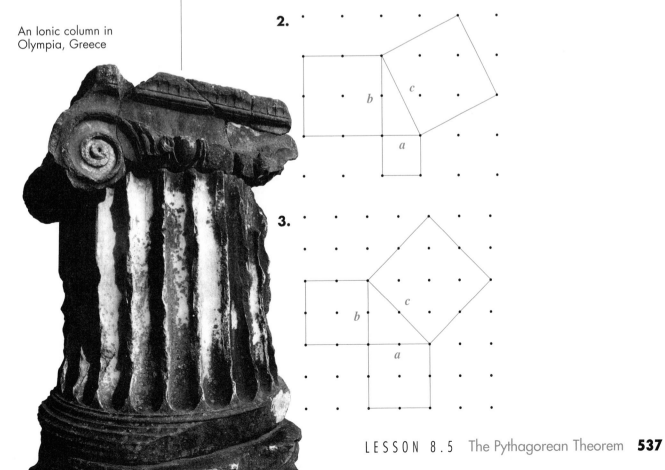

1.

2.

3.

4.

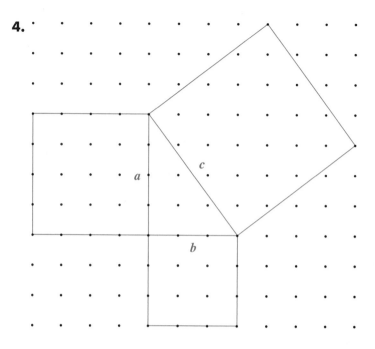

5. Look for a pattern in your table. For the cases you considered, what is the relationship among the areas of the three squares?

6. Draw your own right triangle on dot paper. Does the relationship you described in Problem 5 hold for your triangle as well?

The problems in Problem Set A illustrate the Pythagorean Theorem.

The Pythagorean Theorem

In a right triangle, the area of the square built on the hypotenuse of the triangle is equal to the sum of the areas of the squares built on the legs.

The Pythagorean Theorem is often stated this way:

If c is the length of the hypotenuse of a right triangle and a and b are the lengths of the legs, then $a^2 + b^2 = c^2$.

The diagram below illustrates this idea.

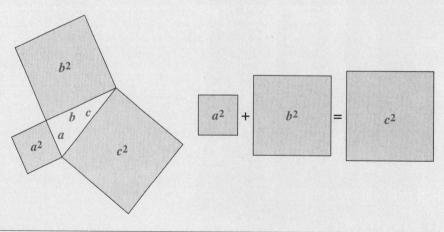

Just the facts

Pythagoras lived 2,500 years ago, but the famous theorem that bears his name was known even earlier. Records show that the Babylonians understood the theorem around 1500 B.C., more than 900 years before Pythagoras was born!

Throughout history, people have found new ways to *prove* the Pythagorean Theorem—that is, to show that it is always true. In Problem Set B, you will explore one such proof.

MATERIALS
* 8 copies of the triangle
* 1 copy of each square
* scissors

Problem Set B

In this problem set, you will use paper triangles and squares to construct a proof of the Pythagorean Theorem.

* Start with a right triangle with squares drawn on its sides.

* Carefully cut out eight copies of the triangle and one copy of each square.

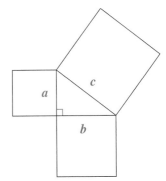

* Use four copies of the triangle, and the square from Side *c*, to make this square:

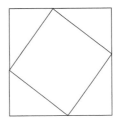

* Use four copies of the triangle, and the squares from Sides *a* and *b*, to make this square:

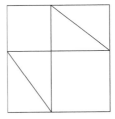

1. The two squares you made have the same area. Explain how you know this is true.

Now take the four triangles away from each square you constructed.

2. Describe what is left of each square.

3. Explain why the area of what is left must be the same for both squares.

4. Explain how your work in this problem set shows that $a^2 + b^2 = c^2$.

State the Pythagorean Theorem in your own words. You might want to
draw a picture to illustrate what you mean.

Investigation 2 Using the Pythagorean Theorem

In this investigation, you will have a chance to practice using the
Pythagorean Theorem.

Problem Set C

Find each missing area. Then use the areas of the squares to find the side
lengths of the triangle.

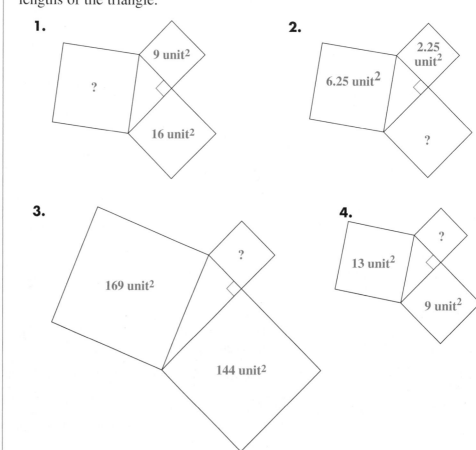

1.
9 unit²
?
16 unit²

2.
2.25 unit²
6.25 unit²
?

3.
?
169 unit²
144 unit²

4.
?
13 unit²
9 unit²

If you know the lengths of any two sides of a right triangle, you can use the Pythagorean Theorem to find the length of the third side.

EXAMPLE

A right triangle has legs of length 1 inch and 2 inches. How long is the hypotenuse?

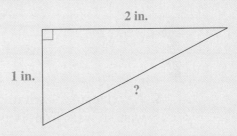

2 in.

1 in.

?

If a and b represent the lengths of the legs, and c represents the length of the hypotenuse, the Pythagorean Theorem says that $a^2 + b^2 = c^2$.

In the triangle shown above, $a = 1$, $b = 2$, and c is the length of the hypotenuse. So,

$$1^2 + 2^2 = c^2$$
$$1 + 4 = c^2$$
$$5 = c^2$$

Therefore, $c = \sqrt{5}$, or about 2.24 inches.

Just the facts

They may not have known the Pythagorean Theorem, but the Egyptians who built the pyramids knew that a triangle with side lengths of 3, 4, and 5 units must be a right triangle. They used a rope device like the one pictured here to check that they made perfect right angles at the corners of the pyramids.

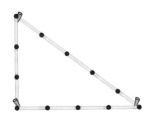

The Great Sphinx in Giza, Egypt

Problem Set D

Find each missing side length. Then find the area of the triangle.

1.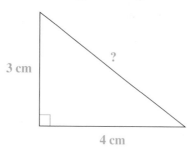

3 cm

?

4 cm

2.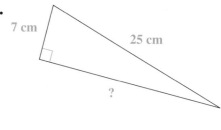

7 cm

25 cm

?

3. A rectangular lawn measures 9 meters by 12 meters. Suppose you want to walk from Point *A* to Point *B*.

A

Keep Off the Grass!

9 meters

12 meters

B

a. If you obey the sign and walk around the lawn, how far will you walk?

b. If you ignore the sign and walk directly across, how far will you walk?

4. A baseball diamond is a square measuring 90 feet on each side. What is the distance from home plate to second base?

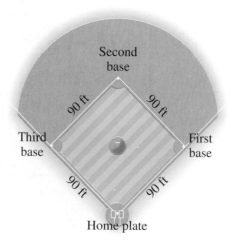

Second base

90 ft

90 ft

Third base

First base

90 ft

90 ft

Home plate

5. Caroline and Marcus are flying a kite. Caroline is holding the kite and has let out 80 feet of kite string. Marcus is standing 25 feet from Caroline and is directly under the kite.

How far above the ground is the kite? Assume Caroline is holding the string 3 feet above the ground. Explain how you found your answer.

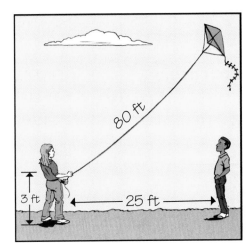

6. Safety regulations say that wheelchair ramps cannot be too steep. Suppose that, for every 1 foot a wheelchair ramp rises, it must cover a horizontal distance of at least 11.5 feet.

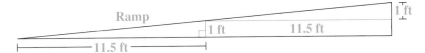

A ramp is being built to a restaurant entrance that is 2.5 feet above the ground.

a. How much horizontal distance does the ramp require?

b. How long must the ramp be?

7. Challenge Find the values of *c* and *h*. Explain the method you used.

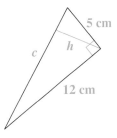

Share & Summarize

Write your own problem that can be solved by using the Pythagorean Theorem. Then show how to solve your problem.

On Your Own Exercises

Find the exact area of each shape. They are drawn on 1-centimeter dot grids.

1.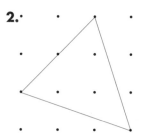

2.

3. In this exercise, you will look at the relationship among the areas of semicircles drawn on the sides of a right triangle.

    a. Find the area, in square units, of each semicircle in this drawing:

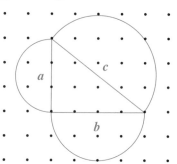

    b. Find the area of each semicircle in this drawing:

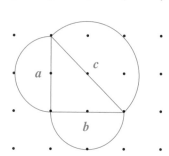

    c. Find the area of each semicircle in this drawing:

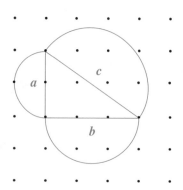

    d. In the cases you looked at, is there a relationship among the areas of the three semicircles? If so, describe it.

 impactmath.com/self_check_quiz

Find each missing area.

4.

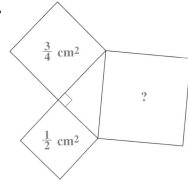

$\frac{3}{4}$ cm^2

?

$\frac{1}{2}$ cm^2

5.

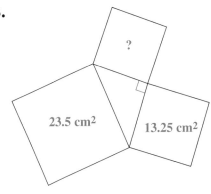

?

23.5 cm^2

13.25 cm^2

Find each missing side length.

6.

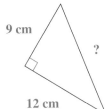

9 cm

?

12 cm

7.

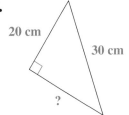

20 cm

30 cm

?

8. The legs of this right triangle are the same length. How long are they?

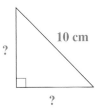

10 cm

?

?

9. To ensure the safety of its workers, a painting company requires that the base of a ladder be 1 foot from a wall for every 4 feet it reaches up the wall.

a. If a ladder reaches 8 feet up a wall, how far should its base be from the wall?

b. How long must the ladder be to reach a height of 8 feet?

c. Challenge What is the highest a 10-foot ladder can reach?

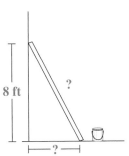

8 ft

?

?

10. Physical Science During its initial climb, an airplane flew 14.4 miles, reaching an altitude of 5 miles. After the initial climb, how far was the airplane from its starting point, as measured along the ground?

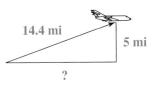

14.4 mi

5 mi

?

Connect & Extend

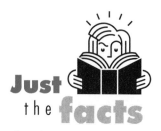

Just the facts

The word *acute* has many meanings. Here are just a few:

- having a sharp point
- shrewd
- sensitive
- severe or intense
- of great importance

11. You know that a *right triangle* has one angle that measures 90°. In an *acute triangle,* all three angles have measures less than 90°. In an *obtuse triangle,* one angle has a measure greater than 90°.

Acute Triangle   Obtuse Triangle

a. Three acute triangles are drawn on the centimeter dot grid below, and squares are drawn on their sides. Find the area of each square, and record your results in a table like this one.

	Acute Triangles		
Triangle	**Area of Square on Side *a***	**Area of Square on Side *b***	**Area of Square on Side *c***
I			
II			
III			

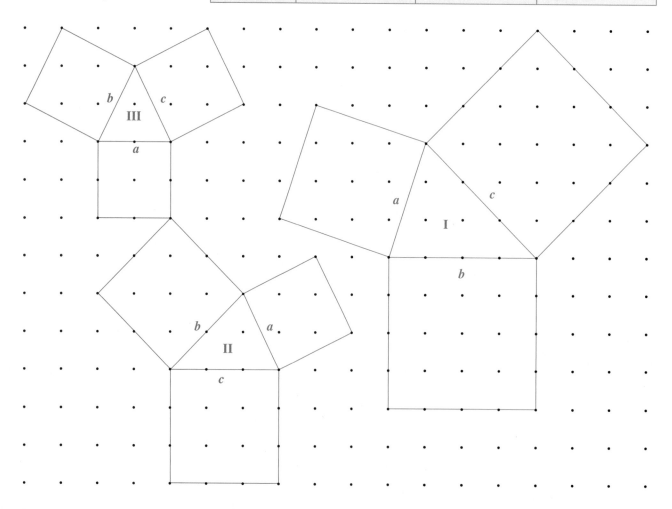

b. Follow the instructions in Part a for these three obtuse triangles.

	Obtuse Triangles		
Triangle	Area of Square on Side a	Area of Square on Side b	Area of Square on Side c
I			
II			
III			

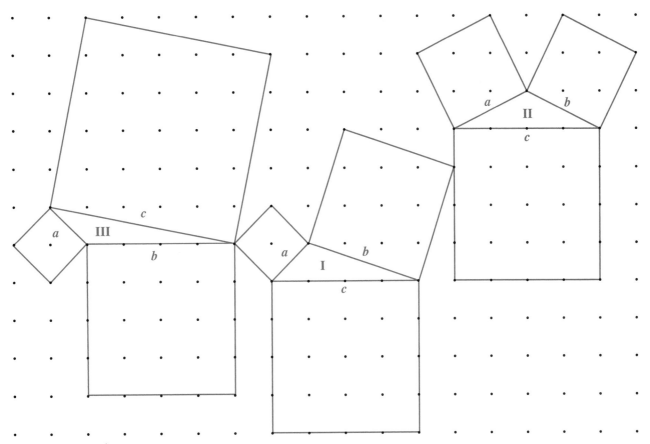

c. Based on your work in this exercise, which of the statements below do you think is true for acute triangles? Which do you think is true for obtuse triangles?

   i. The sum of the areas of the two smaller squares is *less than* the area of the largest square. That is, $a^2 + b^2 < c^2$.

   ii. The sum of the areas of the two smaller squares is *equal to* the area of the largest square. That is, $a^2 + b^2 = c^2$.

   iii. The sum of the areas of the two smaller squares is *greater than* the area of the largest square. That is, $a^2 + b^2 > c^2$.

Just the facts

The word *obtuse* has several other meanings:

- blunt or dull
- slow to understand
- hard to comprehend

In your own words

Explain what the Pythagorean Theorem is and how it is useful for finding lengths.

12. Mr. Mackenzie built a table. He had intended for the table to be rectangular, but he's not sure it turned out that way. He carefully measured the table and found that the side lengths are 60 inches and 45 inches, and the diagonal is 73.5 inches. Is the table a rectangle? Explain how you know.

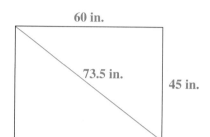

13. Maddie and Jo are building a fence, and they want to make sure each post makes a 90° angle with the ground. Maddie holds one end of a 5-foot piece of rope at a point 4 feet up the post. Jo stretches the rope tight and puts the other end on the ground.

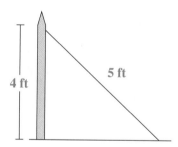

If the post makes a right angle with the ground, how far out will the string reach from the base of the post?

14. Masako and Kai have found the perfect couch for their living room, but they are not sure whether it will fit through the doorway. The doorway measures 37 inches wide and 79 inches high.

They know they can take the legs off the couch. This is a side view of the couch with the legs removed.

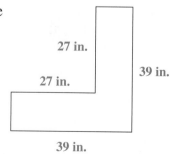

The couch is too wide to fit if they carry it upright, but Marcus thinks it might fit if they tilt it like this:

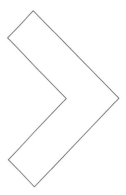

In this exercise, you will use the Pythagorean Theorem to figure out whether the couch will fit through the doorway.

a. Explain why the dashed segments shown here each have length 12 inches. Then find the length of Segment c, and explain how you found it. Round your answer up to the nearest tenth of an inch.

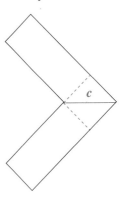

b. Find the length of Segment d to the nearest tenth of an inch.

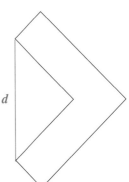

c. Segment b divides Segment d in half. Find the length of Segment b to the nearest tenth.

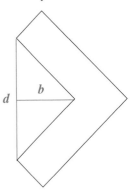

d. Will the couch fit through the doorway? Explain how you know.

Evaluate each expression.

15. $6 \cdot 7 - 5^2$ **16.** $\sqrt{121} - 2^2 + 4 \cdot 3$ **17.** $\frac{4^2 + 6^2}{13}$

18. $\sqrt{6^2 + 8^2}$ **19.** $45 - 3 \cdot 5 + 3^2$ **20.** $(\sqrt{16} + \sqrt{4} + \sqrt{9})^2$

Geometry Find the area of each figure.

21.

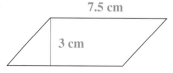

7.5 cm

3 cm

22.

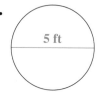

5 ft

23.

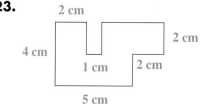

2 cm

4 cm

1 cm 2 cm

2 cm

5 cm

24.

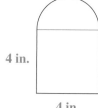

4 in.

4 in.

25. Luke ate 80% of the strawberries he picked. If he picked 30 strawberries, how many did he eat?

26. Ms. Friel's class has 25 students. Of these students, 23 went on the field trip to the museum. What percent of the class did not go to the museum?

27. Statistics Jahmal surveyed students in the cafeteria about their favorite school lunch. He found that $66\frac{2}{3}\%$ like pizza best. If 80 students told Jahmal they like pizza best, how many students did he survey?

28. Preview Devon, Kyle, and Kristi are playing a game with this spinner. They take turns spinning. Each time the spinner lands on blue, Devon scores a point. Each time it lands on green, Kyle scores a point. And each time it lands on white, Kristi scores a point. Do you think the game is fair? Why or why not?

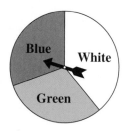

Blue White

Green

VOCABULARY

acute angle
area
base of a parallelogram
base of a triangle
chord
circumference
diameter
height of a parallelogram
height of a triangle
hypotenuse
inverse operations
leg
obtuse angle
parallelogram
perfect square
perimeter
perpendicular
radius
right angle
right triangle
square root

Chapter Summary

In this chapter, you explored ideas about geometry and measurement. You started by working with angles. You measured angles and drew angles with given measures. You looked at relationships among the angles formed by intersecting lines, and you found a rule for finding the angle sum for a polygon based on the number of angles (or sides) it has.

You then found the perimeters of polygons by adding side lengths, and you estimated the perimeters of curved objects by using string and by approximating with polygons. You also learned that the circumference of any circle divided by the diameter is equal to π, and you used this fact to find the formula for the perimeter of a circle.

You learned that the area of a shape is the number of square units that fit inside it. You estimated areas of shapes by counting squares, and you learned formulas for calculating areas of rectangles, parallelograms, triangles, and circles. You learned about the operation of *squaring* and how it is related to the areas of squares. You then learned about the inverse operation *taking the square root,* and you found or estimated the square roots of many numbers.

Finally, you investigated the Pythagorean Theorem, which expresses the relationship among the areas of squares drawn on the sides of a right triangle.

Strategies and Applications

The questions in this section will help you review and apply the important ideas and strategies developed in this chapter.

MATERIALS

- protractor
- ruler
- string

Measuring angles and drawing angles with given measures

1. Victor measured these angles with a protractor. He said both angles have measure of 130°.

   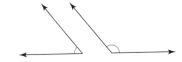

   a. How do you know that Victor is incorrect?

   b. What mistake do you think Victor made?

   c. What advice would you give to help him measure angles correctly?

2. Draw an angle with measure 320°, and explain the steps you followed.

Finding and estimating perimeters

3. Consider this shape:

   a. Describe two methods for estimating the perimeter of the shape.

   b. Use one of the methods you mentioned in Part a to estimate the shape's perimeter.

4. Tell whether the diagram below provides enough information for you to find the perimeter of the shape. If it does, find the perimeter. If it does not, tell what additional information you would need to find the perimeter.

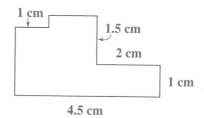

1 cm

1.5 cm

2 cm

1 cm

4.5 cm

Understanding π and the formula for the circumference of a circle

5. Explain what π is and how it is related to the circumference of a circle.

6. Describe how you can find the circumference of a circle if you know its radius.

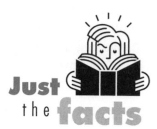

Finding and estimating areas

7. If two shapes have the same perimeter, must they also have the same area? Use words and drawings to help explain your answer.

8. Find the area of this parallelogram in centimeters, and explain the steps you followed:

9. In this chapter, you learned how to find the area of a triangle.

   a. Describe what the base and height of a triangle are.

   b. Explain how to find the area of a triangle if you know the lengths of the base and the height.

   c. How is finding the area of a triangle related to finding the area of a parallelogram?

10. A CD has a diameter of about 12 cm. The hole in the center of a CD has a diameter of about 1.5 cm. Find the area of a CD—not including the hole—to the nearest tenth of a square centimeter. Explain how you found your answer.

Understanding and applying the ideas of squaring and taking the square root

11. In this chapter, you learned about squaring a number and about taking the square root of a number.

  a. Explain what it means to square a number. Give an example.

  b. Explain what it means to take the square root of a number. Give an example.

  c. Explain how you know that squaring and taking the square root are inverse operations. Give an example if it helps you to explain your thinking.

12. How can you predict whether the square of a number will be *greater than, less than,* or *equal to* the original number?

13. Estimate $\sqrt{34}$ to the nearest tenth without using your calculator. Explain each step you take.

Understanding and applying the Pythagorean Theorem

14. The size of a television is given in terms of the length of the diagonal of its screen. For example, a 19-inch television has a screen with diagonal length 19 inches.

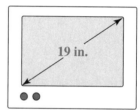

Ms. Perelli's television screen has a length of about 21.75 inches and a width of about 16.25 inches. What is the size of her television? Give your answer to the nearest inch, and explain how you found it.

Demonstrating Skills

Find the measure of each angle.

15.

16.

17.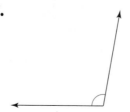

18.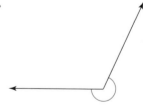

Draw an angle with the given measure.

19. 72°   **20.** 160°   **21.** 210°   **22.** 295°

Find the perimeter and area of each figure.

23.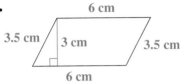

24.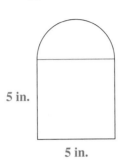

25.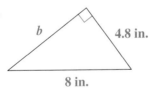

26.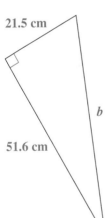

Find the value of each expression.

27. $5 \cdot 7 - 3^2 + 4$ **28.** $(7^2 - 13 \cdot 3)^2$ **29.** $5 + 4 + 6^2$

Approximate each square root to the nearest hundredth.

30. $\sqrt{21}$ **31.** $\sqrt{600}$ **32.** $\sqrt{3}$

In Questions 33 and 34, find the value of b.

33.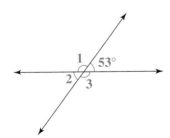

34.

35. Find the measures of Angles 1, 2, and 3.

CHAPTER 9

Solving Equations

It's for the Birds Amy makes and sells birdhouses at her town's craft fair. It costs $10 to rent a table at the fair, and Amy spends an average of $5 for supplies to make each birdhouse.

As she prepares for the event, she can use the equation $c = \$10 + \$5n$ to find the total cost of making the birdhouses. In the equation, c represents the total cost and n represents the number of birdhouses. Suppose Amy has a total of $75 to spend on rent and supplies. To determine how many birdhouses she can make, she can solve the equation $\$75 = \$10 + \$5n$.

Think About It Amy plans to sell her birdhouses for $8 each. So the equation $m = \$8n$ represents the amount of money she'll earn by selling the birdhouses. In the equation, m represents the amount of money she'll earn, and n represents the number of birdhouses. She hopes to earn $120. What equation should she solve?

Family Letter

Dear Student and Family Members,

Our next chapter is about solving equations. Don't worry—you've been working with equations for years. An *equation* is a number sentence that includes an equals sign, which means that two expressions have the same value. Here are three examples:

$$9 + 6 = 15 \qquad 9 + 6 = 5 \times 3 \qquad 7 + 8 = 18 - 3$$

However, in this chapter, you will explore equations with variables (quantities that can change), such as $3 \times n = 18$.

You will learn a method called *backtracking* to solve equations. For example, consider the equation $4 \cdot n + 5 = t$. To find an output (t) with this equation, start with an input (n), multiply it by 4, and add 5. The following flowchart shows these steps.

Input　　　　　　　　　　　　　　**Output**

Here's the flowchart for an input of 3:

Input　　　　　　　　　　　　　　**Output**

If you were given an output of 21, you could use the flowchart to work backward and determine that the input was 4.

You'll discover that some equations cannot be solved using backtracking. So, you'll explore another method, *guess-check-and-improve*.

Vocabulary Here's a list of the new vocabulary words associated with solving equations.

backtracking	guess-check-and-	open sentence
equation	improve	solution
flowchart	inequality	

What can you do at home?

Encourage your student to show you strategies for solving equations. You might even enjoy a game in which you each write a simple equation on a slip of paper, trade papers, and solve one another's equation. Once you have found the solution, talk about what you did to solve it.

Understanding Equations

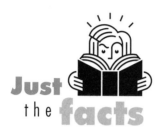

Just the facts

The equals sign was first used by Robert Recorde in his book *Whetstone of Witte*, published in 1557. Recorde states that he chose parallel line segments of equal length because "no two things can be more equal."

You have been working with equations for many years. An **equation** is a mathematical sentence stating that two quantities have the same value. An equals sign, $=$, is used to separate the two quantities. Here are three examples of equations:

$$9 + 6 = 15 \qquad 9 + 6 = 5 \times 3 \qquad 7 + 5 = 15 - 3$$

Explore

In the game *Equation Challenge*, you will see how many equations you can make using a given set of numbers.

Equation Challenge Rules

- Your teacher will call out seven single-digit numbers. Write the numbers on a sheet of paper. (Note: Your teacher may call the same number more than once. For example, the numbers might be 1, 2, 5, 3, 4, 9, and 3.)

- You have 5 minutes to write down as many correct equations as you can, using only the seven numbers, an equals sign, operation symbols, decimal points, and parentheses. Follow these guidelines when writing your equations:

  —Use each of the seven numbers only once in each equation.

  —You don't need to use all the numbers in each equation.

  —You can combine the numbers to make numbers with more than one digit. For example, you could combine 1 and 2 to make 12, 21, 1.2, or $\frac{1}{2}$.

- At the end of the 5 minutes, check your equations to make sure they are correct.

Here are some sample equations for the numbers 1, 2, 5, 3, 4, 9, and 3:

$$9 + 5 = 14 \qquad \frac{35 + 1}{4} = 9 \qquad 9 \cdot 5 = 3 \cdot 3 \cdot (4 + 1)$$

Play *Equation Challenge*. Be creative when writing your equations!

Investigation 1 ▶ Equations and Inequalities

VOCABULARY
inequality

As you know, an *equation* is a mathematical sentence stating that two quantities have the same value. A mathematical sentence stating that two quantities have *different* values is called an **inequality.** Inequalities use the symbols ≠, <, and >. The table below explains what these symbols mean.

Symbol	What It Means	Example
≠	is not equal to	$7 + 2 \neq 5 + 1$
<	is less than	$4 + 5 < 20$
>	is greater than	$6 \cdot 9 > 6 + 9$

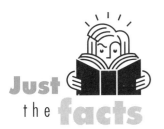

Just as sentences with words can be true or false, so can equations and inequalities. Consider these six sentences:

Texas became the 28th U.S. state in 1845. Alaska became the 49th U.S. state in 1959.

Stop signs are yellow.	$4 = 32 \div 8$
$5 \cdot 6 > 6 \cdot 5$	The sun is hot.
Alaska is south of Texas.	$5 \times 4 = 27 - 3$

Think & Discuss

Which of the sentences above are true, and which are false?

Find a way to make each of the false sentences true by changing or adding just one word, symbol, or number.

Problem Set A

The sentences below are false. Make each sentence true by changing one symbol or number.

1. $17 + 5 < 3^2 + 12$

2. $14 + 5 = 12 + 11$

3. $23 - 11 = 22 \div 2$

4. $6 \cdot 5 > (4 \cdot 7) + 8$

Tell whether each sentence is true or false. If it is false, make it true by replacing the equal sign with < or >.

5. $5 + 13 = 2 + 4^2 + 3$

6. $7 + (2 \times 3) = (6 \times 2) + 1$

7. $24 \div 5 = 2 + 3$

8. $\frac{2}{5} = \frac{1}{2}$

9. $0.25 = \frac{1}{4}$

10. $8 + 12 \div 4 = (8 + 12) \div 4$

Share & Summarize

1. Explain the difference between an equation and an inequality, and give an example of each.

2. Give an example of an equation or an inequality that is false. Then explain how you could change the sentence to make it true.

Investigation 2 ▶ Equations with Variables

You can determine whether the equations in Investigation 1 are true by finding the value of each side. But what if an equation contains a variable? For example, consider this equation:

$$3 \times n = 18$$

VOCABULARY
open sentence

You can't tell whether this equation is true or false unless you know the value of n. An equation or inequality that can be true or false depending on the value of the variable is called an **open sentence.**

Think & Discuss

For each open sentence, find a value of n that makes it true and a value of n that makes it false.

$$3 \times n = 18 \qquad n \div 2 = 2.5 \qquad n + 5 = 25$$

VOCABULARY
solution

Finding the values of the variable or variables that make an equation true is called *solving* the equation. A value that makes an equation true is a **solution** of the equation.

Consider this equation:

$$6 \cdot n - 1 = 29$$

Finding a solution of $6 \cdot n - 1 = 29$ is the same as answering this question:

For what value of n is $6 \cdot n - 1$ equal to 29?

Remember

These are all ways to write "6 times p":

$6 \times p \quad 6 \cdot p \quad 6p$

Jing tried several values for n. Here is what she found:

n	$6 \cdot n - 1$	Test	Solution?
3	17	$6 \cdot n - 1 < 29$	no
4	23	$6 \cdot n - 1 < 29$	no
5	29	$6 \cdot n - 1 = 29$	yes
6	35	$6 \cdot n - 1 > 29$	no
7	41	$6 \cdot n - 1 > 29$	no
8	47	$6 \cdot n - 1 > 29$	no

Jing found that 5 is a solution of $6 \cdot n - 1 = 29$, since $6 \cdot 5 - 1 = 29$. Jing noticed that the results kept increasing as n increased, so she concluded that 5 must be the only solution.

Problem Set B

Each of these equations has one solution. Solve each equation.

1. $6 \cdot n - 1 = 41$

2. $6 \cdot n - 1 = 11$

3. $2p + 7 = 19$

4. $4 + 4 \cdot b = 20$

5. $\frac{5}{4} = \frac{25}{d}$

6. $m + 3 = 2m$

7. Try the numbers 1, 2, 3, 4 in the equation $2 \cdot d + 3 = 15 - 2 \cdot d$ to test whether any of them is a solution.

8. Write three equations with a solution of 13. Check that your equations are correct by substituting 13 for the variable.

All of the equations you have looked at so far have one solution. It is possible for an equation to have more than one solution or to have no solution at all.

For some equations, *every* number is a solution. Such equations are *always* true, no matter what the values of the variables are.

Think & Discuss

Equations that include a squared variable often have two solutions. Find two solutions for the equation $m^2 - 3m = 0$.

Explain why each equation below is always true.

$$n + 3 = 3 + n \qquad a \times 5 = a + a + a + a + a$$

Remember

Equations with squared variables, like $m^2 - 3m = 0$ in the Think & Discuss, are called *quadratic equations*. They can be used to describe the path that a projectile—an object that is tossed, thrown, or shot into the air—travels along.

Problem Set C

1. Try the values 1, 2, 3, and 4 to test whether any are solutions of this equation:

$$t^2 + 8 = 6 \cdot t$$

Tell whether each equation below is *always true, sometimes true,* or *never true,* and explain how you know.

2. $m - m = 0$

3. $\frac{r}{3} = r$

4. $q + 7 = q - 7$

5. $p \div 7 = \frac{1}{7} \times p$

6. $n \times 2 = n + 1$

7. $(a + 3) \cdot 2 = 2a + 6$

8. Challenge Tell whether each equation has a whole-number solution. Explain how you know.

a. $2 \cdot n - 1 = 37$

b. $2n + 1 = 18$

c. $3 \cdot n + 5 = n + 7$

d. $n^2 + 2 = 1$

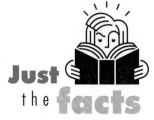

Just the facts

In *meteorology,* the science of weather, an equal sign represents fog.

Share & Summarize

1. Solve the equation $3p + 5 = 11$, and explain how you found the solution.

2. Give an example of an equation that is always true and an example of an equation that is never true.

Investigation ▶ Just Undo It!

In this lab investigation, you will practice "undoing" sets of instructions. In Lesson 9.2, you will see how the strategies you use in this lab can help you solve equations.

MATERIALS

- 1 red block and 1 blue block
- 1 yellow counter and 1 green counter
- Lompoc, California, street map

Undoing Instructions

Starting with a blank sheet of paper, follow these instructions:

- Draw a small X (in pencil) in the center of the paper.
- Put the red block on the X.
- Put the yellow counter on the red block.
- Put the blue block on the yellow counter.
- Put the green counter on the blue block.

1. Write a list of steps you think would undo these instructions, leaving you with a blank sheet of paper. Don't touch any of the items on the paper until you have finished writing your instructions.

2. Follow your steps from Question 1. Did your steps undo the instructions above? If not, rewrite them until they do.

3. How do your steps compare with the original set of steps?

Reversing Directions

Madeline lives in Lompoc, California, on the corner of Nectarine Avenue and R Street. Today she met her friend T.J. at the town pool. T.J. had given her these directions to get to the pool:

- Start at the corner of Nectarine Avenue and R Street.
- Walk 2 blocks east along Nectarine Avenue.
- Turn right from Nectarine Avenue onto O Street.
- Walk 4 blocks south on O Street.
- Turn left from O Street onto Maple Avenue.
- Walk 5 blocks east on Maple Avenue.
- Turn right from Maple Avenue onto J Street.
- Walk 4 blocks south on J Street.
- Turn left from J Street onto Ocean Avenue.
- Walk 7 blocks east on Ocean Avenue.
- The pool is at the corner of Ocean Avenue and C Street.

Now Madeline must reverse the directions to get home.

4. Without looking at the map, write a set of directions Madeline could follow to get home from the pool.

5. On the street map, carefully follow the directions you wrote in Question 4. Do you end up at the corner of Nectarine Avenue and R Street? If not, make changes to your directions until they work.

6. Write a set of directions to get from one place on the map to another. Word your steps like those on page 563.

   • When you describe a turn, mention the street you start on, whether you turn left or right, and the street you end on.

   • When you describe a walk along a street, mention the number of blocks you walk, the direction you walk, and the street name.

   When you are finished, try your directions to make sure they are accurate.

7. Exchange directions with your partner. Without looking at the map, write the steps that reverse your partner's directions. Then use the street map to test your directions.

8. Describe some general strategies you find useful when reversing a set of directions.

What Did You Learn?

9. In this lab investigation, you undid two types of instructions:

   • Steps for stacking blocks and counters

   • Directions for getting from one place to another

   Describe how the methods you used to undo the instructions in each case were similar.

Just the **facts**

It is believed that humans have been making maps since prehistoric times. Archeologists have discovered systems of lines drawn on cave walls and bone tablets that may be maps of hunting trails made by prehistoric peoples.

On Your Own Exercises

1. In a round of the game *Equation Challenge,* the following numbers were called out: 1, 2, 2, 4, 5, 5, and 9. Make at least four equations using these numbers.

In Exercises 2 and 3, the sentence is not true. Change or add one number or symbol to make it true.

2. $5 + 16 = 3 \times 8$

3. $8 \cdot 5 \neq 17 + 16 + 7$

In Exercises 4–9, tell whether the sentence is true or false. If it is false, make it true by replacing the equal sign with $<$ or $>$.

4. $3 \times 11 = 42 - 9$

5. $(3 \times 5) + 4 = 4 + 5 + 1$

6. $\frac{1}{3} + \frac{4}{6} = \frac{3}{7} + \frac{16}{28}$

7. $0.95 = \frac{9}{10}$

8. $3 \times 13 = 54 - 16$

9. $16 - 8 \div 4 = (16 - 8) \div 4$

Solve each equation.

10. $x \cdot 12 = 48$

11. $56 + m = 100$

12. $6p + 10 = 28$

13. $50 - 4 \cdot z = 30$

14. Consider the equations $s + 13 = 20$ and $p + 13 = 20$.

    a. Solve each equation.

    b. How do the solutions to the two equations compare? Explain why this makes sense.

15. Test the values 0, 1, 3, 4, and 6 to see whether any are solutions of $7m - m^2 + 10 = 16$.

Just the facts

The solution to the equation in Exercise 10, $x \cdot 12 = 48$, is the number of 12-egg cartons needed to hold four dozen eggs.

Tell whether each equation is *always true, sometimes true,* or *never true,* and explain how you know.

16. $5 \cdot m = \frac{25 \cdot m}{5}$

17. $5s = 25$

18. $t - 1 = t + 1$

19. $p^2 = p \cdot p$

20. $n + 6 = 7 \div n$

21. $7p = p \div 7$

22. Write three equations with a solution of 3.5.

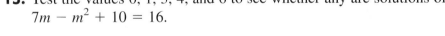

23. Of these three equations, one has no solution, one has one solution, and one has two solutions. Decide which is which, and find the solutions.

    a. $p^2 + 6 = 5 \times p$

    b. $3p + 5 = 3p - 5$

    c. $4 \cdot p + 5 = 7$

24. Isabela and Jada were playing a game of *What's My Rule?* Here is Isabela's secret rule:

$$o = 37 - 4 \times i, \text{ where } o \text{ is the output and } i \text{ is the input}$$

    a. What input value gives an output of 17? Check your answer by substituting it into the rule.

    b. What input value gives an output of 5? Check your answer.

    c. What input value gives an output of 0? Check your answer.

25. Pretend you are playing a game of *What's My Rule?* Make up a secret rule for calculating an output value from an input value. Your rule should use one or two operations.

    a. Write your rule in symbols.

    b. Find the input value for which your rule gives an output of 25.

    c. Find the input value for which your rule gives an output of 11.

26. Paul and Katarina were playing a game of *What's My Rule?* Here is Katarina's secret rule:

$$m = \frac{n}{10}, \text{ where } n \text{ is the input and } m \text{ is the output}$$

    a. Write an equation you could solve to find the input value that gives an output of 1.5.

    b. Solve your equation from Part a to find the input value *n*.

    c. Write an equation you could solve to find the input value that gives an output of 20.

    d. Solve your equation from Part c to find the input value *n*.

Connect & Extend

In your own words

Explain the meaning of each of these words:

- equation
- inequality
- open sentence
- solution

In Exercises 27–31, use this idea:

You can think of an equation as a balanced scale. The scale at right represents the equation $3 \times 5 = 9 + 6$. The scale is balanced because both sides have the same value.

The second scale represents the equation $4 + n = 10$. To solve this equation, you need to find the value of n that will make the scale balance.

In Exercises 27–31, you will solve puzzles involving scales. Thinking about these puzzles may give you some ideas for solving equations.

27. This scale is balancing bags of peanuts and boxes of popcorn:

a. How many bags of peanuts will balance one box of popcorn?

b. If a bag of peanuts weighs 5 ounces, how much does a box of popcorn weigh?

28. Consider this scale:

a. Write the equation this scale represents.

b. What number will balance one n?

In Exercises 29–31, refer to the information on page 567.

29. Consider this scale:

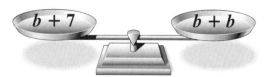

a. Write the equation this scale represents.

b. What number will balance one b?

30. These two scales hold jacks, marbles, and a block:

a. How many jacks will balance the block?

b. How many jacks will balance one marble?

c. If the block weighs 15 grams, how much does a jack weigh? How much does a marble weigh?

31. Challenge These scales hold blocks, springs, and marbles:

a. How many marbles will balance one spring?

b. If a marble weighs 1 ounce, how much does a spring weigh? How much does a block weigh?

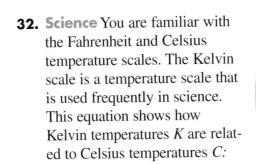

32. Science You are familiar with the Fahrenheit and Celsius temperature scales. The Kelvin scale is a temperature scale that is used frequently in science. This equation shows how Kelvin temperatures K are related to Celsius temperatures C:

$$K = C + 273$$

a. The mean surface temperature on Mercury is about 180°C. Express this temperature in Kelvins.

b. The maximum surface temperature on Mars is about 290 Kelvin. Express this temperature in degrees Celsius.

c. Which planet is hotter, Mercury or Mars? Explain your answer.

33. Economics The unit of currency in South Africa is the rand. On September 4, 2000, one U.S. dollar was worth about 6.96 rand. This relationship can be expressed as an equation, where R stands for the number of rand and D stands for the number of dollars:

$$6.96 \times D = R$$

a. On that day, Jacob exchanged $75 for rand. How many rand did he receive?

b. Jacob wanted to buy a small statue that cost 20 rand. He thought this was about $2.85. His sister thought it was about $139.20. Who was right? Explain your answer.

Mixed Review

Find each product or quotient without using a calculator.

34. $\frac{5}{6} \cdot \frac{7}{12}$

35. $\frac{5}{6} \div \frac{7}{12}$

36. $23.7 \div 1{,}000$

37. $3.1 \cdot 50.7$

38. $5\frac{3}{8} \cdot 2\frac{5}{9}$

39. $4\frac{1}{4} \div \frac{3}{4}$

Number Sense Order each set of numbers from least to greatest.

40. $\frac{2}{3}$, 0.6, $-\frac{5}{6}$, $\frac{3}{4}$, $-\frac{2}{3}$, -0.1

41. $\frac{5}{7}$, $\frac{31}{50}$, $\frac{3}{5}$, $\frac{50}{68}$, $\frac{7}{11}$, $\frac{17}{21}$

Geometry Find the missing length in each triangle.

42.

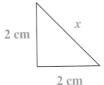

2 cm
2 cm
x

43.

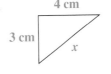

4 cm
3 cm
x

44.

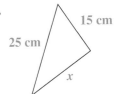

15 cm
25 cm
x

Backtracking

Remember

$3 \times n$, $3 \cdot n$, and $3n$ are all ways to write "3 times n."

Jay was playing *What's My Rule?* with his friend Marla. He figured out that this was the rule:

> *To find the output, multiply the input by 3 and then add 7.*

He wrote the rule in symbols, using n to represent the input and t to represent the output.

$$t = 3n + 7$$

Jay wanted to find the input that would give an output of 43. This is the same as solving the equation $3n + 7 = 43$.

3 times the number plus 7 is equal to 43.

So there must have been 36 before the 7 was added.

3 times a number gives 36—so the number must have been 12.

If I let n be 12, then $3n + 7$ is $3 \times 12 + 7$, which is 43. So I'm right—the input number must be 12.

Think & Discuss

Using Jay's rule, what input gives an output of 40?

Explain the reasoning you used to find the input. Check your answer by substituting it into the equation $3n + 7 = 40$.

Jay's method of solving $3n + 7 = 43$ involves working backward from the output value to find the input value. In this lesson, you will learn a technique for working backward called *backtracking*.

Investigation 1 Learning to Backtrack

Hannah's class was playing *What's My Rule?* Hannah found that the rule was $t = 4n + 5$, where n is the input and t is the output.

To find an output with this rule, you do the following:

> *Start with an input.*
> *Multiply it by 4.*
> *Add 5.*

Hannah drew a diagram, called a **flowchart,** to show these steps.

The oval at the left side of the flowchart represents the input. Each arrow represents a *mathematical operation*. The oval to the right of an arrow shows the result of a mathematical operation. The oval at the far right represents the output.

Here is Hannah's flowchart for the input value 3:

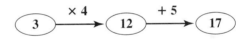

In the next problem set, you will practice working with flowcharts.

Problem Set A

Copy each flowchart, and fill in the ovals.

1.

$$8 \xrightarrow{\times 3} \bigcirc \xrightarrow{-7} \bigcirc$$

2.

$$5 \xrightarrow{\times 3} \bigcirc \xrightarrow{-1} \bigcirc \xrightarrow{\times 2} \bigcirc$$

3. In the rule $j = 7m - 2$, you can think of m as the input and j as the output.

  a. Create a flowchart for the rule, but don't fill in the ovals.

  b. Use your flowchart to find the value of j when the value of m is $\frac{5}{3}$.

4. In the rule $d = 3.2 + a \div 10$, you can think of a as the input and d as the output.

    a. Create a flowchart for this rule, but don't fill in the ovals. (Be sure to think about order of operations.)

    b. Use your flowchart to find the value of d when the value of a is 111.

In Problem Set A, you used flowcharts to *work forward*—starting with the input and applying each operation to find the output. You can also use flowcharts to *work backward*—starting with the output and undoing each operation to find the input. This process, called **backtracking,** is useful for solving equations.

VOCABULARY
backtracking

EXAMPLE

When playing a game of *What's My Rule?*, Hannah figured out that the secret rule was

$$t = \frac{n + 1}{2}$$

Hannah wanted to find the input that gives an output of 33. That is, she wanted to solve this equation:

$$\frac{n + 1}{2} = 33$$

To find the solution, she first made a flowchart:

Then she found the input by backtracking:

"Since 33 is the output, I'll put it in the last oval."

"Since the number in the second oval was divided by 2 to get 33, it must be 66."

"1 was added to the input to get 66, so the input must be 65."

Hannah checked her solution, 65, by substituting it into the original equation:

$$\frac{65 + 1}{2} = \frac{66}{2} = 33$$

Remember

Operations that undo each other are called inverse operations. To undo the division, Hannah used the inverse operation, multiplication. To undo the addition, she used the inverse operation, subtraction.

Problem Set B

1. Marcus used Hannah's rule, $t = \frac{n+1}{2}$, and got the output 53.

Use backtracking to find Marcus' input. Explain each step in your solution.

Tyrone solved three equations by backtracking. Below are the flowcharts he started with. For each flowchart, write the equation he was trying to solve. (Use any letter you want to represent the input variable.) Then backtrack to find the solution. Check your solutions.

2.

3.

4.

5. Terry drew this flowchart:

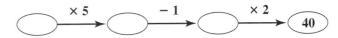

a. Copy Terry's flowchart, and fill in the ovals.

b. Which of these equations can be represented by Terry's flowchart? Explain how you know.

$$5 \cdot k - 1 \cdot 2 = 40 \qquad (k \cdot 5 - 1) \cdot 2 = 40$$
$$2 \cdot (5k - 1) = 40 \qquad 5k - 1 \cdot 2 = 40$$

Share & Summarize

1. Explain what a flowchart is. Demonstrate by making a flowchart for the rule $t = 5n - 3$.

2. Use your flowchart from Question 1 to find the output for input $\frac{7}{10}$.

3. Explain what backtracking is. Demonstrate how backtracking can be used to solve an equation by solving $5n - 3 = 45$. Check your solution.

Investigation 2 ▶ Practicing Backtracking

In this investigation, you will practice backtracking so you can use it to solve equations quickly and easily.

Problem Set C

1. A group of students was playing *What's My Rule?* Miguel and Althea both thought they knew what the rule was. Both students used K to represent the input and P to represent the output.

   Miguel's rule: $P = 14 \cdot (K + 7)$

   Althea's rule: $P = 14 \cdot K + 7$

   a. Make a flowchart for each rule, but don't fill in the ovals.

   b. For each rule, use backtracking to find the input that gives the output 105.

   c. Are these two rules equivalent? Explain why or why not.

2. Gabriela and Erin were playing a game called *Think of a Number.*

Think of a number. Triple it. Subtract 6. Multiply your result by 5. What do you get?

60.

   Gabriela must figure out Erin's starting number.

   a. Draw a flowchart to represent this game.

   b. What equation does your flowchart represent?

   c. Use backtracking to solve your equation. Check your solution by following Gabriela's steps.

Think & Discuss

Luke wanted to solve this equation by backtracking:

$$\frac{2 \times (n + 1)}{3} - 1 = 5$$

He made this flowchart:

Does Luke's flowchart correctly represent the equation? Why or why not?

Solve the equation, and explain how you found the solution.

The equation in the Think & Discuss involves several operations. You can often solve equations like this by backtracking, but you need to pay close attention to order of operations as you draw the flowchart.

Problem Set D

1. Conor, Althea, and Miguel each made a flowchart to represent this equation:

$$\frac{1 + n \cdot 3}{4} - 11 = 10$$

Tell whose flowchart is correct, and explain the mistakes the other students made.

Conor's flowchart:

Althea's flowchart:

Miguel's flowchart:

In Problems 2 and 3, draw a flowchart to represent the equation. Then use backtracking to find the solution. Be sure to check your solution.

2. $\frac{n-13}{2} + 6 = 15$ **3.** $7\left(\frac{n+4}{7} + 1\right) = 84$

Solve each equation. Be sure to check your solutions.

4. $\frac{7z+2}{15} = 2$

5. $(n \cdot 12 + 8) \cdot 100 = 2{,}100$

6. $\frac{q-36}{6} + 16 = 83$

7. $4 \cdot \left(\frac{b}{2} - 3\right) + 1 = 97$

Share & Summarize

Give an example to demonstrate why it is important to pay close attention to order of operations when you make a flowchart.

Investigation ▶3 Using Backtracking to Solve Problems

These "ladders" are made from toothpicks:

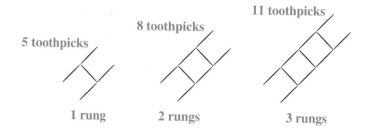

5 toothpicks   8 toothpicks   11 toothpicks

1 rung   2 rungs   3 rungs

The rule for the number of toothpicks n in a ladder with r rungs is $n = 3r + 2$.

Think & Discuss

Can you explain why the rule for the number of toothpicks n in a ladder with r rungs is $n = 3r + 2$?

Suppose you have 110 toothpicks. What size ladder can you make? How can you use backtracking to help you find the answer?

Problem Set E

1. Write and solve an equation to find the number of rungs on a toothpick ladder made with 53 toothpicks.

2. Look at this pattern of toothpick shapes:

1 trapezoid   2 trapezoids   3 trapezoids

   a. Write a rule for finding the number of toothpicks n you would need to make a shape with t trapezoids.

   b. Write and solve an equation to find the number of trapezoids in a shape with 125 toothpicks.

3. Look at the pattern in this table:

n	0	3	6	9	12
y	19	20	21	22	23

   a. Write a rule that relates n and y.

   b. Write and solve an equation to find the value of n when y is 55.

In the next problem set, you will see that backtracking can also be used to solve everyday problems.

Problem Set F

1. Leong makes candy apples to sell at the farmer's market on Saturday. He makes a profit of 35¢ per apple.

   a. Write a rule Leong could use to calculate his profit if he knows how many candy apples he sold. Tell what each letter in your rule stands for.

   b. Leong wants to earn $8 so he can see a movie Saturday night. Write an equation Leong could solve to find the number of candy apples he must sell to earn $8.

   c. Use backtracking to solve your equation. How many candy apples does Leong need to sell?

2. The plumbers at DripStoppers charge $45 for a house call, plus $40 for each hour of work.

    a. Write a rule for the cost of having a DripStoppers plumber come to your home and do *n* hours of work.

    b. DripStoppers sent Mr. Valdez a plumbing bill for $105. Write and solve an equation to find the number of hours the plumber worked at Mr. Valdez's home. Check your solution.

3. When you hire a taxi, you are usually charged a fixed amount of money when the ride starts, plus an amount that depends on how far you travel. Suppose a taxi charges $2 plus $0.75 for every quarter mile.

    a. Write an equation you could solve to find out how far you can travel for $20.

    b. Solve your equation. How far could you travel for $20? Check your answer.

4. Caroline and Althea are making a kite. The materials for the main part of the kite cost $4.50, and the string costs 9¢ per yard.

    a. Write an equation you could solve to find how long the string could be if the friends have $30 to spend on their project.

    b. Solve your equation to find how long the string could be.

Share & Summarize

Jing used toothpicks to create this pattern:

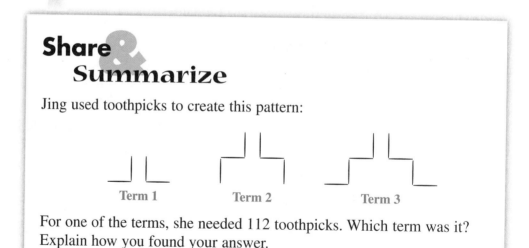

For one of the terms, she needed 112 toothpicks. Which term was it? Explain how you found your answer.

On Your Own Exercises

Practice & Apply

Copy each flowchart, and fill in the ovals.

1.
$$12 \xrightarrow{\times 2} \bigcirc \xrightarrow{+\ 12} \bigcirc$$

2.
$$2.7 \xrightarrow{\times 3} \bigcirc \xrightarrow{-\ 1} \bigcirc \xrightarrow{\times 4} \bigcirc$$

Hannah solved three equations by backtracking. Below are the flowcharts she started with. For each flowchart, write the equation she was trying to solve. Then backtrack to find the solution. Check your solutions.

3.
$$\bigcirc \xrightarrow{\times 16} \bigcirc \xrightarrow{-\ 4} 28$$

4.
$$\bigcirc \xrightarrow{\div 3} \bigcirc \xrightarrow{+\ 11} 41$$

5. Challenge
$$\bigcirc \xrightarrow{+\ 1} \bigcirc \xrightarrow{\times 4} \bigcirc \xrightarrow{-\ 5} 15$$

6. In a game of *What's My Rule?*, Rosita wrote the rule $b = 3a \div 4$, where a is the input and b is the output.

   a. Make a flowchart for Rosita's rule.

   b. Use your flowchart to find the output when the input is 18.

   c. Backtrack to solve the equation $3a \div 4 = 101$.

Make a flowchart to represent each equation, and then use backtracking to solve the equation. Be sure to check your solutions.

   7. $4k + 11 = 91$

   8. $4 \cdot (m - 2) = 38$

9. Neva and Jay were playing *Think of a Number.* Neva said:

Think of a number. Subtract 1 from your number. Multiply the result by 2. Then add 6. What number do you get?

Jay said he got 10. Neva must figure out Jay's starting number.

a. Draw a flowchart to represent this game.

b. What equation does your flowchart represent?

c. Use backtracking to find the number Jay started with. Check your solution by following Neva's steps.

10. For a game of *What's My Rule?,* Mia and Desmond wrote the rule $y = 9(2x + 1) + 1$, where x is the input and y is the output.

a. Draw a flowchart for Mia and Desmond's rule.

b. Use your flowchart to solve the equation $9(2x + 1) + 1 = 46$.

Draw a flowchart to represent each equation. Then use backtracking to find the solution. Be sure to check your solutions.

11. $\frac{3 \cdot m \cdot 2}{6} = 1$

12. $\frac{8p + 2}{5} - 5 = 19$

13. $3\left(\frac{5 + n}{3} - 4\right) = 15$

14. Look at the toothpick sequence below. The rule for the number of toothpicks t needed to make Term n is $t = 2n + 3$.

Term 1   Term 2   Term 3

a. Explain why this rule works for every term.

b. Write and solve an equation to find the number of the term that requires 99 toothpicks.

15. Look at the pattern in this table:

x	0	5	10	15	20
y	100	102.5	105	107.5	110

a. Write a rule that relates x and y.

b. Write and solve an equation to find the value of x when y is 197.5.

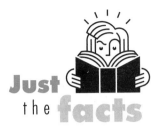

Connect & Extend

16. Economics At Marshall Park, you can rent a canoe for $5, plus $6.50 per hour.

   a. Write a rule for calculating the cost C of renting a canoe for h hours.

   b. Conor, Jing, and Miguel paid $27.75 to rent a canoe. Write and solve an equation to find the number of hours the friends used the canoe. Check your solution.

17. Economics Avocados cost $1.89 each. Hannah plans to make a large batch of guacamole for a party, and she wants to buy as many avocados as possible. She has $14.59 to spend.

   a. Write an equation you could solve to find how many avocados Hannah can buy for $14.59.

   b. Solve your equation. How many avocados can Hannah buy?

18. Evita wants to make a fence out of wooden poles. She drew a diagram to help her figure out how many poles she would need.

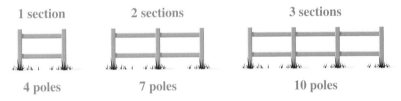

  1 section   2 sections   3 sections

  4 poles   7 poles   10 poles

   a. Write a rule connecting the number of poles p to the number of sections s.

   b. The lumberyard has 100 poles in stock. Write an equation you could solve to find the number of fence sections Evita can build with 100 poles.

   c. Solve your equation. How many fence sections can Evita build?

   d. If each pole is 2 yards long, how long will a 100-pole fence be?

19. Physics A bus is traveling at an average speed of 65 miles per hour.

a. Copy and complete the table to show the distance the bus would travel in the given numbers of hours.

Time (hours), t	1	2	3	4	5
Distance (miles), d					

b. On a grid like the one below, plot the points from your table. If it makes sense to do so, connect the points with line segments.

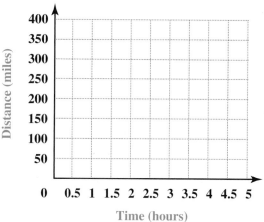

c. Use your graph to estimate how long it would take the bus to travel 220 miles.

d. Write a rule that relates the time traveled t (in hours) to the distance traveled d (in miles).

e. Write and solve an equation to find how long it would take the bus to travel 220 miles. How does the solution compare to your estimate from Part c?

20. Julie and Noah were playing *Think of a Number.* Noah said:

Think of a number. Square it. Add 3. Divide your result by 10. What number do you get?

Julie said she got 8.4. Noah must figure out Julie's starting number. He drew this flowchart to represent the game:

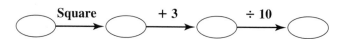

a. What equation does Noah need to solve to find Julie's starting number?

b. Use backtracking to solve your equation. Check your solution by following Noah's steps.

21. **Economics** Althea wants to buy a jacket at Donovan's department store. The store is having a "20% off" sale. In addition, Althea has a coupon for $5 off any item in the store.

   a. Write a rule Althea could use to calculate the price *P* she would pay for a jacket with an original price of *d* dollars. (Note: The $5 is subtracted *after* the 20% discount is calculated.)

   b. Althea pays $37.80 for a jacket. Write and solve an equation to find the original price of the jacket.

22. **Preview** This equation gives the height *h* of a baseball, in feet, *t* seconds after it is thrown straight up from ground level:

$$h = 40 \cdot t - 16 \cdot t^2$$

   a. How high is the ball after 0.5 second?

   b. How high is the ball after 1 second?

   c. Based on what you learned in Parts a and b, estimate how long it would take the ball to reach a height of 20 feet. Explain your estimate.

   d. Substitute your estimate for *t* in the equation to find out how high the ball would be after that number of seconds. Were you close? Was your estimate too high or too low?

23. **Sports** Lana is making fishing lures. Each lure requires 2¢ worth of fishing line. Lana uses feathers and weights to make the lures. The feathers cost 17¢ apiece, and the weights cost 7¢ apiece.

   a. Write a rule for the cost *C* of a lure made with *f* feathers and *w* weights.

   b. Lana doesn't want to spend more than 65¢ on each lure. Write an equation you could solve to find how many feathers she could use on a lure made with two weights.

   c. Solve your equation to find how many feathers Lana can use on a lure with two weights.

24. Economics Jordan has a part-time job as a telemarketer. He earns $14.00 per hour plus 50¢ for every customer he calls.

  a. Write a rule for computing how much Jordan will earn on a three-hour shift if he calls c customers.

  b. Jordan would like to earn $100 on his 3-hour shift. How many customers must he call?

Mixed Review

Find the value of each expression without using a calculator.

25. $27 - 3^2 - 10$

26. $4 + 4 \cdot 3 - 2$

27. $\sqrt{49} + 3^2$

28. $5^2 \div 5$

29. $\frac{2}{3} \cdot \left(\frac{1}{3}\right)^2 - \frac{1}{27}$

30. $\sqrt{3^2 + 4^2}$

31. Sports In a track-and-field competition, 10 women participated in the 100-meter run. Here are their times in seconds:

   12.2   11.3   13.5   11.5   11.7   12.6   15.5   11.8   13.4   11.5

  a. Find the mean and the median time.

  b. Make a stem-and-leaf plot of the times.

Geometry Find each value of x.

32. Area = 4.5 ft²

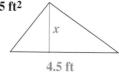

4.5 ft

33. Area ≈ x

6 in.

34. Area = 9.8 cm²

3.5 cm

x

35. Statistics In a landfill, bulldozers spread solid waste (trash and garbage) into layers, alternating with layers of dirt. The list below shows the number of solid-waste landfills that were in use in each region of the United States in a recent year.

Northeast	South Atlantic	South	Midwest	West
Connecticut, 11	Delaware, 3	Alabama, 28	Illinois, 61	Alaska, 217
Maine, 27	Florida, 67	Arkansas, 67	Indiana, 32	Arizona, 59
Massachusetts, 106	Georgia, 159	Kentucky, 12	Iowa, 77	California, 278
New Hampshire, 33	Maryland, 25	Louisiana, 29	Kansas, 58	Colorado, 72
New Jersey, 14	North Carolina, 114	Mississippi, 14	Michigan, 54	Hawaii, 10
New York, 42	South Carolina, 37	Oklahoma, 94	Minnesota, 26	Idaho, 37
Pennsylvania, 47	Virginia, 152	Tennessee, 81	Missouri, 30	Montana, 82
Rhode Island, 4	West Virginia, 22	Texas, 678	Nebraska, 21	Nevada, 56
Vermont, 61			North Dakota, 12	New Mexico, 79
			Ohio, 63	Oregon, 88
			South Dakota, 13	Utah, 54
			Wisconsin, 46	Washington, 25
				Wyoming, 59

Source: United States Environmental Protection Agency Web site, www.epa.gov.

a. Find the region in which your state is located, and choose one of the other regions. Find the mean, median, and mode of each of those two regions.

b. Make a back-to-back stem plot of the two data sets.

c. Using the measures you found and the stem plot you made, write a paragraph comparing the numbers of municipal landfills in use in the two regions.

9.3

Guess-Check-and-Improve

Backtracking is useful for solving many types of equations. However, as you will see in this lesson, some equations are difficult or impossible to solve by backtracking.

Johanna and Rosita were playing *What's My Rule?* Johanna made this table to keep track of her guesses:

From her table, Johanna figured out that Rosita's secret rule was

Input	Output
10	110
20	420
30	930
40	1,640

$$n = m \cdot (m + 1)$$

where m is the input and n is the output.

Now the two friends want to figure out what input gives an output of 552. That is, they want to solve the equation $m \cdot (m + 1) = 552$.

Think & Discuss

Rosita suggests they solve the equation by backtracking. Try to solve the equation this way. Are you able to find the solution? Why or why not?

What advice would you give Rosita and Johanna to help them solve the equation?

Investigation 1 ▶ Using Guess-Check-and-Improve

VOCABULARY
guess-check-and-improve

As you have seen, backtracking does not work for every equation. In this lesson, you will learn another solution method. This method is called **guess-check-and-improve** because that is exactly what you do.

The Example on page 587 shows how Rosita and Johanna used guess-check-and-improve to solve the equation $m \cdot (m + 1) = 552$.

From the table Johanna had made during the game, she could see that the output for $m = 20$ was too low and the output for $m = 30$ was too high.

Input	Output
10	110
20	420
30	930
40	1,640

Using this information, the friends decided to try 25, the number halfway between 20 and 30. They checked their guess by substituting it for m in the expression $m \cdot (m + 1)$:

$$m \cdot (m + 1) = 25 \cdot (25 + 1)$$
$$= 25 \cdot 26$$
$$= 650$$

The output 650 is too high. Johanna recorded the guesses and the results in the table.

m	$m \cdot (m + 1)$	Comment
20	420	too low
30	930	too high
25	650	too high, but closer

The friends now decided the solution must be between 20 and 25. The table below shows their next two guesses.

m	$m \cdot (m + 1)$	Comment
20	420	too low
30	930	too high
25	650	too high, but closer
22	506	too low, but close
23	552	23 is the solution!

The solution to $m \cdot (m + 1) = 552$ is 23.

Review the process that Johanna and Rosita used:

- They *guessed* the solution.
- They *checked* their solution by substituting it into the equation.
- They used the result to *improve* their guess.

Now it's your turn to try guess-check-and-improve.

Problem Set A

1. Conor, Marcus, and Jing are playing *What's My Rule?* Here is Jing's secret rule:

   $$d \cdot (d + 3) = J, \text{ where } d \text{ is the input and } J \text{ is the output}$$

   a. Conor gave Jing an input, and Jing calculated the output as 8,554. Write an equation you could solve to find Conor's input.

   b. Find a solution of your equation using guess-check-and-improve.

   c. Marcus gave Jing an input, and Jing calculated the output as 32.56. Write an equation you could solve to find Marcus' input.

   d. Find a solution of your equation from Part c using guess-check-and-improve.

For each equation, use guess-check-and-improve to find a solution.

2. $2n + 6 = 20$ **3.** $19 = \frac{4}{q} + 3$ **4.** $s^2 + 2s = 19.25$

5. Miguel is trying to solve this equation using guess-check-and-improve: $25 - 3 \cdot d = 17.8$. The table shows his first two guesses. Miguel asked:

   "Why was the output for 8 lower than the output for 7? Shouldn't a greater input give a greater output?"

d	$25 - 3 \cdot d$	Comment
7	4	too low
8	1	still too low

   a. Answer Miguel's questions.

   b. What input do you think Miguel should try next? Explain.

6. Hannah and Luke were trying to solve $7.25t - t^2 = 12.75$. They made this table using guess-check-and-improve:

   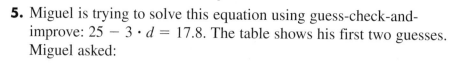

t	$7.25t - t^2$	Comment
5	11.25	too low
6	7.5	too low
4	13	too high
2	10.5	too low

   Hannah thinks the solution must be between 2 and 4. Luke thinks it must be between 4 and 5.

   a. Is there a solution between 2 and 4? If so, find it. If not, explain why not.

   b. Is there a solution between 4 and 5? If so, find it. If not, explain why not.

Just the facts

Imagine that a garage contains two tricycles and several bicycles, and that there are 20 bicycle wheels in all. You can solve the equation in Problem 2, $2n + 6 = 20$, to find the number of bicycles in the garage.

Investigation 2 Solving Problems Using Guess-Check-and-Improve

In this investigation, you will solve problems by writing equations and then using guess-check-and-improve. As you work, you will find that you can't always give an exact decimal value for a solution. In such cases, you can approximate the solution.

Problem Set B

1. The floor of Mr. Cruz's basement is shaped like a rectangle. The length of the floor is 2 meters greater than the width.

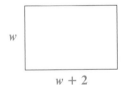

w

$w + 2$

 a. Write a rule to show the connection between the floor's area A and the width w.

 b. The area of the basement floor is 85 square meters. Write an equation you could solve to find the floor's width.

 c. Use guess-check-and-improve to find an approximate solution of your equation. Give the solution to the nearest tenth.

 d. What are the dimensions of the basement floor?

2. A *cube* is a three-dimensional shape with six identical square faces.

The *surface area* of a cube is the sum of the areas of its six faces.

This cube has edges of length *L*.

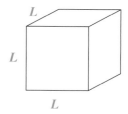

a. What is the area of one face of this cube?

b. Write a rule for finding the surface area *S* of the cube.

c. Suppose the cube has a surface area of 100 square centimeters. Write an equation you could solve to find the cube's edge length.

d. Use guess-check-and-improve to find the edge length of the cube to the nearest 0.1 centimeter.

In all the equations you have solved so far, the variable appears on only one side of the equation. You can also use guess-check-and-improve to solve equations in which the variable appears on *both* sides.

EXAMPLE

Solve this number puzzle: *12 more than a number is equal to 3 times the number. What is the number?*

If you let *m* stand for the number, you can write the puzzle as an equation:

$$m + 12 = 3m$$

To solve the equation, substitute values for *m* until the two sides of the equation are equal.

m	*m* + 12	3*m*	Comment
1	13	3	not the same
2	14	6	a bit closer
5	17	15	very close
6	18	18	Got it!

So, 6 is the solution of the equation $m + 12 = 3m$.

Check that 6 is the solution of the original puzzle: 12 more than 6 is 18, which is equal to 3 times 6.

In the next problem set, you will practice solving equations in which the variable appears on both sides.

Problem Set C

In Problems 1–3, write an equation to represent the number puzzle. Then find the solution using guess-check-and-improve.

1. 5 added to a number is the same as 1 subtracted from twice the number. What is the number?

2. 4 times a number, plus 1, is equal to 4 added to twice the number. What is the number?

3. 2 added to the square of a number is equal to 5 times the number, minus 4. What is the number? (There are two solutions. Try to find them both.)

4. Consider the equation $2m = 5m - 18$.

   a. Make up a number puzzle that matches the equation.

   b. Solve the equation, and check that the solution is also the answer to your puzzle.

5. Peta and Ali went to the pet store to buy fish for their tanks. Peta bought three black mollies for b dollars each and one peacock eel for $12. Ali bought seven black mollies for b dollars each and an Australian rainbow fish for $2. Peta and Ali spent the same amount of money.

   a. How much money did Peta spend? Your answer should be an expression containing the variable b.

   b. How much money did Ali spend? Your answer should be an expression containing the variable b.

   c. Using your two expressions, write an equation that states that Peta and Ali spent the same amount of money.

   d. Solve your equation to find the value of b. How much did each black molly cost?

   e. How much did each friend spend?

Use guess-check-and-improve to find a solution of each equation.

6. $3n = 9 - 2n$ 7. $p + 1 = 5(p - 4)$

Share & Summarize

Make up a number puzzle like those you have worked with in this investigation. Exchange puzzles with your partner, and solve your partner's puzzle.

Investigation Choosing a Method

Now you have two methods for solving equations: *backtracking* and *guess-check-and-improve*. The problems in this investigation will help you decide which solution method is more efficient for a particular type of equation.

In the cartoon, Marcus and Rosita are trying to solve some equations. Marcus uses backtracking for each equation, and Rosita uses guess-check-and-improve for each equation.

Think & Discuss

For which equation in the cartoon is backtracking a better method than guess-check-and-improve? Why?

For which equation is guess-check-and-improve a better method than backtracking? Why?

For which equation do both methods seem to work well?

Problem Set D

In Problems 1–4, do Parts a and b.

a. Find a solution of the equation using one method while your partner finds a solution using the other. Switch methods for each equation so you have a chance to practice both.

b. Discuss your work with your partner. Indicate whether both methods work, and tell which method seems more efficient.

1. $3p - 8 = 25$ **2.** $1.6r + 3.96 = 11$

3. $k + 4 = 6k$ **4.** $(j - 2) \cdot j = 48$

Solve each equation. Tell which solution method you used, and explain why you chose that method.

5. $4w + 1 = 2w + 8$ **6.** $5 \cdot \frac{2k + 4}{6} = 10$

7. Luke says, "When the variable appears only once in an equation, I use backtracking. If it occurs more than once, I use guess-check-and-improve." Discuss Luke's strategy with your partner. Would it be effective for the equations in Problem 1–6? Explain.

Share & Summarize

Tell which solution method you would use to find a solution of each equation. Explain your choice.

1. $n^2 + n = 30$ **2.** $4 + (v - 3) = 8$

3. $3g = 6g - 7$ **4.** $2e - 7 = 4$

On Your Own Exercises

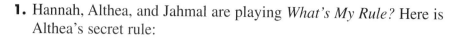

Practice & Apply

1. Hannah, Althea, and Jahmal are playing *What's My Rule?* Here is Althea's secret rule:

$$3 \cdot p + p^2 = q, \text{ where } p \text{ is the input and } q \text{ is the output}$$

   a. Hannah gives Althea an input, and Althea calculates the output as 24.79. Write an equation you could solve to find Hannah's input.

   b. Find a solution of your equation using guess-check-and-improve.

   c. Jahmal gives Althea an input, and Althea calculates the output as 154. Write an equation you could solve to find Jahmal's input.

   d. Find a solution of your equation in Part c using guess-check-and-improve.

For each equation, use guess-check-and-improve to find a solution.

   2. $16 - 5k = 2$

   3. $h \cdot (5 + h) = 26.24$

   4. $y^2 + 72 = 17y$ (There are two solutions. Try to find both.)

   5. Geometry Reina is planning her summer garden.

   a. Reina wants to plant strawberries in a circular plot covering an area of 15 square meters. Write an equation you could solve to find the radius Reina should use to lay out the plot.

   b. Use guess-check-and-improve to find the strawberry plot's radius to the nearest tenth of a meter.

   c. Reina also wants to plant tomatoes in five identical circular plots with a total area of 25 square meters. Write an equation you could solve to find the radius of one of the tomato plots.

   d. What should the radius of each tomato plot be, to the nearest tenth of a meter?

Remember

The area of a circle is given by the formula $A = \pi \cdot r^2$, where r is the radius of the circle.

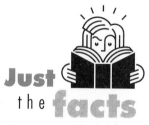

Just the facts

Tomatoes are the world's most popular fruit—in terms of tons produced each year—followed by bananas, apples, oranges, and watermelons.

impactmath.com/self_check_quiz

In Exercises 6–8, write an equation to represent the number puzzle. Then find the solution using guess-check-and-improve.

6. 3 times a number, plus 5, is equal to 5 times the number. What is the number?

7. 10 plus the square of a number is equal to 6 times the number, plus 2. What is the number? (This puzzle has two solutions. Find both of them.)

8. 3 times a number, plus 1, is equal to 9 plus the number. What is the number?

9. Marjorie said, "If you double my macaw's age and then subtract 21.75, your answer will be half my macaw's age."

   a. Write an equation you could use to find how old Marjorie's macaw is.

   b. Use guess-check-and-improve to find the macaw's age. Check your answer in Marjorie's original statement.

10. Consider the equation $3m - 11 = m + 3$.

   a. Make up a number puzzle that matches the equation.

   b. Solve the equation, and check that the solution is also the answer to your puzzle.

Find a solution of each equation.

11. $3.3h - 7 = 2.801$

12. $2l = 4l - 20$

13. $j \cdot (3 - 2j) = 1$

14. $2 = m^2 - m$

15. $9 \times \frac{g \div 2 + 1}{5} = 12$

16. $143 = (q + 1) \cdot (q - 1)$

In Exercises 17–19, tell whether you would use backtracking or guess-check-and-improve to find a solution, and explain your choice.

17. $n^2 + n = 30$ **18.** $47(2v - 3.3) = 85$ **19.** $s = 17.5s - 0.5$

20. Aisha and Terrell were playing *Think of a Number.* Terrell said:

Think of a number. Multiply the number by 1 less than itself. What number do you get?

Aisha said she got 272. Terrell must figure out Aisha's starting number.

a. What equation does Terrell need to solve to find Aisha's number?

b. Use guess-check-and-improve to find Aisha's number. Check your answer by following the steps to verify that you get 272.

21. Geometry The elevator in Rafael's apartment building has a square floor with an area of 6 square meters.

a. Write an equation you could solve to find the dimensions of the elevator floor.

b. Use guess-check-and-improve to find the dimensions of the elevator floor to the nearest tenth of a meter.

22. Physical Science When an object is dropped, the relationship between the distance it has fallen and the amount of time it has been falling is given by the rule

$$d = 4.9 \cdot t^2$$

where *d* is the distance in meters and *t* is the time in seconds.

a. Jing dropped a ball from a pier. It took the ball 1.1 seconds to hit the water. How many meters did the ball travel?

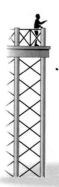

b. A bolt falls 300 meters down a mine shaft. Write an equation you could solve to find how long it took the bolt to fall.

c. Find how long it took the bolt to fall, to the nearest tenth of a second.

23. Nutrition Three blueberry muffins and a plain bagel have the same number of calories as two blueberry muffins and one bagel with cream cheese. A bagel has 150 calories, and cream cheese adds 170 calories. How many calories are in a blueberry muffin?

24. Number Sense This formula can be used to find the sum S of the whole numbers from 1 to n:

$$S = \frac{n \cdot (n + 1)}{2}$$

For example, you can use the formula to find the sum of the whole numbers from 1 to 100:

$$S = \frac{n \cdot (n + 1)}{2}$$
$$= \frac{100 \cdot (100 + 1)}{2}$$
$$= \frac{100 \cdot 101}{2}$$
$$= \frac{10,100}{2}$$
$$= 5,050$$

a. Use the formula to find the sum of the whole numbers from 1 to 9. Check your answer by calculating $1 + 2 + 3 + 4 + 5 + 6 + 7 + 8 + 9$.

b. If the sum of the numbers from 1 to n is 6,670, what must n be?

c. If the sum of the numbers from 1 to n is 3,003, what must n be?

For each equation, use guess-check-and-improve to find a solution.

25. $x^3 + x = 130$

26. $4n^4 = 48$

27. $2a^4 - 4a^2 = 16$

28. $c^4 = \frac{1}{100}$

29. Geometry The *volume* of a three-dimensional shape is the amount of space inside of it. Volume is measured in cubic units, such as cubic centimeters and cubic inches. You can calculate the volume V of a cylinder with radius r and height h using this formula: $V = \pi \cdot r^2 \cdot h$.

a. Calculate the volume of a cylinder with radius 2 centimeters and height 5 centimeters.

b. A soft drink can has a volume of 350 cubic centimeters and a height of 15 centimeters. Write an equation you could solve to find its radius.

c. Solve your equation to find the can's radius to the nearest tenth of a centimeter.

Write each fraction as a decimal without using a calculator.

30. $\dfrac{99}{150}$ **31.** $\dfrac{73}{90}$ **32.** $\dfrac{7}{18}$

33. $\dfrac{78}{110}$ **34.** $\dfrac{4}{11}$ **35.** $\dfrac{63}{125}$

36. Of the 80 acres on Ms. McDonald's farm, 28 acres are devoted to growing corn. What percent of the farm's area is devoted to corn?

37. Of the animals on Ms. McDonald's farm, 12.5% are goats. If Ms. McDonald has 7 goats, what is the total number of animals on her farm?

38. At the farmer's market, Ms. McDonald sold 60% of the 42 pounds of tomatoes she picked last week. How many pounds of tomatoes did she sell at the market?

39. **Physical Science** The graph shows the level of water in a bathtub over a one-hour period. Write a story that explains all the changes in the water level.

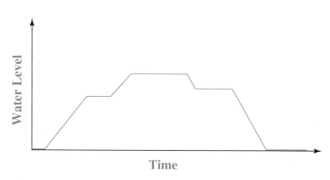

Chapter Summary

You started this chapter by looking at numeric equations and inequalities. Then you turned your attention to equations containing variables. You learned that a value of a variable that makes an equation true is called a *solution* of the equation. You saw that while many equations have only one or two solutions, some have every number as a solution, and others have no solution at all.

You then created flowcharts to represent rules, and you saw how *backtracking*—working backward from an output using a flowchart—can be used to solve some equations. You were then faced with equations that could not be solved by backtracking, and you learned how to use *guess-check-and-improve* to solve them. You also learned some strategies for determining which solution method to use for a given equation.

Strategies and Applications

The questions in this section will help you review and apply the important ideas and strategies developed in this chapter.

Understanding equations and inequalities

1. Explain what the symbols $=$, $>$, $<$, and $\neq$ mean. For each symbol, write a true mathematical sentence using that symbol.

2. Give an example of an open sentence, and give a value of the variable that makes your sentence true.

3. Explain why the sentence $P + 5 = P$ is never true.

4. Explain why the sentence $2 \cdot (x + 3) = 2 \cdot x + 2 \cdot 3$ is always true.

Solving equations by backtracking

5. Solve this equation by creating a flowchart and backtracking:

$$10 \cdot \frac{4 + 9x}{7} - 25 = 45$$

Explain each step in your solution.

6. Admission to the town carnival is $4.50 per person. Rides cost 75¢ each. Russ has $10 to spend at the carnival, and he wants to go on as many rides as possible.

    a. Write an equation you could solve to find the number of rides r Russ can go on.

    b. Solve your equation by backtracking. How many rides can Russ go on?

Solving equations using guess-check-and-improve

7. The Smallville community garden is made up of 15 identical square plots with a total area of 264.6 m^2.

    a. Write an equation you could solve to find the side length of each plot.

    b. Solve your equation using guess-check-and-improve. Make a table to record your guesses and the results. What is the side length of each plot?

8. A number multiplied by 2 more than the number is 9 times the number, plus 8.

    a. Write an equation to represent the number puzzle.

    b. Solve your equation using guess-check-and-improve.

Choosing a solution method for an equation

9. Explain why you could not use backtracking to solve this equation: $(n + 3.5) \cdot n = 92$.

10. Tell which solution method you would use to solve the equation $6.34 + 10.97 \cdot y = 208.188$. Give reasons for your choice.

Demonstrating Skills

Tell whether each sentence is true or false. If it is false, make it true by changing one number or symbol.

11. $(4 + 5) \cdot 6 = 24 + 30$

12. $4^2 = 4 \cdot 2$

13. $30 \div (3 + 2) = 30 \div 3 + 30 \div 2$

14. $7 + 4 - 1 > 5 + 6$

Tell whether each equation is *always true*, *sometimes true*, or *never true*, and explain how you know.

15. $c^3 = c \cdot c \cdot c$

16. $n - 5 = n + 1$

17. $x - 9 = 0$

18. $2m - 2m = 0$

19. $\frac{13}{6 + p} = 1$

20. $5 \cdot (4 + y) = 20$

Find a solution of each equation.

21. $4.7x + 12.3 = 42.85$

22. $m \cdot (m + 5) = 336$

23. $5n + 7 = 14$

24. $\frac{5(x - 1)}{3} = 7.5$

25. $6z - 29 = 7 + 3z$

26. $7 \cdot \frac{3b + 4}{10} = 11.2$

CHAPTER 10

Understanding Probability

Real-Life Math

Fat Chance! A probability is a number between 0 and 1 that tells you the chance that an event will occur. The closer a probability is to 0, the less likely the event is to happen.

Just how likely is it that some everyday events happen?

- If you toss a coin to determine the answers of a 10-question true-false test, the probability you will get all the answers correct is $\frac{1}{1,024}$ or about 0.001.

- If you randomly dial a three-number combination on a dial lock with numbers from 0 to 29, the probability you will open the lock is $\frac{1}{27,000}$ or about 0.00004.

Think About It To better understand how unlikely these events are, compare these probabilities to the probability that it will rain in your community tomorrow.

Family Letter

Dear Student and Family Members,

We close our exciting year of mathematics by exploring probability. Probability tells you that it is very unlikely for you to win the grand prize in a state lottery. Suppose you have to pick six different numbers from 1 to 54. To win the grand prize, all six numbers must match those selected in a random drawing. The probability that you will win is only 1 in 25,827,165 or 0.00000004.

The probability that some event will happen can be described by a number between 0 and 1.

- A probability of 0 means that the event has no chance of happening. (So the probability of winning a state lottery is very close to 0.)

- A probability of 1 means that the event is certain to happen.

- A probability of $\frac{1}{2}$ or 50% means that the event is just as likely to happen as not to happen.

For example, if a weather forecaster says that the probability of rain tomorrow is 90%, it's probably a good idea to take your umbrella, although it might not rain after all. If the probability of rain is 10%, it is unlikely that it will rain.

In this chapter, you'll use mathematical reasoning to calculate probabilities in simple situations like tossing coins or drawing names from a hat. You will also do some experiments in which you actually toss coins or draw names and then compare the results with the calculated probabilities.

Vocabulary Along the way, you'll learn these new vocabulary terms:

equally likely	**simulation**
experimental probability	**theoretical probability**
probability	

What can you do at home?

Look for situations in everyday life that involve probability, such as the chance of rain or the odds in sports games. You might encourage your student's exploration by playing games of chance together. Have your student teach you the games we play in class and ask him or her to describe what part probability plays in each game.

10.1 The Language of Chance

People often make comments like these:

- "It probably won't rain tomorrow."
- "I expect to have a lot of homework this week."
- "Our team has a good chance of winning the game."
- "It's not likely she will eat that entire cake!"
- "The chances are 50/50 that we'll go to the movies tonight."
- "There's a 40% chance of rain tomorrow."

The words *probably, expect, chance,* and *likely* are used when someone is making a prediction.

Just the facts

Some gorillas have been taught to communicate with humans by using sign language. One such gorilla, Koko, has a vocabulary of over 1,000 words!

Think & Discuss

What are some other words or phrases people use when predicting the chances of something happening?

Listed below are six events. How likely do you think each event is? Talk about them with your class, and come to an agreement about whether each event

- has no chance of happening
- could happen but is unlikely
- is just as likely to happen as not to happen
- is likely to happen
- is certain to happen

Event 1: Our class will have homework tonight.

Event 2: It will snow tomorrow.

Event 3: If I toss a penny in the air, it will land heads up.

Event 4: A gorilla will eat lunch with us today.

Event 5: You choose a name from a hat containing the names of all the students in your class, and you get a girl's name.

Event 6: You draw a number from a hat containing the numbers from 1 to 5, and you get 8.

Investigation 1 ▶ Probability in Everyday Life

VOCABULARY
probability

Remember

The probability *P* of an event *E* is the ratio that compares the number of favorable outcomes *f* to the number of possible outcomes *n*.

$P(E) = \frac{f}{n}$

The **probability,** or chance, that an event will happen can be described by a number between 0 and 1:

- A probability of 0, or 0%, means the event has no chance of happening.

- A probability of $\frac{1}{2}$, or 50%, means the event is just as likely to happen as not to happen.

- A probability of 1, or 100%, means the event is certain to happen.

For example, the probability of a coin landing heads up is $\frac{1}{2}$, or 50%. This means you would expect a coin to land heads up $\frac{1}{2}$, or 50%, of the time.

The more likely an event is to occur, the greater its probability. If a weather forecaster says the probability of rain is 90%, it's a good idea to take your umbrella when you go outside. Of course, it *might* not rain after all! On the other hand, if the forecaster says the probability of rain is 10%, you might want to leave your umbrella at home. Still, you *might* get wet!

You can represent the probability of an event by marking it on a number line like this one:

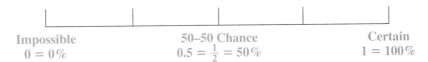

Impossible	50–50 Chance	Certain
0 = 0%	$0.5 = \frac{1}{2} = 50\%$	1 = 100%

For example, the next number line shows the probabilities of tossing a coin and getting heads, of a goldfish walking across a room, and of Alaska getting snow this winter.

Goldfish Walking	Tossing Heads	Snow in Alaska
0%	50%	100%

Problem Set A

1. Describe an event you think has the given chance of happening.

a. The event has no chance of happening.

b. The event could happen but is unlikely.

c. The event is just as likely to happen as not to happen.

d. The event is likely to happen.

e. The event is certain to happen.

2. Copy this number line. Label the number line with your events from Problem 1. You can use the letters of the events for the labels.

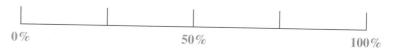

0%　　　　　　　　50%　　　　　　　100%

In Problem Set A, you used your experience to estimate the chances that certain events would occur. For example, you know from experience that when you toss a coin, it lands tails up about half the time. In some situations, you can use data to help estimate probabilities.

Problem Set B

On Saturday, Caroline's baseball team, the Rockets, is playing Jahmal's team, the Lions. Caroline decided to look at the scores from the last six times their teams played each other.

Lions	3	8	6	4	4	5
Rockets	5	2	4	5	7	6

1. How many times did the Rockets win?

2. Which team do you think is more likely to win the next game?

3. Caroline can estimate the probability that her team will win by dividing the number of times the Rockets won by the number of games played. What probability estimate would she get based on the results of the six games? Give your answer as a fraction and as a percent.

4. Suppose Caroline knew the results of only the first three games. What would her probability estimate be? If she knew the results of only the last three games, what would her probability estimate be?

5. These two teams have played six games against each other. Suppose they had played eight games and each team had won one more game. What probability estimate would you give for the Rockets winning the next game?

The probabilities you found in Problem Set B are examples of **experimental probabilities.** Experimental probabilities are always estimates, and they can vary depending on the particular set of data you use.

Suppose you want to find an experimental probability when you have no data available. In such cases, you might perform *experiments* to create some data.

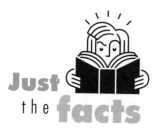

Just the facts

Little League baseball was started in 1939 in Williamsport, Pennsylvania, with three teams and 45 players. Today, more than 2.5 million children worldwide play Little League baseball.

VOCABULARY
experimental probability

Problem Set C

Much like a science experiment, a probability experiment involves trying something to see what happens. You may have some idea of what will occur, but the actual results can be surprising.

1. Toss a paper cup so that it spins in the air. Record how it lands: right side up, upside down, or on its side. This is one *trial* of the experiment.

| Right Side Up | Upside Down | On Its Side |

Toss the cup 29 more times, for a total of 30 tosses. Record the landing position each time. You may want to use tally marks as shown below.

Right Side Up	Upside Down	On Its Side
II	I	IIII

2. How many trials did you perform in your experiment?

3. How many times did the cup land right side up? Upside down? On its side?

4. Find the portion of the trials for which the cup landed right side up, stating your answer as a fraction or a percent. Your answer is an experimental probability that the cup lands right side up when tossed.

5. Now find an experimental probability that the cup lands upside down and an experimental probability that it lands on its side.

6. Share your results with the class, and consider the results found by your classmates. Suggest at least one way you might use them to find a class experimental probability for the cup landing right side up.

Share & Summarize

1. What does it mean to say that the probability of an event is 1?

2. Conor is in a basketball league. He was practicing free throws one afternoon, and he made 32 out of 50 shots. Estimate the probability that he makes a free throw, expressing it as a fraction and a percent.

3. Suppose you conducted y trials of an experiment, and a particular event happened x times. Find an experimental probability that the event will occur.

Investigation 2 Theoretical Probability

VOCABULARY
equally likely
theoretical
probability

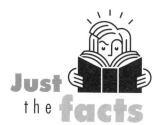

Just the facts

In 2001, the most popular names for girls born in the U.S. were Emily, Hannah, Madison, Samantha, Ashley, Sarah, Elizabeth, Kayla, Alexis, and Abigail.

In some situations, all of the possibilities for a situation—called the *outcomes*—have the same probability of occurring. For example, a coin toss has two possible outcomes: heads or tails. If the coin is fair, about half of the tosses will come up heads and half will come up tails. In situations like these, the outcomes are **equally likely.**

When outcomes are equally likely, you can calculate probabilities by reasoning about the situation. Since these **theoretical probabilities** do not depend on experiments, they are always the same for a particular event.

EXAMPLE

In a class competition, five students—Althea, Conor, Hannah, Luke, and Rosita—are tied for first place. To break the tie, they will write their names on slips of paper and place them in a bowl. A judge will choose one slip without looking, and the student whose name is on that slip will receive first prize. What is the probability that the name chosen has three syllables?

| Althea | Conor | Hannah | Luke | Rosita |

There are five names. The judge chooses a name *at random*—that is, in a way that all five names have the same chance of being selected. *Althea* and *Rosita* are the only names with three syllables.

Since there are five equally likely outcomes, and two of them are three-syllable names, you would expect a three-syllable name to be selected $\frac{2}{5}$ of the time. So, the probability of choosing a name with three syllables is $\frac{2}{5}$, or 40%.

Problem Set D

Think about the situation described in the Example. Decide how likely each of the following events is, and determine its theoretical probability. Explain your answers.

1. The name chosen does not begin with *R*.

2. The name chosen begins with *J*.

3. The name chosen has four letters or more.

4. The name chosen has exactly four letters.

5. The name chosen ends with *A*.

6. The name chosen does *not* end with *A*.

7. Add the probabilities you found in Problems 5 and 6. Why does this sum make sense?

When people talk about probabilities involved in games, rolling dice, or tossing coins, they usually mean *theoretical* rather than experimental probabilities. In the rest of this investigation, you will consider the relationship between these two types of probability.

Think & Discuss

Discuss what the students are saying. Which students do you agree with?

Just the facts

In 2001, the most popular names for boys born in the U.S. were Jacob, Michael, Joshua, Matthew, Andrew, Joseph, Nicholas, Anthony, Tyler, and Daniel.

Problem Set E

1. If you toss a coin 12 times, how many times would you expect to get heads? Explain.

2. Conduct an experiment to find an experimental probability of getting heads. Toss a coin 12 times and record the results, writing H for each head and T for each tail.

   a. How many heads did you get?

   b. Use your results to find an experimental probability of getting heads when you toss a coin.

3. Compare the theoretical results with your experimental results.

   a. Is your result in Part a of Problem 2 the same as the answer you computed in Problem 1?

   b. Is your experimental probability the same as the theoretical probability?

Just the facts

The United States produced half-cent coins from 1793 to 1857.

4. Now combine your experimental results with those of other students. Make a table like this one, showing how many times out of 12 each student's coin came up heads.

Student	Number of Heads
James	7
Ali	5

5. Calculate the total number of tosses your class made. How many heads would you expect for that number of tosses?

6. Now add the entries in the "Number of Heads" column. How closely does the result agree with your expectations?

7. What percent of the total number of coin tosses came up heads? In other words, what is the experimental probability of getting heads based on the data for your entire class?

8. Which experimental probability is closer to the theoretical probability: the one you calculated for Part b of Problem 2, or the one you calculated for Problem 7?

Just the facts

Lincoln is the only U.S. president on a coin who faces to the right.

It's normal for experimental probabilities to be different from theoretical probabilities. In fact, when you repeat an experiment a small number of times, the experimental and theoretical probabilities may not be close at all.

However, when you repeat an experiment a large number of times—for example, by combining all the coin tosses of everyone in your class or by performing more tosses—the experimental probability will usually grow closer to the theoretical probability.

The theoretical probability tells what is likely to happen *in the long run*—that is, if you try something a large number of times. It doesn't reveal exactly what will happen each time.

Problem Set F

A standard die has six faces, each indicating a different number from 1 to 6. When you roll a die, the *outcome* is the number facing up.

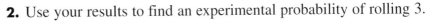

1. Roll a die 12 times, and record each number you roll. How many 3s did you get?

2. Use your results to find an experimental probability of rolling 3.

3. Collect results from others in your class, and make a table with these columns:

Student	Number of 3s

4. Find the total number of rolls and the total number of 3s for all the students in your class. Then compute an experimental probability of rolling 3.

5. Now think about the *theoretical* probability of rolling 3.

a. How many possible outcomes are there for a single roll? Are they all equally likely?

b. On each roll of a die, what is the probability that you will roll 3?

c. If you roll a die 12 times, how many 3s would you expect?

6. Compare your theoretical results from Problem 5 with your own experimental results from Problem 2 and to your class experimental results from Problem 4. Which experimental result is closer to the theoretical result?

Share & Summarize

1. At Jenna's birthday party, her mother assigned each of the ten partygoers—including Jenna and her two sisters—a number from 1 to 10. To see who would play *Pin the Tail on the Donkey* first, Jenna's father pulled one of ten balls, numbered 1 to 10, from a box. The person whose number was selected would play first.

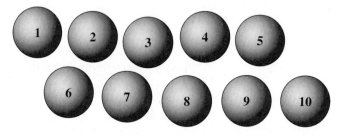

   a. What is the probability that the number chosen was Jenna's or one of her sisters?

   b. What is the probability that the number chosen was *not* Jenna's or one of her sisters—that is, what is the probability that the number belonged to one of the seven guests? Explain how you found your answer.

2. Chris has a spinner divided into five same-sized sections, numbered from 1 to 5. He spun the arrow 100 times, recording the result each time.

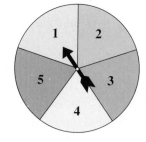

   a. How many times would you expect the arrow to land on Section 4?

   b. If the actual number of 4s Chris recorded was different from the number you answered in Part a, does that mean your calculation was wrong? Explain.

3. Suppose you perform an experiment 20 times and a friend performs the same experiment 200 times. You both use your results to calculate an experimental probability. Whose experimental probability would you expect to be closer to the theoretical probability? Explain.

Investigation ▶ The Spinning Top Game

The Spinning Top game is played with a four-cornered top that contains four symbols. There are many variations of this game.

In this investigation, you will look at the probabilities for one of the most common versions of the Spinning Top game.

Play the Game

You will play this game in a group of four. Here are the rules:

MATERIALS

- 4-cornered top with sides marked DN, TA, TH, and P1.
- counters

- Each player begins with ten counters.

- Each player puts a counter in the center of a table.

- Players take turns spinning the top. One of these four symbols will land face up:

DN TA TH P1

The letter facing up tells the player what to do:

DN   Do nothing.
TA   Take all the counters in the center.
TH   Take half of the counters in the center. (Round down. For example, if the number of counters is five, take two.)
P1   Pay one by placing counter in the center.

- Before each turn, each player puts another counter in the center. A player with no counters left is out of the game.

- The game continues until only one player has counters left or your teacher says time is up. The player with the most counters wins.

1. Play the game with your group, keeping a tally of the symbols the players spin..

DN	TA	TH	P1
‖				

2. How many times did each symbol land face up in the whole game?

3. Draw a bar graph showing the number of times each symbol landed face up.

Calculate Probabilities

Now that you have some experience playing the Spinning Top game, you can calculate the probabilities of certain outcomes.

4. Begin by calculating an experimental probability that each letter lands face up, based on the results of your game.

Assume each symbol is equally likely to land face up.

5. Calculate the theoretical probability of each symbol landing face up.

6. How do the theoretical probabilities you calculated in Question 5 compare to your experimental probabilities from Question 4?

7. What is the (theoretical) probability of winning counters in a turn? Explain.

8. What is the probability of losing a counter in a turn? Explain.

9. Jahmal said the first player has a better chance to win all the counters in the center than the other three players do. Do you agree with Jahmal? Explain your answer.

10. Suppose the first player wins all the counters on his or her turn. What is the probability that the second player will also win all the counters? Explain your answer.

What Have You Learned?

11. If you are equally likely to get each of the four sides of the top, why are you not equally likely to win counters as to lose counters?

12. Change the rules of the Spinning Top game so that the probability of winning counters on each turn is $\frac{1}{4}$ and the probability of losing counters is $\frac{1}{2}$.

On Your Own Exercises

Practice & Apply

1. Copy this number line. In Parts a–d, add a label to your number line indicating how likely you think the event is.

```
|_____|_____|_____|
0%              50%           100%
```

   a. I will listen to the radio tonight.

   b. I will go to a movie sometime this week.

   c. Everyone in my math class will get a perfect score on the next test.

   d. I will wake up before 7:00 A.M. tomorrow.

2. Estimate the probability of each event, and explain your reasoning.

   a. The school lunch will taste good tomorrow.

   b. Everyone in our class will come to school next Monday.

   c. A giraffe will come to school next Monday.

3. **Sports** Jahmal is practicing his archery skills. He hit the bull's-eye with 3 of the first 12 arrows he shot. Use these results to find an experimental probability that Jahmal will hit the bull's-eye on his next shot.

4. Get a spoon (preferably plastic) and conduct this experiment:

   For each trial, drop the spoon and record how it lands: right side up (so it would hold water) or upside down. Conduct 30 trials for your experiment. Use your results to find an experimental probability that the spoon will land right side up.

5. A word is chosen at random from *book, paper, pencil,* and *eraser.*

   a. What is the probability that the word has only one syllable?

   b. What is the probability that the word begins with *P*?

   c. What is the probability that the word ends with *P*?

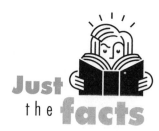
6. Lupe tossed a coin 10 times and got 6 heads. Jing tossed a coin 1,000 times and got 530 heads.

a. Based on Lupe's results, what is an experimental probability of getting heads? Express your answer as a percent.

b. Using theoretical probabilities, how many heads would you expect to get in 10 coin tosses? How far was Lupe's result from that number?

c. Based on Jing's results, what is an experimental probability of getting heads? Express your answer as a percent.

d. Using theoretical probabilities, how many heads would you expect to get in 1,000 coin tosses? How far was Jing's result from that number?

e. Challenge The difference between the actual number of heads and the expected number of heads is much greater for Jing than for Lupe. How is it possible that Jing's experimental probability is closer to the theoretical probability?

7. Marika has a spinner divided into 10 equal sections, numbered from 1 to 10. Think about a single spin of the arrow.

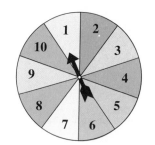

a. What is the probability that the arrow will point to Section 1?

b. What is the probability that the arrow will point to an odd number?

c. What is the probability that the arrow will point to an even number?

d. What is the probability that the arrow will point to a prime number?

Connect & Extend

8. Earth Science Imelda and a group of her friends planned a beach outing for a certain day. The local weather service said there was a 20% chance of rain on that day. When the day came, it rained and the trip was canceled.

Imelda said the weather service had been wrong when they gave the 20% rain prediction. They said it wasn't going to rain, but it did.

a. Do you agree with Imelda? Why or why not?

b. If the weather service prediction didn't mean that it wouldn't rain, what do you think it meant?

9. Miguel's radio alarm clock goes off at 6:37 every morning. He complained that, almost every morning, he wakes up to commercials rather than music. Describe an experiment he could conduct to estimate the probability that he will wake up to a commercial. Explain how you would use the result to find an experimental probability.

10. Describe a situation for which the probability that something will occur is $\frac{1}{6}$.

11. A whole number is chosen at random from the numbers 1 to 10.

a. What is the probability that the number is odd?

b. What is the probability that the number is prime?

c. What is the probability that the number is a perfect square?

d. What is the probability that the number is a factor of 36?

12. A whole number is chosen at random from the numbers 10 to 20.

a. What is the probability that the number is odd?

b. What is the probability that the number is prime?

c. What is the probability that the number is a perfect square?

d. What is the probability that the number is a factor of 36?

13. **Sports** Two sixth graders and two seventh graders are having a checkers tournament. They decide to choose randomly who the first two players will be and who will use which color. Consider the possible arrangements. For example, the first game might be a seventh grader playing black and a sixth grader playing red.

a. What other possible arrangements are there?

Assume each of the arrangements is equally likely to occur.

b. What is the probability that the seventh graders will play each other in the first game?

c. What is the probability that at least one seventh grader will play in the first game?

d. What is the probability that exactly one sixth grader will play in the first game?

14. Ruben's class went to a carnival. There was a game of chance he wanted to play, but his teacher told him that only one player out of four wins a prize at that game. Ruben got in line anyway, behind three other people. As Ruben waited, he noticed that none of the three people ahead of him won a prize. He was very excited, because he was sure that meant he would win. Was he right? Explain.

15. **Preview** Cut or tear a sheet of paper to create four slips of paper, as identical to each other as possible. Number the slips of paper from 1 to 4, fold them once, and put them in a hat or bag.

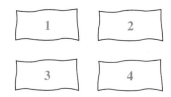

Then conduct this experiment at least 40 times:

- Without looking, draw a slip of paper and look at the number you drew.

- Put the slip back in the hat or bag, mix them up, and draw another.

Keep a tally of how many times the numbers were the same and how many times they were different.

a. Based on your results, find the probability that two chosen numbers will be the same.

b. Is the probability you found an experimental probability or a theoretical probability?

Find a solution of each equation.

16. $6 + 7m = 41$ **17.** $1.9z + 14.3 = 37.1$ **18.** $3 + 3p = 5p - 12$

19. $x^2 - 5x = 14$ **20.** $4(v - 2) = 20$ **21.** $\frac{2(3c + 12)}{6} = 8$

Find the prime factorization of each number.

22. 3,740   **23.** 19,551   **24.** 1,872

Use the fact that $783 \cdot 25 = 19,575$ to find each product without using a calculator.

25. $7.83 \cdot 25$ **26.** $78.3 \cdot 2.5$ **27.** $7,830 \cdot 250$

Use the fact that $7,848 \div 12 = 654$ to find each quotient without using a calculator.

28. $7,848 \div 0.12$ **29.** $7.848 \div 12$ **30.** $78.48 \div 1.2$

Just the facts

The contiguous United States are all those states that border on another, which means all U.S. states are contiguous except for Alaska and Hawaii.

31. Life Science The table shows the number of bald eagle pairs in the contiguous United States in even-numbered years between 1982 and 1998.

a. Graph the data on a grid like the one below.

b. Describe the overall change in the population of bald eagle pairs.

c. Use your graph to predict the number of bald eagle pairs in 1995 and in 2006.

Year	Bald Eagle Pairs
1982	1,480
1984	1,757
1986	1,875
1988	2,475
1990	3,035
1992	3,749
1994	4,449
1996	5,094
1998	5,748

Source: U.S. Fish and Wildlife Service

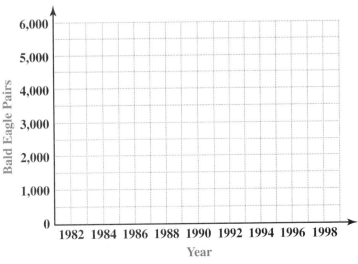

U.S. Bald Eagle Pairs

In Lesson 10.1, you considered *events* such as rolling a particular number on a die or getting heads when you flip a coin. Sometimes more than one outcome can cause an event to occur.

For example, when you roll a die, there are three outcomes for which the event *rolling a prime number* occurs: 2, 3, and 5. To calculate the probability of an event, you have to know the number of possible outcomes and which outcomes cause the event to happen.

For a die roll, you can easily see that there are six possible outcomes. For some situations, it is more difficult to identify the number of possibilities.

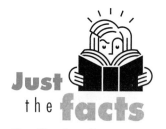

Just the **facts**

The Booker T. Washington commemorative half dollar was the first U.S. coin to feature an African American.

Explore

When you flip two coins, one possible outcome is HH. List all the possible outcomes. How many are there?

When you roll two dice, one possible outcome is 1-1. List all the possible outcomes. How many are there?

In the case of the dice, how can you be sure you have found all the possible outcomes?

For some experiments, figuring out all the possible outcomes can be a real problem in organization! You might forget one possibility or count the same one twice. Making a table can help you systematically count all the outcomes.

This table shows the possible outcomes when you flip two coins:

	Coin A	
	H	**T**
Coin B **H**	HH	TH
T	HT	TT

Investigation Who's Greater?

You have probably played games of chance with standard dice. In this investigation, you will play a game with different kinds of dice.

Just the facts

Cube-shaped dice, with markings similar to those on modern dice, have been found in Egyptian tombs dating from 2000 B.C.

Problem Set A

Who's Greater? is a game for two players. To play, you need two dice. Die A should be marked 5 on four of its faces and 0 on the other two. Die B should have 1, 2, and 3 on three of its faces and 4 on the remaining faces.

Each player uses one of the dice. A player uses the same die for the entire game. On a turn, each player rolls his or her die, and the player who rolls the greater number scores 1 point. The first person to score 10 points wins.

1. Play *Who's Greater?* with your partner. Record the die each player used and the final score of the game.

2. Which die did the winner use?

3. Use your results to calculate an experimental probability that, on a single turn, the player using Die A will score a point.

4. Which die would you prefer to play this game with? Give reasons for your choice.

- die with faces labeled 5, 5, 5, 5, 0, 0
- die with faces labeled 1, 2, 3, 4, 4, 4

Problem Set B

You can use this table to analyze the results of *Who's Greater?* The numbers on the faces of Die A are listed across the top; those of Die B are listed down the side. The entry "A" means that when Die A shows 5 and Die B shows 1, the roller of Die A scores a point.

Die A

	5	5	5	5	0	0
1	A					
2					B	
3						
4						
4						
4						

Die B (label at left of table)

1. What combination does the cell with "B" in it represent?

2. Copy and complete the table to show the winning die for each roll.

3. Calculate the theoretical probability that Die A will score a point on a turn. Do this by counting the number of outcomes in which Die A scores and dividing by the total number of outcomes.

4. Based on the theoretical probabilities, which die would you prefer to play *Who's Greater?* with? Explain.

5. Will the player who uses the die in your answer to Problem 4 always win the game? Explain.

6. Play the game again a few times with your partner. Trade dice so that each of you has a chance to play with both Die A and Die B.

    a. For each game, record which die the winner used. Combine your results with the other pairs in your class.

    b. How do your results compare with what you expected would happen?

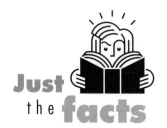

Just the facts

The ancient Egyptian game senet is one of the earliest known dice games. The dice first used to play senet were made of sticks or the knucklebones of animals.

Share & Summarize

Design a pair of dice so the probability that the player using Die A scores a point on a turn is $\frac{5}{6}$. Use a table to show that your dice give the desired probability.

Investigation 2 ▶ Dice Sums

Three and Seven is a game two people play with a pair of standard dice, with faces numbered from 1 to 6. Each player rolls one die, and the two numbers are added.

- If the sum is 3, Player 1 scores a point.
- If the sum is 7, Player 2 scores a point.
- If the sum is any other number, neither player scores.

The player with the most points after 40 turns wins.

Problem Set C

1. What is the least sum you can get in this game? What number would be on each die to get this sum?

2. What is the greatest sum you can get? What number would be on each die to get this sum?

3. What other sums are possible?

4. Do you think one player has a better chance of winning the game, or are the players' chances the same? Explain.

If you play the game, you can collect evidence to test your answer to Problem 4.

Problem Set D

1. Play *Three and Seven* with a partner. If you can't agree who will be Player 1 and who will be Player 2, toss a coin to decide. Keep track of all the sums you roll. Record your results in a table like this one:

Turn	Die A	Die B	Sum
1			
2			

2. Is there a sum that appears more often than any other?

3. Use your results to calculate an experimental probability for each of the possible sums.

4. Compare your results with other students. How many games did Player 1 win? How many games did Player 2 win?

5. Which player seems to have the better chance of winning?

MATERIALS
2 standard dice

Think & Discuss

To find all the possible rolls of two dice that will produce a sum of 3 or a sum of 7, it is helpful to use a table. Describe or create a table that might be useful to you.

Problem Set E

Answer these questions about the game *Three and Seven*.

1. On a single turn, what is the probability that the sum will be 3?

2. On a single turn, what is the probability that the sum will be 7?

3. Which player has a better chance of winning? Explain.

Share & Summarize

Hannah thinks there are 11 equally likely outcomes in the *Three and Seven* game because there are 11 possible sums. How would you help her understand the mistake in her thinking?

Investigation ▶3 Rolling Differences

Rolling Differences is a dice game played by two people. To play, you need a pair of standard dice and a game board like the one below. Each player also needs 20 counters of a single color. Your counters should be a different color from your opponent's.

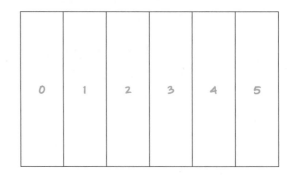

0	1	2	3	4	5

Here are the rules for *Rolling Differences:*

- Each player distributes 20 counters in the six sections of the game board in any way he or she chooses. (It's fine to leave some sections blank or even to put all counters in the same section.)

- On a player's turn, he or she rolls the dice and finds the difference of the numbers rolled. If the player has one or more counters in the section labeled with that difference, he or she removes one.

- The first player to remove all his or her counters from the board wins.

▶ **M A T E R I A L S**
- *Rolling Differences game board*
- *2 dice*
- *40 counters (20 in each of 2 colors)*

Problem Set F

1. Play *Rolling Differences* one or two times, keeping a record of the differences rolled. Which difference was rolled least often?

2. What do you think is a good strategy for distributing the counters on the game board? Why?

Problem Set G

Just the facts

Dominoes were invented in China in the 12th century. Each domino was created to represent one of the possible results of throwing two dice.

In this problem set, you will analyze the probability of rolling each difference with two dice.

1. Make a table that will help you find the number of ways to get each possible difference on a roll of two dice.

2. Copy and complete the table to show the probability of rolling each difference.

3. Using your table, determine what you think is the best way to arrange the counters.

Difference	Ways to Get the Difference	Probability
0		
1		
2		
3		
4		
5		

Share & Summarize

In the game *Rolling Differences,* do you think putting all the counters on the most likely difference is a good strategy? Explain.

Investigation 4 Geometric Probability

To calculate theoretical probabilities for the situations you have worked with so far, you divided the number of outcomes that mean an event occurred by the total number of outcomes. In this investigation, you will look at a situation for which you need to use a different strategy.

MATERIALS

• *Rice Drop* game board
• grain of uncooked rice

Problem Set H

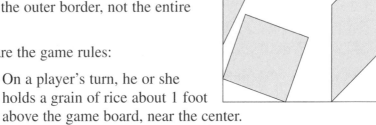

Rice Drop is a game of chance. To play, a larger version of the game board at right should be placed on a hard, flat surface such as a table or desk. The game board is the area inside the outer border, not the entire page.

Here are the game rules:

- On a player's turn, he or she holds a grain of rice about 1 foot above the game board, near the center.

- The player drops the grain of rice and watches where it lands. If it bounces outside the outer border, the drop does not count—the player must try again.

- If the grain of rice lands in one of the four figures (square, circle, triangle, or parallelogram), the player scores 1 point. A grain that lands on the edge of a figure should be counted as inside the figure if half or more of the grain is inside.

- Each player gets 10 chances to score. (Remember, if the grain of rice bounces off the board, the drop does not count.) The player with the greatest score wins.

Just the facts

Rice is the staple food (providing more than one-third of a person's caloric intake) for about 60% of the world's population.

1. Play the game with your group. On your turn, record the results of your drops by making tally marks in a chart like this one:

Circle	Square	Triangle	Nonsquare Parallelogram	No Figure

2. Use the results of your 10 drops to estimate the chances that the grain of rice will land in

   a. the circle   **b.** the square

   c. the triangle   **d.** the nonsquare parallelogram

   e. no figure   **f.** any figure

3. Now combine the results of your group, and calculate new estimates for the probabilities in Problem 2.

4. Which set of probabilities do you think are more reliable: those from Problem 2 or Problem 3? Explain.

Think & Discuss

Do you think a grain of rice is as likely to land on the circle as it is to land outside any of the figures?

Can you think of a way you might calculate the theoretical probability of scoring a point on a single drop?

Just the facts

Hundreds of thousands of varieties of rice are thought to exist.

For the rest of this investigation, assume that the rice lands in a completely random spot on the game board. That is, assume the rice is just as likely to land in one spot on the game board as in another.

Problem Set I

1. Suppose you use a game board divided into four equal rectangles, like this one. What is the probability that the rice lands on the shaded rectangle?

2. The game board at right is also divided into four equal sections. What is the probability that the rice lands on the shaded rectangle?

Just the facts

The average Cambodian eats more than 350 pounds of rice each year. The average American eats about 20 pounds of rice each year.

3. To create this game board, a square was removed from the shaded rectangle of the game board in Problem 2. Is the probability that the rice lands in the shaded figure *less than, greater than,* or *the same as* your answer to Problem 2? Explain your reasoning.

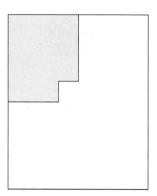

4. Now the square that was removed from the rectangle in Problem 2 has been returned, but in a different place. Is the probability that the rice lands in a shaded figure on this board *less than, greater than,* or *the same as* your answer to Problem 2? Explain your reasoning.

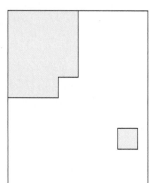

Problem Set J

Rosita thought she could find the theoretical probabilities for the original *Rice Drop* game board using the areas of the figures and the area of the board.

1. Find the area of the game board and the area of the square.

2. Use your answer to Problem 1 to find the probability that the rice lands in the square. Express your answer as a percent.

3. Find the probability that the rice lands in each of the remaining figures. Express your answers as percents.

4. What is the probability that a player will score on a single drop? Explain how you found your answer.

5. **Challenge** The theoretical probabilities you found in this problem set assume that the rice lands randomly. Compare the theoretical probabilities you found in this problem set to the experimental probabilities you found in Problem Set H.

   a. Do you think the rice lands in a completely random spot, or are some spots more likely than others? Explain why you think so.

   b. What could you do to test whether your answer to Part a is correct?

Share & Summarize

1. Suppose this was the game board for a game of *Rice Drop*. The measurements are given in inches.

   Assuming that the rice lands in a completely random place, what is the probability of scoring a point on a single drop?

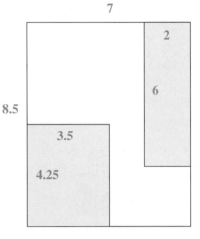

2. Miguel and Althea were playing *Rice Drop* using a checkerboard. They decided that a drop scored a point if the rice landed on a green square. A drop that doesn't land on the board doesn't count.

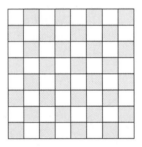

   Assuming that the rice lands in a completely random place, what is the probability that a drop scores a point? Explain how you found your answer.

3. Luke likes playing darts. He throws a dart at the dartboard shown here. The points earned for each ring are shown.

   a. Assuming Luke's dart hits the dartboard in a random place, what is the probability that Luke scores at least 3 points? (The radius of the inner circle is equal to the width of each ring.)

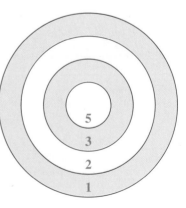

   b. Do you think the assumption in Part a is reasonable? Explain.

On Your Own Exercises

Practice & Apply

1. Suppose two standard dice are rolled. What is the probability that one die shows 5 and the other shows 1, 2, 3, or 6?

2. Althea and Rosita are playing a game. Each girl rolls one standard die, and the girl with the greater number scores a point. If the numbers are the same, neither girl scores.

   a. How many possible combinations of dice rolls are there?

   b. On a single turn, what is the probability that the girls will roll the same number?

   c. What is the probability that the girls will roll different numbers?

   d. What is the probability that Althea will score a point?

3. **Technology** Some calculators have a *random number generator.* María uses her calculator to choose a whole number from 3 to 7. Rashid uses his to choose a whole number from 1 to 9. In both cases, each possible outcome is equally likely. Which of the two friends has a better chance of getting the greater number? Explain. You may want to create a chart to help you.

4. Ramón says the probability of getting a sum of 12 when rolling two dice is equal to the probability of getting a sum of 6. To explain his reasoning, he pointed out that each number has the same probability to be rolled on each die.

   a. Explain why Ramón is incorrect.

   b. What is the probability of rolling a sum of 6? A sum of 12? A sum of 1?

5. Suppose you have two sets of eight cards. The cards in each set are numbered from 1 to 8. You pick one card at random from each set and find the sum of the numbers.

   a. What is the probability that the sum is 1?

   b. What is the probability that the sum is 3?

   c. Suppose you play a game in which Player A scores a point when the sum of the two cards is 3 and Player B scores a point when the sum is 9. Which player would you rather be? Explain.

6. Suppose someone rolls two standard dice.

   a. Find the probability that the difference rolled is even. Explain.

   b. Find the probability that the difference is odd. Explain.

   c. Find the probability that the difference is less than 3. Explain.

   d. Find the probability that the difference is 3 or more. Explain.

   e. Find the probability that the difference is 6 or more. Explain.

7. Suppose someone rolls two dice, each with faces numbered 2, 4, 6, 8, 10, and 12. Consider the differences between two rolled numbers.

   a. How many possible differences are there? What are they?

   b. Are the probabilities for all the possible differences the same? Explain.

   c. Suppose you were playing a game of *Rolling Differences* with these dice instead of standard dice. The spaces on the game board would be labeled with the possible differences, rather than 0 to 5. How would you arrange your counters in the spaces? Explain.

8. Imagine a game called *Rolling Sums* that is just like *Rolling Differences,* but in which you *add* the numbers to determine which counter to remove.

   a. How many sections should there be on the game board?

   b. What is the greatest number that should appear on the board?

   c. What is the least number that should appear on the board?

   d. Describe a strategy you would use to arrange your counters on the board.

9. Darnell and Camila were playing *Rice Drop* using this game board. The dimensions are in inches. Assume the rice will land on a completely random spot.

   a. Find the probability that the rice lands in a triangle.

   b. Find the probability that the rice lands in a shaded quadrilateral.

   c. Find the probability that the rice lands in a shaded figure.

   d. Find the probability that the rice does not land in a shaded figure.

10. A small area created by buildings built close together is often called a *courtyard.* Suppose a group of six friends are standing in a courtyard like the one shown here when it starts to rain.

What is the probability that the first raindrop hits one of the friends? (Assume each person occupies a circle about 50 cm in diameter.)

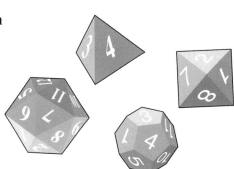

Connect Extend

11. Marcus and Miguel played *Who's Greater?* with two different dice. Marcus' die had 100 on one face and 0 on each of the other five faces. Miguel's was numbered from 1 to 6.

Before playing, Marcus said: "I will probably win because the greatest number on my die is 100, which is much greater than any of the numbers on your die."

Miguel said: "I have the better chance of winning because my die has more faces with numbers that are greater than the numbers on your die."

Who is right? Explain your answer.

12. Charo and Irene have dice with 4 sides (numbered 1 to 4), 8 sides (numbered 1 to 8), 12 sides (numbered 1 to 12), and 20 sides (numbered 1 to 20). With each of the dice, all the possible outcomes are equally likely.

The friends want to play *Who's Greater?* with these dice. To play, each chooses a die and rolls it. Whoever rolls the greater number scores a point. If the numbers are the same, no one scores.

Irene chose a 12-sided die. Which die should Charo choose? Explain.

13. Design two *different* six-sided dice that have an equal probability of beating each other at *Who's Greater?*

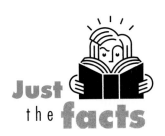

Just the facts

A 4-sided die is called a *tetrahedron.* An 8-sided die is an *octahedron.* A 12-sided die is a *dodecahedron.* And a 20-sided die is an *icosahedron.*

14. Economics Omry and Shaked are twin brothers. For their birthday, they each received gift certificates from the local candy store, worth 25¢, 50¢, $1, $5, and $10.

a. Suppose both boys have only one gift certificate left, but neither can remember which one. They want to buy and share a chocolate that costs $2.25. Assuming the gift certificates are equally likely to be remaining, what is the probability that they have enough to buy the chocolate? Explain your answer.

b. What is the probability that each boy has a $5 gift certificate left?

15. Suppose you roll three standard dice and add the results. There are 216 equally likely outcomes.

a. What is the least possible sum? What is the probability of rolling this sum?

b. What is the greatest possible sum? What is the probability of rolling this sum?

c. Find all the ways to get a sum of 4, and find the probability of rolling that sum.

16. Mr. Shu's suitcase has a lock consisting of two number wheels with the numbers 0, 2, 4, 5, 6, and 8 on each. To open the lock, the wheels must be set so the product of the two numbers is 32.

Locked

Unlocked

a. Suppose a stranger finds the suitcases and attempts to open it by trying different combinations. What is the probability that he will open the suitcase on the first try?

b. Mr. Shu can reset the lock to open with a different product. Which product would give a more secure lock than 32? Explain.

c. Which product would you suggest Mr. Shu *not* use? Why?

d. Mrs. Shu's suitcase has two number wheels identical to those on Mr. Shu's suitcase. Her suitcase opens when the numbers form a particular pair, such as 0-5 or 6-2, rather than a particular product. Assuming Mr. Shu sets his lock as you suggested in Part b, whose suitcase will be more secure, Mr. Shu's or Mrs. Shu's? Explain.

17. Avril has a die numbered from 1 to 6. Chelsea has a spinner divided into ten identical sections numbered from 1 to 10. Avril rolls her die, and Chelsea spins her spinner.

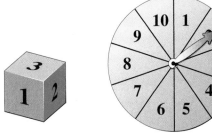

a. One possible outcome is 1-10, meaning Avril rolled a 1 on the die and Chelsea spun a 10 on the spinner. How many possible outcomes are there?

b. What is the probability that the difference of the two numbers is 10? Explain your answer.

c. What is the probability that the difference is 7? Explain your answer.

d. What is the probability that the difference is greater than 7? Explain your answer.

18. Avril has a die numbered from 1 to 6. Chelsea has a spinner divided into ten identical sections, numbered from 1 to 10. They decide to play the following game: On each turn, they spin the spinner and roll the die. Avril scores a point if the difference is 5, and Chelsea scores a point if the difference is 7.

a. Who has the better chance of scoring a point? Explain.

b. How could the girls change the scoring rules so that they have the same chance of winning?

c. Suppose you were playing the original game and could choose any number as the difference you earn a point on. Which number would you choose? Why?

19. Technology Marcus videotaped all six episodes of his favorite hour-long television show. His friend missed an episode, and Marcus said he would loan her the tape. The number line illustrates where on the tape the show had been recorded.

Hours

Marcus often rewinds the tape and watches different parts of previous episodes. He doesn't remember where on the tape he last stopped watching the show. Before giving the tape to his friend, Marcus put it in his VCR and pushed the play button. What is the probability that the tape started somewhere within the show his friend wanted to see?

20. While visiting a friend, Carla parked her car under a large oak tree. She left her moon roof open. (A *moon roof* is a small window in the roof.)

The tree drops acorns in a circular area around its trunk, as shown in the diagram. Assume the acorns fall randomly within the circle, which has a radius of 20 feet. No acorns fall where the trunk is. The trunk has a radius of 5 feet.

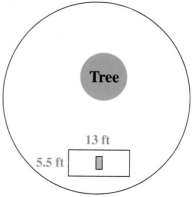

Tree

13 ft

5.5 ft

a. Find the area of the region in which an acorn might fall.

b. While Carla was at her friend's house, an acorn fell from the tree. Find the probability that the acorn hit Carla's car (including the moon roof).

c. The dimensions of the moon roof are 32 inches by 16 inches. Find the probability that the acorn fell through the moon roof.

d. Challenge Suppose the acorn hit Carla's car. Find the probability that it fell through the moon roof. (Hint: Should this probability be *more than* or *the same as* the probability you answered in Part c?)

Statistics Find the mean, median, and mode of each set of test scores. Then tell which measure you think best represents the data.

21. 85, 99, 73, 64, 99, 80, 69, 72, 70

22. 0, 90, 93, 6, 85, 97, 84

23. 52, 94, 73, 81, 65, 88

Geometry Find the value of each lettered angle without measuring.

24.

c
b
$24°$
a

25.

$233°$
x

26.

y
$125°$
$125°$
$45°$

27. Economics Mr. Morales hired Althea to distribute takeout menus for his new restaurant to homes in the area. He told Althea he would pay her $14.50, plus 4¢ for every menu she delivered.

   a. Write a rule you can use to calculate the amount Althea will earn, A, if you know the number of menus she delivers, m.

   b. Althea wants to earn $23 to buy a pair of earrings for her mother. Write an equation you could solve to find the number of menus should would need to deliver to earn enough money.

   c. Solve your equation. How many menus does Althea need to deliver?

28. Geometry Make sketches to show how you can arrange seven identical square tiles to form three shapes with different perimeters. You must use all the tiles in each of your shapes. Give the perimeter of each shape.

Making Matches

In the probability games you have considered so far, the result of one round or trial does not affect the result of another. For example, if you toss a coin and get heads, your chances of getting heads when you flip the coin again are still 50%. These are called *independent events*.

In this lesson, you will work with situations in which what happens in one case *does* affect what can happen in the next, or *dependent events*.

Think & Discuss

Suppose you roll a die several times.

- What is the probability of getting 6 on the first roll? What is the probability of getting 4?

- Suppose the first time you roll the die, you get 6. You roll a second time. What is the probability of getting 6 on the second roll? What is the probability of getting 4?

Now imagine that you and some friends are making a poster for a school party. Your teacher gives you six markers, in six different colors, and each of you chooses one without looking. Two of the colors are red and green.

- If you choose first, what is the probability that you will get the red marker? What is the probability that you will get the green marker?

- Suppose you chose first and got the red marker. What is the probability that the second person will get the red marker? What is the probability that the second person will get the green marker?

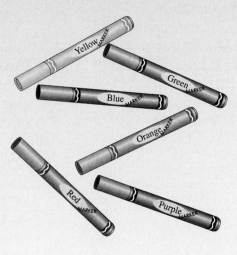

- Why are the probabilities for the dice situation different from those for the marker situation?

Investigation Matching Colors

VOCABULARY
simulation

A **simulation** is an experiment in which you use different items to represent the items in a real situation. For example, to simulate choosing markers and looking at their colors, you can write the color of each marker on a slip of paper and put all the slips into a bag. You can simulate choosing markers by drawing slips from the bag. Mathematically, the situations are identical.

Using a simulation can help with some of the problems in this investigation.

MATERIALS
- counters or blocks (2 in each of 2 colors) or slips of paper
- bucket or bag

Problem Set A

Ken woke up early and found that a storm had knocked out the power in his neighborhood. He has to dress in the dark. Ken has four socks in his drawer, two black and two brown. Color is the only difference between them. As long as both socks are the same color, Ken doesn't care which he wears. He takes two socks out of the drawer.

1. Simulate this situation, using counters, blocks, or slips of paper with colors written on them. If you use counters or blocks, you may have to let other colors stand for the sock colors. For example, a red block might represent a brown sock, and a blue block might represent a black sock. Use a bucket or bag to represent Ken's sock drawer.

   a. Without looking, pick two "socks," one at a time, from the "drawer." Record whether the socks match. Then put the socks back, mix them up, and try again. Repeat this process 16 times, and record the results.

   b. Use your results to find an experimental probability that Ken will choose matching socks.

2. If the first sock Ken picks is brown, what is the (theoretical) probability that the second sock will also be brown? Explain.

3. If the first sock is black, what is the (theoretical) probability that the second sock will also be black?

4. Ken says that since he has two colors of socks, he has a 50% chance of getting a matching pair. Do you think he is correct? If not, what do you think the actual probability is? Explain.

Just the facts

On August 14, 2003, a massive power failure caused the largest blackout in U.S. history. The blackout affected over 50 million people in eight states and parts of Canada.

In Lesson 10.2, you saw how you could use a table to keep track of the possible outcomes for two coin tosses. You can also draw a *tree diagram* to show all the possibilities. The possible results for the first coin can be shown like this:

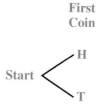

The possibilities for the second coin can be shown as branches from each of the first two branches:

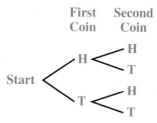

You can read off the possible outcomes by following the branches, beginning with Start. For example, following the top set of branches gives the outcome HH.

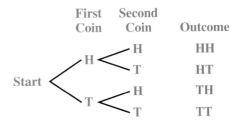

Problem Set B

1. Suppose you choose one of six markers from a bag: red, orange, yellow, green, blue, or purple. Draw a tree diagram showing the possible colors for the marker.

2. Now consider what happens when you choose a second marker.

   a. Suppose the first marker you chose was red. What are the possible choices for the second marker?

   b. Add a new set of branches to the "red" branch of your tree diagram for Problem 1, showing the possibilities for your second choice.

   c. Complete your tree diagram by adding branches that show the possibilities when each of the other colors are picked first.

   d. What is the probability that, if you choose the two colors at random, you will get red and green (chosen in either order)?

3. Draw a tree diagram to show the possible choices of sock colors for Ken if he chooses two socks from a drawer containing two brown and two black socks. Since there are two socks of each color, label the socks as brown 1, brown 2, black 1, and black 2.

4. How many possible sock pairs are there? How many of them have matching colors?

5. What is the probability that Ken will choose a matching pair?

Just the facts

The knitting machine was invented in 1589 by William Lee of England. Queen Elizabeth I refused to give Lee a patent for his machine because she felt the stockings it produced were too coarse. (It may also be that she didn't want to put people who knit by hand out of business.)

Problem Set C

After choosing socks, Ken has to choose pants and a shirt. His school requires uniforms of blue, tan, or green. He has two pairs of pants, one blue and one tan. He has two shirts, also one blue and one tan. Now he takes one shirt and one pair of pants.

1. Suppose the shirt is blue. What is the probability that the pants will also be blue? Explain how you found your answer.

2. Draw a tree diagram showing the possible choices of shirts and pants.

3. Ken says the probability that he will choose matching shirt and pants is 50%. Is he right? How do you know?

4. Suppose Ken has a third shirt and a third pair of pants, both green. Now what is the probability that he will chose a shirt and pants of the same color? Explain.

5. Suppose Ken has one tan shirt and two blue shirts, and one tan pair of pants and two blue pairs of pants. Find the probability that the shirt and pants will match, and show how you found your answer.

Share & Summarize

1. When Ken is choosing socks in the dark, does the color of the first sock affect the chance of choosing a particular color next? Why or why not?

2. When Ken is choosing shirts and pants in the dark, does the color of the shirt chosen affect the chance of choosing pants of a particular color? Why or why not?

Investigation 2 Matching Cards

In an ordinary deck of playing cards, there are four suits:

| Clubs | Diamonds | Spades | Hearts |

There are 13 cards in each suit, one for each of the numbers from 1 to 10, and three *face cards:* jack, queen, and king. Clubs and spades are black while diamonds and hearts are red.

Many kinds of games—involving various combinations of chance and skill—are played with decks of cards. In this investigation, you will work with some simple games of chance that involve choosing cards from a deck.

Problem Set D

In the first game, a deck of cards is shuffled and placed on a table face down. For one round of the game, players do the following:

- Player 1 chooses a card from the deck without looking and writes down its suit (spades, hearts, diamonds, or clubs).
- Player 1 puts the card back and shuffles the deck.
- Player 2 chooses a card without looking. If it has the same suit as the first card, Player 1 scores a point. Otherwise, Player 2 scores a point.
- Player 2 returns the card and shuffles the deck.

The winner is the player with more points at the end of 20 rounds.

1. What is the probability that Player 1 will choose a heart?

2. If Player 1 chooses a heart, what is the probability that Player 2 will also choose a heart? Explain.

3. What is the probability that Player 2 chooses a card of the same suit as Player 1's card, no matter what that suit was? How do you know?

4. What would you expect the score to be after 20 rounds?

5. Think of a way to change the scoring rules to give both players the same chance of winning.

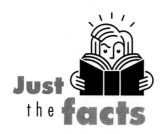

Just the **facts**

Problem Set E

The second card game is similar to the first. The only difference is that Player 1 does not put the card back before Player 2 chooses. After both players have chosen, the cards are returned to the deck. Player 1 scores a point if the two cards have the same suit, and Player 2 scores a point if they have different suits.

1. What is the probability that Player 1 will choose a heart?

2. Suppose Player 1 chooses a heart.

    a. What is the probability that Player 2 also chooses a heart?

    b. Is your answer to Part a different from your answer to Problem 2 of Problem Set D? Why or why not?

3. What is the probability that Player 2 chooses a card of the same suit as Player 1, no matter what suit Player 1 chooses?

4. Is this game *more fair, less fair,* or *just as fair* as the game in Problem Set D? Explain.

5. Challenge Find a way to assign points so that the game will be fair. Explain how you devised your point system.

Problem Set F

Suppose you want to draw a tree diagram to show the possible choices for the first card game, in which Player 1 replaces the card before Player 2 chooses.

1. How many branches would you need to show the possibilities for the first card?

2. How many branches would you have to add to show the possibilities for the second card?

3. How many total branches would your tree diagram have?

As you have probably realized, this tree diagram would be very large! Since the game concerns only the suits of the cards, and since the four suits are equally likely for each draw, you can draw a simplified tree diagram showing the four possible suits for each draw.

For example, suppose the first card chosen is a heart. Here is the part of the tree diagram showing the possible suits for the second card:

First Card Second Card

Heart
- heart
- diamond
- spade
- club

4. Draw a tree diagram showing all the possible suit combinations for the first game.

5. Hearts and diamonds are red while clubs and spades are black. What is the probability that the two cards have the same color?

6. Can you use a simplified tree diagram for the second game, in which Player 1 keeps the card instead of returning it to the deck before Player 2 chooses? Explain.

Share & Summarize

In some probability situations, one event can affect the probability of another.

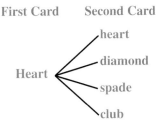

1. For each pair of events, decide whether the first event affects the probability of the second. If your answer is "yes," explain why.

First Event	Second Event
a. getting heads on the flip of a coin	getting heads on the second flip of the coin
b. getting a king when choosing a card from a deck	getting a king when choosing a second card without returning the first
c. drawing a certain name from names written on slips of paper and chosen at random from a hat	drawing a second name if the first slip is returned to the hat before the second choice

2. Make up your own sequence of two events for which the first event affects the probability of the second.

3. Now make up your own sequence of two events for which the first event *does not* affect the probability of the second.

On Your Own Exercises

1. Suppose there are two cards, numbered 1 and 2. The cards are mixed, and placed face down.

   a. You arrange the cards in a row and then turn them over. What is the probability that the first card will be a 2?

   b. What is the probability that the cards will form the number 21? Explain.

2. Suppose you have three cards, numbered 1, 2, and 3. The cards are shuffled and placed face down in a row.

   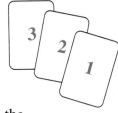

   a. List all the three-digit numbers that can be created from these three cards.

   b. What is the probability that the cards will form the number 213 when they are turned over?

   c. What is the probability that the cards will form a number between 200 and 300?

   d. What is the probability that the cards will form an even number?

   e. What is the probability that the cards will form a number less than 300?

3. Manuel and Leila are splitting a box of marbles. The box contains three red, two green, and one orange marble. Each friend chooses a marble at random.

   a. If Leila chooses first and gets a red marble, what is the probability that Manuel's marble will also be red?

   b. Draw a tree diagram showing all the possible combinations when each friend chooses one marble. Label the red marbles R1, R2, and R3; the green marbles G1 and G2; and the orange marble O.

   c. What is the probability that the two marbles will be the same color?

impactmath.com/self_check_quiz

4. Jahmal and Hannah are playing a game with an ordinary deck of playing cards. For each turn, Hannah chooses a card and returns it to the deck. She shuffles the deck, and then Jahmal chooses a card.

   a. If Hannah picks the 5 of clubs, what is the probability that Jahmal will pick the 5 of clubs?

   b. If Hannah picks a black card (either a spade or a club), what is the probability that Jahmal will pick a red card (either a heart or a diamond)? How do you know?

   c. If Hannah picks the 6 of spades, what is the probability that Jahmal will pick a king?

   d. If Hannah picks a red queen, what is the probability that Jahmal will pick a red queen?

5. Jahmal and Hannah are playing a game with an ordinary deck of playing cards. For each turn, the deck is shuffled and the cards are spread out face down. At the same time, Hannah and Jahmal each choose a card.

   a. If Hannah picks the 5 of clubs, what is the probability that Jahmal also picks the 5 of clubs?

   b. If Hannah picks a black card (either a spade or a club), what is the probability that Jahmal picks a red card (either a heart or a diamond)? How do you know?

   c. If Hannah picks the 6 of spades, what is the probability that Jahmal picks a king?

   d. If Hannah picks a red queen, what is the probability that Jahmal picks a red queen?

6. Ramesh has created a game using these six cards:

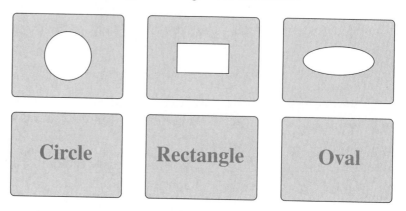

The three shape cards are placed face up. The three word cards are shuffled, and one word card is placed face down next to each shape card. A player scores 1 point for each word card that matches a shape card.

a. Write all the possible arrangements of the three word cards. Use C to stand for the circle card, R to stand for the rectangle card, and O to stand for the oval card. How many possibilities are there?

b. What is the probability that a player will match all three word cards correctly?

c. What is the probability that a player will match at least one word card correctly?

7. Shaunda has written letters to four friends: Caroline, Raul, Jing, and Ernest. She has four envelopes, each with the name and address of one of the friends. Shaunda's little brother wants to help, so he puts one letter in each envelope. Since he can't read yet, he puts the letters in the envelopes at random.

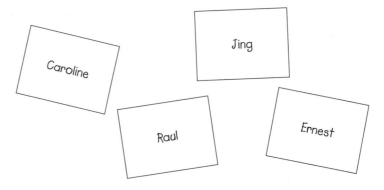

a. How many ways can the letters and envelopes be paired?

b. What is the probability that everyone receives the right letter?

8. Life Science A person's *genes* determine many things about that person, including how he or she looks. For example, a person has two genes that determine eye color. The gene for blue eyes is *recessive* and the gene for brown eyes is *dominant*. This means that if a person has one blue-eye gene and one brown-eye gene, he or she has brown eyes. A person with two brown-eye genes also has brown eyes. To have blue eyes, both genes must be blue.

A child gets one eye-color gene from each parent. Assume the chances of passing either gene to a child are equal. For example, a father with one blue-eye gene and one brown-eye gene has a 50% chance of passing the blue-eye gene to his child.

a. Suppose two people are having a child. One has a blue-eye gene and a brown-eye gene, and the other has two brown-eye genes. What is the probability that the child will have blue eyes? (Hint: You can find the possible gene combinations for the child by making a table or a tree diagram.)

b. Suppose the two parents both have one blue-eye gene and one brown-eye gene. What is the probability that the child will have blue eyes?

c. Now suppose one of the parents has two blue-eye genes and the other has one blue-eye gene and one brown-eye gene. What is the probability that the child will have blue eyes?

In your
own
words

Describe a situation in which one event affects the probability of another.

9. Maria and David are playing a game with an ordinary deck of playing cards. For each turn, the deck is shuffled and placed face down. Maria chooses a card and records it. She returns the card to the deck and shuffles the cards, and then David chooses a card. The player with the higher card scores a point. (Aces are the lowest cards, and kings are the highest.) If the cards have the same value, neither player scores.

a. Suppose Maria chooses a king. What is the probability that she will score a point?

b. Suppose Maria chooses the 6 of diamonds. What is the probability that she will *not* score a point?

10. Maria and David are playing a game with an ordinary deck of playing cards. For each turn, the deck is shuffled and placed face down. David chooses a card and keeps it. Then Maria chooses a card. The player with the higher card earns a point. (Aces are the lowest cards, and kings are the highest.) If the cards have the same value, neither player scores.

a. Suppose David picks a king. What is the probability that he will score a point?

b. Suppose David picks the 6 of diamonds. What is the probability that he will *not* score a point?

11. Althea shuffled a standard deck of playing cards and turned over the first two cards.

a. What is the probability that the first card was an ace?

b. Suppose the first card was an ace. What is the probability that the second card was a 2 with the same suit as the ace?

c. How many possible combinations of two cards are there in a standard deck?

d. How many of those combinations include an ace and then a 2 (in that order) of the same suit?

e. What is the probability of getting an ace and then a 2 of the same suit?

f. What is the probability of getting an ace and a 2 in *any* order, but of the same suit?

Find the value of each expression in simplest form.

12. $\frac{5}{6} + \frac{4}{9}$ **13.** $\frac{19}{26} - \frac{17}{39}$ **14.** $\frac{45}{56} \cdot \frac{32}{35}$

15. $\frac{14}{15} \div \frac{2}{5}$ **16.** $11\frac{19}{21} + 6\frac{1}{7}$ **17.** $5\frac{1}{4} - 2\frac{5}{12}$

18. $3\frac{5}{8} \cdot 1\frac{3}{4}$ **19.** $9\frac{1}{3} \div \frac{5}{9}$ **20.** $\frac{11}{14} \div 1\frac{3}{14}$

21. Geometry A rectangle has an area of 48 square feet and a perimeter of 32 feet. What are the dimensions of the rectangle?

22. Which has the greater area: a circle with diameter 11 meters, or a square with side length 10 meters?

23. Economics The Book Bin is having a clearance sale.

a. All dictionaries are marked $33\frac{1}{3}\%$ off. Ramesh bought a French dictionary with a sale price of $18. What was the dictionary's original price?

b. Novels are on sale for 20% off. Althea bought a novel with an original price of $11.95. What was the sale price?

c. Travel books are all marked down by a certain percent. Miguel bought a book about African safaris. The book was originally priced at $27.50, but Miguel paid only $16.50. What percent did Miguel save?

Chapter Summary

Probability is useful in many areas of life, from playing games to making plans based on weather predictions. In this chapter, you learned how to find *experimental* and *theoretical probabilities* for events in which the possible outcomes are *equally likely*.

You examined probabilities in several types of situations, including some in which the possible outcomes were not easy to determine. For certain games of chance, you came up with strategies for play based on your knowledge of probabilities. You also used *simulation* and tree diagrams to examine situations in which the number of possible outcomes were affected by what had happened before.

Strategies and Applications

The questions in this section will help you review and apply the important ideas and strategies developed in this chapter.

Understanding probability

Althea took the king of hearts and the king of clubs from a standard deck of cards, leaving only 50 cards. She told Leah that the probability of selecting a queen was now 8%, but she didn't tell her how many or what cards she had removed.

1. What does it mean that the probability was 8%?

2. Leah selected a card from Althea's deck, looked at it, and then put it back. Althea shuffled the cards. They repeated this process until Leah had chosen a card 100 times.

   a. How many times would you expect Leah to have picked a queen?

   b. Leah chose a queen 7 times. She said that this means Althea was wrong and that the actual probability is 7%. Althea and Leah both calculated the probabilities they gave. Is either incorrect in her calculation? Explain.

   c. Whose is the more accurate probability, Leah's or Althea's?

d. Leah kept selecting cards until she had 1,000 trials. She chose a queen 88 times. Althea said, "The difference between the 88 queens you selected and the 80 you should have expected was 8, but the difference was only 1 when you drew 100 cards. Your experimental probability will be less accurate for the 1,000 draws than it was for the 100 draws."

Is Althea correct? Explain.

Identifying outcomes

3. Name two strategies for identifying the outcomes of a probability situation. Illustrate each strategy by using them to find the number of outcomes for spinning this spinner twice:

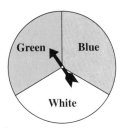

Finding probabilities of events

4. Explain the difference between a theoretical probability and an experimental probability. You may want to illustrate with an example.

5. Josh said, "Suppose you roll a standard die. To calculate the probability of getting a prime number, you have to divide 3 by 6, giving 0.5."

a. Why did Josh choose 6 for the divisor?

b. Why did Josh choose 3 for the dividend?

c. Consider this *Rice Drop* game board. Explain why the procedure for calculating the probability that the rice lands in a shaded square is the same as the one Josh used for getting a prime number on a die roll.

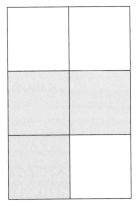

6. A bag contains five slips of paper numbered 1 to 5. In the game *Find the Difference,* each player chooses one of the cards below. Players take turns drawing two numbers from the bag. If the difference of the numbers is on the player's card, the player covers that difference. The numbers are returned to the bag after each turn. The first player to cover all his numbers wins.

Card A

| 1 | 2 |
| 3 | 4 |

Card B

| 1 | 2 |
| 2 | 1 |

Card C

| 3 | 4 |
| 4 | 3 |

Which card gives a player the best chance of winning? Explain.

Working with situations in which the probabilities depend on previous results

7. Craig and Kenna were playing a board game in which they rolled two dice. Rolling doubles (that is, rolling the same number on both dice) lets you take an extra turn. Kenna rolled two 3s and then two 5s. As she was getting ready to take another extra turn, Craig said, "The chances of you getting doubles again are next to nothing!" Is Craig correct? Explain your answer.

8. Describe a probability experiment in which the result of one trial changes the probabilities for the next trial's result. You might want to use dice, cards, spinners, or slips of paper drawn from a bag in your experiment.

Demonstrating Skills

9. At a fund-raising carnival, Marcus operated a game in which each player spun a wheel. The section on which the wheel stopped would indicate what prize, if any, the player won.

The table shows how many equal-sized spaces listed each type of prize as well as how many people won each prize by the end of the day.

	Key Chain	Troll Doll	Baseball Cap	Stuffed Animal	Beach Ball	No Prize
Number of Spaces	5	4	3	2	1	45
Number of Winners	14	16	13	6	3	148

a. Find an experimental probability of winning each prize.

b. Find the theoretical probability of winning each prize.

Use this information for Questions 10 and 11:

At the beginning of a computer game called *Geometry Bug,* players take turns choosing circles, squares, and triangles on the screen. After all the shapes have been chosen, a small "bug" appears and flies over the shapes. The bug lands on a random place on the screen. If it lands on one of the players' shapes, that player scores a point. The winner is the player with the most points after 50 landings.

Rosa and Cari were playing with this screen. Rosa's shapes are green and Cari's are white. The screen is 8 inches wide and 5 inches high. The circles have a radius of $\frac{1}{2}$ inch. The squares have a side length of 1 inch. The triangles are right triangles with legs 1 inch long.

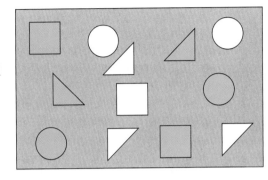

10. Consider probabilities for a single bug landing. Write your answers as decimals to the nearest thousandth.

    a. What is the probability that the bug lands in Cari's square?

    b. What is the probability that the bug lands in one of Rosa's circles?

    c. Find each player's probability of scoring on a single bug landing.

11. The girls decide to play with an optional rule. When the bug lands on a shape, the shape is removed from the board. For example, if the bug lands on Cari's square, Cari scores a point but the square disappears.

    a. Find the probability that the bug lands on one of Cari's triangles.

    b. Suppose the first shape the bug landed on was Cari's triangle in the bottom right. Now what is the probability that the bug lands on any of Cari's shapes?

12. A group of six friends wanted to play a game that requires three teams. To decide which two people would play on each team, Luke put six cubes in a bag. Two of the cubes were white, two were black, and two were red. Each person took a cube without looking, and the two with the same color formed a team.

    a. Draw a tree diagram to show the first *two* drawings of the cubes. To make the labeling easier, use W1 and W2 for the white cubes, B1 and B2 for the black cubes, and R1 and R2 for the red cubes.

    b. Jing drew a cube first and then Jahmal. Use your tree diagram to find the probability that Jing and Jahmal will be on the same team.

English	**Español**

absolute value The *absolute value* of a number is its distance from 0 on the number line, and is indicated by drawing a bar on each side of the number. For example, |⁻20| means "the absolute value of ⁻20." Since ⁻20 and 20 are each 20 units from 0 on the number line, |20| = 20 and |⁻20| = 20. [page 143]

valor absoluto El *valor absoluto* de un número es su distancia desde 0 en la recta numérica, lo cual se indica trazando una barra en cada lado del número. Por ejemplo, |⁻20| significa "el valor absoluto de ⁻20". Como ⁻20 y 20 se encuentran a 20 unidades de 0 en la recta numérica, |20| = 20 y |⁻20| = 20.

acute angle An angle that measures less than 90°. Each of the angles shown below is an *acute angle*. [page 469]

ángulo agudo Ángulo que mide menos de 90°. Cada uno de los siguientes es un *ángulo agudo*.

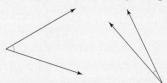

angle Two rays with the same endpoint. For example, the figure below is an *angle*. [page 47]

ángulo Dos rayos que parten del mismo punto. Por ejemplo, la siguiente figura muestra un *ángulo*.

area The amount of space inside a two-dimensional shape. [page 494]

área La cantidad de espacio dentro de una figura bidimensional.

axes The horizontal line and vertical line that are used to represent the variable quantities on a graph. For example, in the graph below, the horizontal axis represents width and the vertical axis represents length. [page 279]

ejes La recta horizontal y la recta vertical que se usan para representar las cantidades variables en una gráfica. Por ejemplo, en la siguiente gráfica el eje horizontal representa longitud.

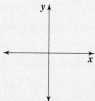

backtracking The process of using a flowchart to work backward, starting with the output and undoing each operation to find the input. [page 572]

vuelta atrás El proceso de usar un flujograma para trabajar en sentido inverso, comenzando con la salida y anulando cada operación hasta llegar a la entrada.

base of a parallelogram Any of the sides of a parallelogram. The *base of a parallelogram* is used in computing its area. (See the figures in the glossary entry for height of a parallelogram.) [page 515]

base de un paralelogramo Cualquiera de los lados de un triángulo. La *base de un paralelogramo* se usa para calcular su área. (Ver las figuras en el inciso del glosario para *altura de un paralelogramo*.)

base of a triangle Any of the sides of a triangle. The *base of a triangle* is used in computing its area. (See the figures in the glossary entry for *height of a triangle*.) [page 518]

base de un triángulo Cualquiera de los lados de un triángulo. La *base de un triángulo* se usa para calcular su área. (Ver las figuras en el inciso del glosario para *altura de un triángulo*.)

English	Español

chord A segment connecting two points on a circle. (See the figure in the glossary entry for radius. This figure shows a chord of a circle.) [page 486]

cuerda Segmento que conecta dos puntos en un círculo. (Ver la figura en el inciso del glosario para *radio*. Dicha figura muestra una cuerda de un círculo.)

circumference The perimeter of a circle (distance around a circle). [page 486]

circunferencia El perímetro de un círculo (distancia alrededor del círculo).

common factor A common factor of two or more numbers is a number that is a factor of all the numbers. For example, 5 is a *common factor* of 15, 25, and 40. [page 82]

factor común Un factor común de dos o más números es un número que es un factor de todos los números. Por ejemplo, 5 es un *factor común* de 15, 25 y 40.

common multiple A *common multiple* of a set of numbers is a multiple of all the numbers. For example, 24 is a common multiple of 3, 8, and 12. [page 85]

múltiplo común Un *múltiplo común* de un conjunto de números es un múltiplo de todos los números. Por ejemplo, 24 es un múltiplo común de 3, 8 y 12.

composite number A whole number greater than 1 with more than two factors. For example, 12 is a *composite number* since it is greater than 1 and has more than two factors. In fact, it has six factors: 1, 2, 3, 4, 6, and 12. [page 80]

número compuesto Un número entero mayor que 1 con más de dos factores. Por ejemplo, 12 es un *número compuesto* dado que es mayor que 1 y tiene más de dos factores. De hecho, tiene seis factores: 1, 2, 3, 4, 6 y 12.

concave polygon A polygon that looks like it is "collapsed" or has a "dent" on one or more sides. Any polygon with an angle measuring more than 180° is concave. The figures below are *concave polygons*. [page 50]

polígono cóncavo Polígono que parece que se hubiera "hundido" o que tiene una hendidura en uno más de sus lados. Cualquier polígono con un ángulo mayor que 180° es cóncavo. Las siguientes figuras son *polígonos cóncavos*.

coordinates Numbers that represent the location of a point on a graph. For example, if a point is 3 units to the right and 7 units up from the origin, its *coordinates* are 3 and 7. [page 302]

coordenadas Números que representan la posición de un punto en una gráfica. Por ejemplo, si un punto se encuentra a 3 unidades a la derecha y 7 unidades hacia arriba del origen, sus *coordenadas* son 3 y 7.

diameter A chord that passes through the center of a circle. *Diameter* also refers to the distance across a circle through its center. (See the figure in the glossary entry for radius. This figure shows a diameter of a circle.) [page 486]

diámetro Cuerda que pasa por el centro de un círculo. El *diámetro* también se refiere a la distancia a través de un círculo, pasando por su centro. (Ver las figuras en el inciso del glosario para radio. Dicha figura muestra un diámetro de un círculo.)

distribution The *distribution* of a data set shows how the data are spread out, where there are gaps, where there are lots of values, and where there are only a few values. [page 351]

distribución La *distribución* de un conjunto de datos muestra la extensión de los datos, las brechas entre los datos, los lugares donde hay muchos valores y donde hay pocos valores.

equally likely Outcomes of a situation or experiment that have the same probability of occurring. For example, if one coin is tossed, coming up heads and coming up tails are *equally likely* outcomes. [page 608]

equiprobables Resultados de una situación o experimento que tienen la misma posibilidad de ocurrir. Por ejemplo, si se lanza una moneda, es *equiprobable* que la moneda caiga mostrando cara o escudo.

English	Español

equation A mathematical sentence stating that two quantities have the same value. An equal sign, =, is used to separate the two quantities. For example, $5 + 8 = 3 + 10$ is an *equation*. [page 558]

ecuación Enunciado matemático que establece que dos cantidades tienen el mismo valor. Se usa un signo de igualdad, =, para comparar las dos cantidades. Por ejemplo, $5 + 8 = 3 + 10$ es una *ecuación*.

equivalent fractions Fractions that describe the same portion of a whole, or name the same number. For example, $\frac{3}{4}$, $\frac{9}{12}$, and $\frac{30}{40}$ are *equivalent fractions*. [page 100]

fracciones equivalentes Fracciones que describen la misma parte de un todo o que representan el mismo número. Por ejemplo, $\frac{3}{4}$, $\frac{9}{12}$ y $\frac{30}{40}$ son *fracciones equivalentes*.

experimental probability A probability based on experimental data. An *experimental probability* is always an estimate and can vary depending on the particular set of data that is used. [page 606]

probabilidad experimental Probabilidad que se basa en datos experimentales. Una *probabilidad experimental* es siempre una estimación y puede variar según el conjunto de datos en particular que se usen.

exponent A small, raised number that tells how many times a factor is multiplied. For example, in 10^3, the exponent 3 tells you to multiply 3 factors of 10: $10 \cdot 10 \cdot 10 = 1,000$. [page 81]

exponente Número pequeño y elevado que indica cuántas veces se multiplica un factor. Por ejemplo, en 10^3, el exponente 3 te indica que multipliques 3 factores de 10: $10 \cdot 10 \cdot 10 = 1,000$

factor A *factor* of a whole number is another whole number that divides into it without a remainder. For example, 1, 2, 3, 4, 6, 8, 12, and 24 are factors of 24. [page 78]

factor Un factor de un número entero es otro número entero que lo divide sin que quede un residuo. Por ejemplo, 1, 2, 3, 4, 6, 8, 12 y 24 son factores de 24.

factor pair A *factor pair* for a number is two factors whose product equals that number. For example, the factor pairs for 24 are 1 and 24, 2 and 12, 3 and 8, and 4 and 6. [page 78]

par de factores Un *par de factores* para un número consta de dos factores cuyo producto es igual al número. Por ejemplo, los pares de factores para 24 son 1 y 24; 2 y 12; 3 y 8; y 4 y 6.

flowchart A diagram, using ovals and arrows, that shows the steps for going from an input to an output. For example, the diagram below is a *flowchart*. [page 571]

flujograma Diagrama que usa óvalos y flechas para mostrar los pasos a seguir desde un dato de entrada hasta uno de salida. Por ejemplo, el siguiente diagrama es un *flujograma*.

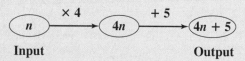

Input   **Output**

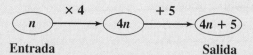

Entrada   **Salida**

greatest common factor, GCF The *greatest common factor* (often abbreviated *GCF*) of two or more numbers is the greatest of their common factors. For example, the greatest common factor (or GCF) of 24 and 36 is 12. [page 83]

máximo común divisor, MCD El *máximo común divisor* (a menudo abreviado, MCD) de dos números es el mayor de sus factores comunes. Por ejemplo, el máximo común divisor (o MCD) de 24 y 36 es 12.

guess-check-and-improve A method for solving an equation that involves first guessing the solution, then checking the guess by substituting into the original equation, and then using the result to improve the guess until the correct solution is found. [page 586]

conjetura, verifica y mejora Método para resolver una ecuación que implica primero hacer una conjetura, verificar la conjetura y sustituirla en la ecuación original y luego usar el resultado para mejorar la conjetura hasta hallar la solución correcta.

English	**Español**

height of a parallelogram The distance from the side opposite the base of a parallelogram to the base. The *height of a parallelogram* is always measured along a segment perpendicular to the base (or to the line containing the base). The figures below show a base and the corresponding height for two parallelograms. [page 515]

altura de un paralelogramo La distancia desde el lado opuesto a la base de un paralelogramo, hasta la base. La *altura de un paralelogramo* se mide siempre a lo largo de un segmento perpendicular a la base (o a la recta que contiene la base). Las siguientes figuras muestran una base y la altura correspondiente de dos paralelogramos.

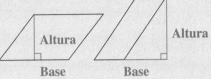

height of a triangle The distance from the base of a triangle to the vertex opposite the base. The *height of a triangle* is always measured along a segment perpendicular to the base (or the line containing the base). The figures below show a base and the corresponding height for two triangles. [page 518]

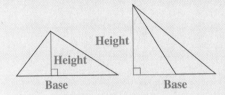

altura de un triángulo La distancia desde la base de un triángulo hasta el vértice opuesto a la base. La *altura de un triángulo* se mide siempre a lo largo de un segmento perpendicular a la base (o de la recta que contiene la base). Las siguientes figuras muestran una base y la altura correspondiente de dos triángulos.

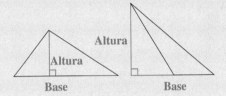

histogram A bar graph in which data are divided into equal intervals, with a bar for each interval. The height of each bar shows the number of data values in that interval. [page 350]

histograma Gráfica de barras en la cual los datos se dividen en intervalos iguales, con una barra para cada intervalo. La altura de cada barra muestra el número de valores de los datos en ese intervalo.

hypotenuse The side opposite the right angle in a right triangle. The *hypotenuse* is the longest side of a right triangle. (See the figure in the entry for *leg* below.) [page 536]

hipotenusa El lado opuesto al ángulo recto en un triángulo rectángulo. La *hipotenusa* es el lado más largo de un triángulo rectángulo. (Ver a continuación la figura en el inciso del glosario para cateto.)

inequality A mathematical sentence stating that two quantities have different values. For example, $5 + 9 > 12$ is an *inequality*. The symbols $\neq$, $<$, and $>$ are used in writing inequalities. [page 559]

desigualdad Enunciado matemático que establece que dos cantidades tienen distintos valores. Por ejemplo, $5 + 9 > 12$ es una *desigualdad*. Los símbolos $\neq$, $<$ y $>$ se usan para escribir desigualdades.

inverse operations Two operations that "undo" each other. For example, addition and subtraction are inverse operations, and multiplication and division are inverse operations. [page 504]

operaciones inversas Dos operaciones que se "anulan" entre sí. Por ejemplo, la adición y la sustracción son operaciones inversas y la multiplicación y la división son operaciones inversas.

least common multiple, LCM The *least common multiple* (often abbreviated *LCM*) of two or more numbers is the smallest of their common multiples. For example, the least common multiple (or LCM) of 6 and 15 is 30. [page 85]

mínimo común múltiplo, mcm El *mínimo común múltiplo* (a menudo abreviado mcm) de dos o más números es el menor de sus múltiplos comunes. Por ejemplo, el mínimo común múltiplo de 6 y 15 es 30.

English

leg One of the sides of a right triangle that is not the hypotenuse, or one of the shorter two sides of a right triangle. The figure at the right shows the two legs and hypotenuse of a right triangle. [page 536]

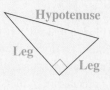

line graph A graph in which points are connected with line segments. [page 289]

line plot A number line with X's indicating the number of times each data value occurs. [page 363]

line symmetry A polygon has *line symmetry* (or reflection symmetry) if you can fold it in half along a line so that the two halves match exactly. The polygons below have *line symmetry*. The lines of symmetry are shown as dashed lines. [page 51]

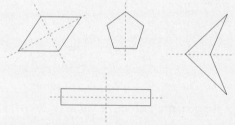

lowest terms A fraction is in *lowest terms* if its numerator and denominator are relatively prime. For example, $\frac{5}{6}$ is in lowest terms because the only common factor of 5 and 6 is 1. [page 100]

mean The number you get by distributing the total of the values in a data set among the members of the data set. You can compute the *mean* by adding the values and dividing the total by the number of values. For example, for the data 5, 6, 6, 8, 8, 8, 9, 10, 12, the total of the values is 72 and there are 9 values, so the mean is $72 \div 9 = 8$. [page 371]

median The middle value when all the values in a data set are ordered from least to greatest. For example, for the data set 4.5, 6, 7, 7, 8.5, 10.5, 12, 12, 14.5, the *median* is 8.5. [page 363]

mixed number A whole number and a fraction. For example $12\frac{3}{4}$ is a *mixed number*. [page 98]

mode The value in a data set that occurs most often. For example, for the data set 4.5, 6, 7, 7, 7, 8.5, 10.5, 12, 12, the *mode* is 7. [page 363]

Español

cateto Uno de los lados de un triángulo rectángulo que no es la hipotenusa o uno de los dos lados más cortos de un triángulo rectángulo. La figura muestra los dos catetos y la hipotenusa de un triángulo rectángulo.

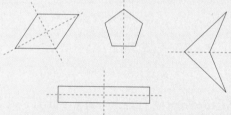

gráfica lineal Gráfica en la cual los puntos se conectan con segmentos de recta.

esquema lineal Recta numérica que contiene equis que indican el número de veces que ocurre cada valor de los datos.

simetría lineal Un polígono tiene *simetría lineal* (o simetría de reflexión) si se puede doblar por la mitad a lo largo de una línea de modo que las dos mitades coincidan exactamente. Los siguientes polígonos tienen *simetría lineal*. Los ejes de simetría se muestran como líneas punteadas.

en términos reducidos o reducida Una fracción está *en términos reducidos* o *reducida* si su numerador y denominador son primos relativos. Por ejemplo, $\frac{5}{6}$ está reducida dado que el único factor común de 5 y 6 es 1.

media El número que se obtiene al distribuir el total de los valores en un conjunto de datos entre los miembros del conjunto de datos. Se puede calcular la *media* sumando los valores y luego dividiendo el total entre el número de valores. Por ejemplo, para los datos 5, 6, 6, 8, 8, 8, 9, 10, 12, el total de los valores es 72 y hay 9 valores, de modo que la media es $72 \div 9 = 8$.

mediana El valor central cuando todos los valores en un conjunto de datos se ordenan de menor a mayor. Por ejemplo, para el conjunto de datos 4.5, 6, 7, 7, 8.5, 10.5, 12, 12, 14.5, la *mediana* es 8.5.

número mixto Un número entero y una fracción. Por ejemplo, $12\frac{3}{4}$ es un *número mixto*.

moda El valor en un conjunto de datos que ocurre con más frecuencia. Por ejemplo, para el conjunto de datos 4.5, 6, 7, 7, 7, 8.5, 10.5, 12, 12, la *moda* es 7.

English	Español

multiple A *multiple* of a whole number is the product of that number and another whole number. For example, 35 is a multiple of 7 since $35 = 7 \times 5$. [page 85]

negative number A number that is less than 0. For example, $^-18$ (read "negative eighteen") is a *negative number*. [page 142]

obtuse angle An angle that measures more than 90° and less than 180°. Each of the angles shown below is an *obtuse angle*. [page 469]

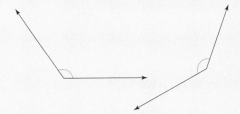

open sentence An equation of inequality that can be true or false depending on the value of the variable. For example, $5 + n = 20$ is an *open sentence*. [page 560]

opposites Two numbers that are the same distance from 0 on the number line, but on different sides of 0. For example 35 and $^-35$ are opposites. [page 143]

ordered pair A pair of numbers that represent the coordinates of a point, with the horizontal coordinate of the point written first. For example, the point with horizontal coordinate 3 and vertical coordinate 7 is represented by the ordered pair (3, 7). [page 302]

order of operations A convention for reading and evaluating expressions. The order of operations says that expressions should be evaluated in this order:

• Evaluate any expressions inside parentheses and above and below fraction bars.
• Evaluate all exponents, including squares.
• Do multiplications and divisions from left to right.
• Do additions and subtractions from left to right.

For example, to evaluate $5 + 3 \times 7$, you multiply first and then add: $5 + 3 \cdot 7 = 5 + 21 = 26$.

To evaluate $10^2 - 6 \div 3$, you evaluate the exponent first, then divide, then subtract:

$10^2 - 6 \div 3 = 100 - 6 \div 3 = 100 - 2 = 98$.

[pages 19 and 501]

múltiplo Un *múltiplo* de un número entero es el producto de ese número y otro número entero. Por ejemplo, 35 es un múltiplo de 7 dado que $35 = 7 \times 5$.

número negativo Un número menor que 0. Por ejemplo, $^-18$ (que se lee "dieciocho negativo") es un *número negativo*.

ángulo obtuso Ángulo que mide más de 90° y menos de 180°. Cada uno de los siguientes ángulos es un *ángulo obtuso*.

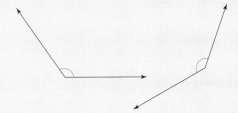

enunciado abierto Ecuación de desigualdad que puede ser verdadera o falsa dependiendo del valor de la variable. Por ejemplo, $5 + n = 20$ es un *enunciado abierto*.

opuestos Dos números equidistantes de 0 en la recta numérica, pero en lados opuestos de 0. Por ejemplo, 35 y $^-35$ son opuestos.

par ordenado Un par de números que representa las coordenadas de un punto, en el cual la coordenada horizontal del punto se escribe primero. Por ejemplo, el punto con la coordenada horizontal 3 y coordenada vertical 7 se representa con el par ordenado (3, 7).

orden de las operaciones Una convención para leer y evaluar expresiones. El orden de las operaciones indica que las expresiones se deben evaluar en el siguiente orden:

• Evalúa cualquier expresión entre paréntesis y sobre y debajo de barras de fracciones.
• Evalúa todos los exponentes, incluyendo los cuadrados.
• Efectúa las multiplicaciones y las divisiones de izquierda a derecha.
• Efectúa las sumas y las restas de izquierda a derecha.

Por ejemplo, para evaluar $5 + 3 \times 7$, multiplica primero y luego suma. $5 + 3 \cdot 7 = 5 + 21 = 26$.

Para evaluar $10^2 - 6 \div 3$, evalúa primero el exponente, luego divide y por último resta.

$10^2 - 6 \div 3 = 100 - 6 \div 3 = 100 - 2 = 98$.

English	**Español**

origin The point where the axes of a graph meet. The *origin* of a graph is usually the 0 point for each axis. [page 279]

origen El lugar donde se encuentran los ejes de una gráfica. El *origen* de una gráfica es por lo general el punto 0 de cada eje.

outlier A value that is much greater than or much less than most of the other values in a data set. For example, for the data set 6, 8.2, 9.5, 11.6, 14, 30, the value 30 is an *outlier*. [page 375]

valor atípico Un valor que es mucho mayor o mucho menor que la mayoría de los otros valores en un conjunto de datos. Por ejemplo, para el conjunto de datos 6, 8.2, 9.5, 11.6, 14, 30, el valor 30 es un *valor atípico*.

parallelogram A quadrilateral with opposite sides that are the same length. The opposite sides of a *parallelogram* are parallel. Each of the figures shown below is a parallelogram. [page 515]

paralelogramo Cuadrilátero cuyos lados opuestos tienen la misma longitud. Los lados opuestos de un *paralelogramo* son paralelos. Cada una de las siguientes figuras es un paralelogramo.

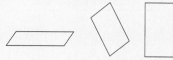

percent *Percent* means "out of 100." A percent represents a number as a part out of 100 and is written with a percent sign. For example, 39% means 39 out of 100, or $\frac{39}{100}$, or 0.39. [page 227]

por ciento, porcentaje *Por ciento* significa "de cada 100". Un por ciento representa un número como una parte de 100 y se escribe con un signo de porcentaje. Por ejemplo, 39% significa 39 de cada 100 ó $\frac{39}{100}$ ó 0.39.

perfect square A number that is equal to a whole number multiplied by itself. In other words, a *perfect square* is the result of squaring a whole number. For example, 1, 4, 9, 16, and 25 are perfect squares since these are the results of squaring 1, 2, 3, 4, and 5, in that order. [page 500]

cuadrado perfecto Número que es igual a un número entero multiplicado por sí mismo. Es decir, un *cuadrado perfecto* es el resultado de elevar al cuadrado un número entero. Por ejemplo, 1, 4, 9, 16 y 25 son cuadrados perfectos, dado que son el resultado de elevar al cuadrado 1, 2, 3, 4 y 5, en ese orden.

perimeter The distance around a two-dimensional shape. [page 482]

perímetro La distancia alrededor de una figura bidimensional.

perpendicular Two lines or segments that form a right angle area are said to be *perpendicular*. For example, see the figures below. [page 469]

perpendicular Se dice que dos rectas o segmentos que forman un ángulo recto son *perpendiculares*. Por ejemplo, observa las siguientes figuras.

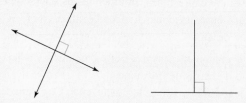

Perpendicular Lines   Perpendicular Segments

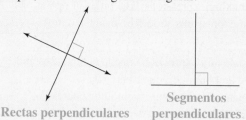

Rectas perpendiculares   Segmentos perpendiculares

polygon A flat (two-dimensional) geometric figure that has these characteristics:

- It is made of straight line segments.
- Each segment touches exactly two other segments, one at each of its endpoints.

The shapes below are *polygons*. [page 42]

polígono Figura geométrica plana (bidimensional) que posee las siguientes tres características:

- Está compuesta de segmentos de recta
- Cada segmento interseca exactamente otros dos segmentos, uno en cada uno de sus extremos.

Las siguientes figuras son *polígonos*.

positive number A number that is greater than 0. For example, 28 is a *positive number*. [page 142]

prime factorization The *prime factorization* of a composite number shows that number written as a product of prime numbers. For example, the prime factorization of 98 is $2 \cdot 7 \cdot 7$. [page 80]

prime number A whole number greater than 1 with only two factors: itself and 1. For example, 13 is a *prime number* since it only has two factors, 13 and 1. [page 80]

probability The chance that an event will happen, described as a number between 0 and 1. For example, the probability of tossing a coin and getting heads is $\frac{1}{2}$ or 50%. A *probability* of 0 or 0% means the event has no chance of happening, and a probability of 1 or 100% means the event is certain to happen. [page 605]

radius (plural: radii) A segment from the center of a circle to a point on a circle. *Radius* also refers to the distance from the center to a point on a circle. The figure below shows a *chord,* a *diameter,* and a *radius* of a circle. [page 486]

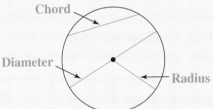

range The difference between the minimum and maximum values of a data set. For example, for the data set 4.5, 6, 7, 7, 7, 8.5, 10.5, 12, 12, the *range* is $12 - 4.5 = 7.5$. [page 363]

reciprocal Two numbers are *reciprocals* if their product is 1. For example, the reciprocal of $\frac{5}{7}$ is $\frac{7}{5}$. [page 188]

regular polygon A polygon with sides that are all the same length and angles that are all the same size. The shapes below are *regular polygons.* [page 50]

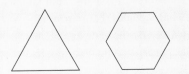

número positivo Número mayor que 0. Por ejemplo, 28 es un *número positivo.*

factorización prima La *factorización prima* de un número compuesto muestra ese número escrito como un producto de números primos. Por ejemplo, la factorización prima de 98 es $2 \cdot 7 \cdot 7$.

número primo Número entero mayor que 1 cuyos únicos factores son 1 y el número mismo. Por ejemplo, 13 es un *número primo* dado que sólo tiene dos factores: 13 y 1.

probabilidad La oportunidad de que un evento ocurra, descrita como un número entre 0 y 1. Por ejemplo, la probabilidad de lanzar una moneda y que ésta caiga mostrando cara es de $\frac{1}{2}$ ó 50%. Una probabilidad de 0 ó 0% significa que el evento no tiene oportunidad de ocurrir y una *probabilidad* de 1 ó 100% significa que el evento ocurrirá con seguridad.

radio Un segmento desde el centro del círculo hasta un punto del mismo. *Radio* también se refiere a la distancia desde el centro hasta un punto del círculo. La siguiente figura muestra una *cuerda,* un *diámetro* y un *radio* de un círculo.

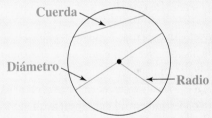

rango La diferencia entre los valores mínimo y máximo en un conjunto de datos. Por ejemplo, para el conjunto de datos 4.5, 6, 7, 7, 7, 8.5, 10.5, 12, 12, el *rango* es $12 - 4.5 = 7.5$.

recíproco Dos números son *recíprocos* si su producto es 1. Por ejemplo, el recíproco de $\frac{5}{7}$ es $\frac{7}{5}$.

polígono regular Polígono cuyos lados son todos de la misma longitud y cuyos ángulos tienen la misma medida. Las siguientes figuras son *polígonos regulares.*

English

relatively prime Two or more numbers are *relatively prime* if their only common factor is 1. For example, 7 and 9 are relatively prime. [page 83]

repeating decimal A decimal with a pattern of digits that repeat without stopping. For example, 0.232323 … is a *repeating decimal*. Repeating decimals are usually written with a bar over the repeating digits, so 0.232323 … can be written as $0.\overline{23}$. [page 132]

right angle An angle that measures exactly 90°. Right angles are often marked with a small square at the vertex. Each angle shown below is a *right angle*. [page 469]

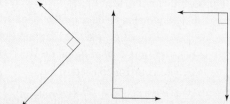

right triangle A triangle that has one 90° angle. [page 536]

sequence An ordered list. For example, 2, 5, 8, 11, … is a *sequence*. [page 6]

simulation An experiment in which you use different items to represent the items in a real situation. For example, to simulate choosing markers and looking at their colors, you can write the color of each marker on a slip of paper and put all the slips into a bag. You can simulate choosing markers by drawing slips from the bag. Mathematically, the situations are identical. [page 639]

solution A value of a variable that makes an equation true. For example, 4 is the *solution* of the equation $3n + 7 = 19$. [page 560]

square root The number you need to square to get a given number. For example, the *square root* of 36 is 6. [page 504]

stem-and-leaf plot A visual display of data that groups data, but also allows you to read individual values. To make a *stem-and-leaf plot* of data that are two-digit numbers, use the tens digits as the stems and the ones digits as the leaves. [page 366]

Español

relativamente primo Dos o más números son *relativamente primos* si su único factor común es 1. Por ejemplo, 7 y 9 son relativamente primos.

decimal periódico Decimal con un patrón de dígitos que se repiten indefinidamente. Por ejemplo, 0.232323 … es un *decimal periódico*. Los decimales periódicos por lo general se escriben con una barra sobre los dígitos que se repiten, de este modo, 0.232323 … se puede escribir como $0.\overline{23}$.

ángulo recto Ángulo que mide exactamente 90°. Los ángulos rectos por lo regular se marcan con un cuadrado en el vértice. Cada ángulo siguiente es un *ángulo recto*.

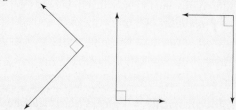

triángulo rectángulo Triángulo que posee un ángulo de 90°.

sucesión Lista ordenada. Por ejemplo, 2, 5, 8, 11, … es una *sucesión*.

simulacro Un experimento que usa diferentes artículos para representar una situación real. Por ejemplo, en un simulacro para elegir marcadores y verificar sus colores, puedes escribir el color de cada marcador en tiras de papel y colocar las tiras en una bolsa. Saca tiras de la bolsa para llevar a cabo el experimento de escoger marcadores. Matemáticamente, las situaciones son idénticas.

solución Un valor de una variable que hace verdadera una ecuación. Por ejemplo, 4 es la *solución* de la ecuación $3n + 7 = 19$.

raíz cuadrada El número que debes elevar al cuadrado para obtener un número dado. Por ejemplo, la *raíz cuadrada* de 36 es 6.

diagrama de tallo y hojas Representación visual de datos, la cual agrupa los datos, pero también permite leer valores individuales. Para trazar un *diagrama de tallo y hojas* de datos de dos dígitos, usa los dígitos de las decenas como el tallo y los dígitos de las unidades como las hojas.

English

term An item in a sequence. For example, 8 is a *term* in the sequence 2, 5, 8, 11, ... [page 6]

theoretical probability Probability calculated by reasoning about the situation. Since *theoretical probabilities* do not depend on experiments, they are always the same for a particular event. [page 608]

triangle inequality The *triangle inequality* states that the sum of the lengths of any two sides of a triangle is greater than the length of the third side. For example, the triangle inequality tells you that it is impossible to have a triangle with sides of lengths 4, 5, and 10, since 4 + 5 is not greater than 10. [page 56]

variable A quantity that varies, or changes. For example, in a problem about the sizes of buildings, the height and width of the buildings would be variables. [page 278]

vertex (plural: vertices) A corner of a polygon, where two sides meet. *Vertices* are usually labeled with capital letters, such as *A, B,* and *C* for the vertices of a triangle. [page 44]

Español

término Un artículo en una sucesión. Por ejemplo, 8 es un *término* de la sucesión 2, 5, 8, 11, ...

probabilidad teórica Probabilidad que se calcula mediante el razonamiento de la situación. Dado que las *probabilidades teóricas* no dependen de experimentos, son siempre idénticas para un evento en particular.

desigualdad del triángulo La *desigualdad del triángulo* establece que la suma de las longitudes de cualquier par de lados de un triángulo es mayor que la longitud del tercer lado. Por ejemplo, la desigualdad del triángulo indica que es imposible tener un triángulo cuyos lados miden 4, 5 y 10 de longitud, dado que 4 + 5 no es mayor que 10.

variable Cantidad que varía o cambia. Por ejemplo, en un problema sobre el tamaño de edificios, la altura y el ancho de los edificios serían variables.

vértice La esquina de un polígono, donde se encuentran dos de sus lados. Por lo general, los *vértices* se designan con letras mayúsculas, como por ejemplo, *A, B* y *C* para los vértices de un triángulo.

INDEX

PHOTO CREDITS

Chapter 1 2–3 (bkgd), PhotoDisc; 2 (tr), PhotoDisc; 2 (bl), Jeff J. Daly/Visuals Unlimited; 7, PhotoDisc; 11, PhotoDisc; 15, Mark Steinmetz; 21, Bischel Studios; 24, Laura Sifferlin; 32, Mali Apple; 35, Digital Vision; 36, Mali Apple; 39, PhotoDisc; 46 (tl), PhotoDisc; 46 (tr), PhotoDisc; 46 (cr), CORBIS; 62, Aaron Haupt; 70, Aaron Haupt

Chapter 2 74–75 (bkgd), CORBIS; 74 (tr), PhotoDisc; 76, PhotoDisc; 82, Matt Meadows; 87, CORBIS; 97, PhotoDisc; 107, Apple.com; 110, CORBIS; 114, CORBIS; 125, Courtesy D. Carr & H. Craighead/Cornell University; 130, Bogart Photography; 138, CORBIS; 141, Aaron Haupt; 143, PhotoDisc; 146, PhotoDisc

Chapter 3 152–153 (bkgd), Ken Frick; 152 (b), Courtesy Habitat for Humanity/Columbus Ohio Chapter; 158, Aaron Haupt; 161, PhotoDisc; 164, Doug Martin; 167, PhotoDisc; 169, CORBIS; 177, PhotoDisc; 179, Matt Meadows; 186, CORBIS; 189, Aaron Haupt; 195, Tim Courlas; 199, PhotoDisc; 202, Doug Martin; 211, CORBIS; 212, PhotoDisc; 215, Aaron Haupt; 216, Tim Courlas; 219, CORBIS; 222, PhotoDisc

Chapter 4 224–225 (bkgd), PhotoDisc; 224 (b), CORBIS; 229, CORBIS; 230, PhotoDisc; 242, PhotoDisc; 245, PhotoDisc; 246, CORBIS; 251, Elaine Shay; 261, Doug Martin; 265, CORBIS; 269, PhotoDisc; 271, Christopher Ena/AP/Wide World Photos; 273, CORBIS

Chapter 5 276–277 (bkgd), AP/Wide World Photos; 276 (l), AP/Wide World Photos; 285, CORBIS; 289, PhotoDisc; 291, CORBIS; 292, PhotoDisc; 298, Ann Summa; 306–307, Mali Apple; 314, PhotoDisc; 316, PhotoDisc; 318, Courtesy ABDO Publishing Co., 329, PhotoDisc; 332, CORBIS; 337, Mali Apple

Chapter 6 340–341 (bkgd), PhotoDisc; 340 (b), Aaron Haupt; 344, PhotoDisc; 345 (b), CORBIS; 352, Aaron Haupt; 360, Aaron Haupt; 374, PhotoDisc; 375, Aaron Haupt; 378, CORBIS; 384, CORBIS; 388, Mark Burnett; 390, CORBIS; 391, Doug Martin; 395, PhotoDisc; 402, Inga Spence/Visuals Unlimited; 404, Chuck Savage/CORBIS

Chapter 7 408–409 (bkgd), CORBIS; 408 (tr), PhotoDisc; 416, Leslye Borden/PhotoEdit, Inc.; 417, PhotoDisc; 419, Aaron Haupt; 422, Tim Courlas; 425, Mali Apple; 440, Mali Apple; 440, Mali Apple; 444, PhotoDisc; 459, CORBIS

Chapter 8 464–465 (bkgd), CORBIS; 464 (l), DUOMO/CORBIS; 471, Doug Martin; 475, Doug Martin; 483, Doug Bryant; 492 (tr), 2002 Cordon Art B.V. -Baarn-Holland. All rights reserved.; 492 (bl), Doug Martin; 493, CORBIS; 500, PhotoDisc; 510, Mali Apple; 514, Jack Demuth; 519, PhotoDisc; 527, Mark Ransom; 528, PhotoDisc; 530, PhotoDisc; 531, Fran Brown; 534, Wesley Treat; 552, Mark Ransom; 553, PhotoDisc

Chapter 9 557–558 (bkgd), CORBIS; 556 (b), CORBIS; 556 (tr), PhotoDisc; 563, Creative Publications; 564, Creative Publications; 569, NASA; 571, Aaron Haupt; 573, Mali Apple; 575, PhotoDisc; 579, Aaron Haupt; 584, PhotoDisc; 591, CORBIS; 598, PhotoDisc; 600, PhotoDisc

Chapter 10 602–603 (bkgd), Brand X Pictures; 602 (bl), Geoff Butler; 602 (tr), PhotoDisc; 604, PhotoDisc; 608, PhotoDisc; 609, PhotoDisc; 610 (tl), Courtesy American Numismatic Society; 610 (bl), PhotoDisc; 615, Aaron Haupt; 626–627, Matt Meadows; 629, CORBIS

Unlisted photographs are property of Glencoe/McGraw-Hill